Parasitic Infections

Parasitic Infections

Immune Responses and Therapeutics

Edited by
Abhay Prakash Mishra
University of the Free State,
Bloemfontein, South Africa

Manisha Nigam
Hemvati Nandan Bahuguna Garhwal University,
Srinagar Garhwal, India

WILEY

Published by John Wiley & Sons, Inc., Hoboken, New Jersey.

Published simultaneously in Canada.

For general information on our other products and services or for technical support, please contact our Customer Care Department within the United States at (800) 762-2974, outside the United States at (317) 572-3993 or fax (317) 572-4002.

Wiley also publishes its books in a variety of electronic formats. Some content that appears in print may not be available in electronic formats. For more information about Wiley products, visit our web site at www.wiley.com.

A catalogue record for this book is available from the Library of Congress

Hardback ISBN: 9781119878025; ePub ISBN: 9781119878049; ePDF ISBN: 9781119878056; oBook ISBN: 9781119878063

Cover image(s): © KATERYNA KON/SCIENCE PHOTO LIBRARY/Getty Images
Cover design: Wiley

Set in 9.5/12.5pt STIXTwoText by Integra Software Services Pvt. Ltd, Pondicherry, India

We dedicate Parasitic Infections: Immune Responses and Therapeutics to frontline corona warriors––doctors and healthcare workers who sacrificed their lives while saving patients during the ongoing COVID-19 pandemic.

The sacrifice of doctors and healthcare workers across the globe will not go to waste. Humanity will always be grateful to them for what they did during these unprecedented times.

First and foremost, Abhay Prakash Mishra dedicates this book to his parents, Mr. Jay Prakash Mishra and Mrs. Kavita Mishra, who gave birth to him and moulded him into the person he is today. Another word of gratitude is extended to his beloved wife Priyanka Mishra, who has stood by him steadfastly over the years.

Manisha Nigam dedicates this book to her late mother, Adarsh Nigam, who has always inspired her with strength and inspiration. She also appreciates her husband Anand Kaushal for his support, encouragement, and love throughout this journey.

Contents

List of Contributors

Ayodeji Mathias Adegoke
Cancer Research and Molecular Biology Laboratories
Department of Biochemistry
Faculty of Basic Medical Sciences
College of Medicine
University of Ibadan
Nigeria

Kayode Olayinka Afolabi
Pathogenic Yeast Research Group
Department of Microbiology and Biochemistry
Faculty of Natural and Agricultural Sciences
University of the Free State
Bloemfontein
South Africa

and

Molecular Epidemiology and Public Health Research Group
Department of Biological Sciences
Anchor University
Lagos
Nigeria

Mustapha Abdullahi Akpaki
Department of Biochemistry
Federal University of Technology
Owerri
Imo
Nigeria

Jacob Kehinde Akintunde
Applied Biochemistry and Molecular Toxicology Research Group
Department of Biochemistry
College of Biosciences
Federal University of Agriculture
Abeokuta
Nigeria

Yahya S. Al-Awthan
Department of Biology
Faculty of Science
Ibb University
Ibb
Yemen

Omar Bahattab
Department of Biochemistry
Faculty of Sciences

University of Tabuk
Tabuk
Saudi Arabia

Luit Moni Barkalita
Department of Animal Biotechnology
College of Veterinary Science
Assam Agricultural University
Assam
India

Maryam Bello-Akinosho
Pathogenic Yeast Research Group
Department of Microbiology and Biochemistry
Faculty of Natural and Agricultural Sciences
University of the Free State
Bloemfontein
South Africa

Munni Bhandari
Department of Microbiology
School of Life Sciences
Hemvati Nandan Bahuguna Garhwal University
Srinagar Garhwal
Uttarakhand
India

Anupam Brahma
SRF
Krishi Vigyan Kendra
Dudhnoi
Assam
India

Harish Chandra
Department of Botany and Microbiology
Gurukul Kangri (Deemed to be University)
Haridwar
Uttarakhand
India

Abdulkadir Mohammed Danyaro
Department of Biochemistry
Bauchi State University
Gadau
Bauchi
Nigeria

Sarmistha Debbarma
Veterinary Officer
Animal Resources Development Department
Govt of Tripura
Prani Sampad Bikash Bhawan
Pandit Nehru Complex
Gurkhabusti
Kunjaban
West Tripura

Aqsa Farooqui
Department of Nursing
Indian Institute of Health and Technology (IIHT)
Deoband
Saharanpur
Uttar Pradesh
India

Arti Gautam
Pharmacology Division
National Botanical Research Institute (NBRI-CSIR)
Lucknow
Uttar Pradesh
India

Seetha Harilal
Faculty of Pharmaceutical Sciences
Kerala University of Health Sciences
Kerala
India

Fida Hussain
Department of Pharmacy
University of Swabi
Swabi
Pakistan

Swinder Jeet Singh Kalra
Department of Chemistry,
D.A.V College,
Kanpur,
India

Salah-Ud-Din Khan
Department of Biochemistry
College of Medicine
Imam Mohammad Ibn Saud Islamic
University (IMSIU)
Riyadh
Saudi Arabia

Shahanavaj Khan
Department of Medical Lab
Technology
Indian Institute of Health and
Technology (IIHT)
Deoband, Saharanpur
Uttar Pradesh
India

and

Department of Health Sciences
Novel Global Community Educational
Foundation
Australia

Rajesh Kumar
Faculty of Pharmaceutical Sciences
Kerala University of Health Sciences
Kerala
India

Vijay Jyoti Kumar
Department of Pharmaceutical
Sciences
H. N. B. Garhwal University
Srinagar Garhwal
Uttarakhand
India

Prabhakar Maurya
Consultant
CEHTRA Chemical Consultants
Pvt. Ltd
B1/A5
Mohan Co-Operative Industrial
Estate
New Delhi
India

Dearikha Karina Mayashinta
Department of Parasitology
Faculty of Medicine
Universitas Brawijaya
Malang
East Java
Indonesia

Abhay Prakash Mishra
Department of Pharmacology
School of Clinical Medicine
Faculty of Health Sciences
University of the Free State
Bloemfontein
South Africa

Tribhuvan Mohan Mohapatra
Department of Microbiology
Institute of Medical Sciences
Banaras Hindu University
Varanasi
India

Adeline Lum Nde
Department of Pharmacology
School of Clinical Medicine
Faculty of Health Sciences
University of the Free State
Bloemfontein
South Africa

Rahul Negi
Department of Microbiology
School of Life Sciences
Hemvati Nandan Bahuguna Garhwal
University
Srinagar Garhwal
Uttarakhand
India

Manisha Nigam
Department of Biochemistry
Hemvati Nandan Bahuguna Garhwal
University
Srinagar Garhwal
Uttarakhand
India

Shakir Mayowa Obidola
Department of Crop Production
Technology
Federal College of Forestry
Jos
Plateau
Nigeria

Olalekan Ogunro
Department of Biological Sciences
Kola Daisi University
Ibadan
Oyo
Nigeria

Ahmed Olatunde
Department of Medical
Biochemistry
Abubakar Tafawa Balewa
University
Bauchi
Nigeria

Carolina Pohl-Albertyn
Pathogenic Yeast Research Group
Department of Microbiology and
Biochemistry
Faculty of Natural and Agricultural
Sciences
University of the Free State
Bloemfontein
South Africa

Abdu Rauf
Department of Chemistry
University of Swabi
Swabi, Anbar
Khyber Pakhtunkhwa
Pakistan

Muhammad Rizwan
Centre for Biotechnology and
Microbiology
University of Swat
Khyber Pakhtunkhwa
Pakistan

Saira
Department of Zoology
University of Swabi
Swabi
Pakistan

Yulia Dwi Setia
Department of Parasitology
Faculty of Medicine
Universitas Brawijaya
Malang
East Java
Indonesia

and

Department of Parasitology
Department of Infectious Disease
Faculty of Medicine
University of Miyazaki
Japan

Syed Muhammad Mukarram Shah
Department of Pharmacy
University of Swabi
Swabi
Pakistan

Rahime Şimşek
Department of Pharmaceutical Chemistry
Faculty of Pharmacy
Başkent University
Ankara
Turkey

Gurdeep Singh
School of Pharmaceutical Sciences
Lovely Professional University
Phagwara
Punjab
India

Mukesh Singh
School of Pharmaceutical Sciences
IFTM University
Moradabad
Uttar Pradesh
India

Nisha Singh
Hemvati Nandan Bahuguna Garhwal University (A Central University)
Srinagar (Garhwal)
Uttarakhand
India

Rahul Kunwar Singh
Department of Microbiology
School of Life Sciences
Hemvati Nandan Bahuguna Garhwal University
Srinagar Garhwal
Uttarakhand
India

Kishawar Sultana
Centre of Biotechnology and Microbiology
University of Peshawar
Khyber Pakhtunkhwa
Pakistan

Pratichi Singh Swetanshu
Department of Biosciences
School of Basic and Applied Sciences
Galgotias University
Greater Noida
Uttar Pradesh
India

Jupi Talukdar
Consultant
Mahisah Perspicientia LLP

Irkuchi Guwahati
Assam
India

Habibu Tijjani
Department of Biochemistry
Bauchi State University
Gadau
Bauchi
Nigeria

Abhishek Tiwari
Pharmacy Academy
IFTM University
Moradabad
Uttar Pradesh
India

Varsha Tiwari
Pharmacy Academy
IFTM University
Moradabad
Uttar Pradesh
India

Archana Yadav
Department of Microbiology
Institute of Biosciences and
Biotechnology
C.S.J.M. University
Kanpur
Uttar Pradesh
India

Shikha Yadav
Department of Pharmacy
School of Medical and Allied Sciences
Galgotias University
Greater Noida
Uttar Pradesh
India

Nasib Zaman
Centre for Biotechnology and
Microbiology
University of Swat
Khyber Pakhtunkhwa
Pakistan

Preface

Parasitic infections remain a significant cause of morbidity and mortality in the world today. Indeed, many of the key parasites found today existed and were widely distributed before written records were created, and our forefathers must have been aware of the presence of the largest and most frequent worms, as well as some parasitic disorders. Often endemic in developing countries, many parasitic diseases are neglected in terms of research funding and much remains to be understood about parasites and the interactions they have with the immune system. These diseases result in more severe consequences in tropical and subtropical countries due to their low economy that makes it even more difficult to design and implement health control programs. This situation opens the door to the emergence and reemergence of these diseases; therefore, it is essential to study and update the immunopathological behavior of parasitic diseases with the objective of the designing of strategy that could eradicate it fully. Remarkable achievements in parasitic infections and immune responses, both basic and translational, have occurred over the last ten years, and we have incorporated these into the this book. We have added around 1,000 references to document these advances. There exists great variation in the degree to which parasites are fastidious with respect to their hosts. This is exemplified by the expression "host range," a descriptive feature of a given parasite used to characterize the variety of species that can be infected by that parasite. Many parasites are highly selective (i.e., host specific), their range of hosts being limited to one, or a few related, species. This book encompasses the status of research concerned with natural resistance of prospective hosts toward animal parasites and indicates profitable directions of future research. Featuring the work and dedication of several world experts, this book emphasizes the immunopathology by delineating the concepts of immunity and inflammation against parasitic infections. It covers almost all the current issues in the concerned area of parasitic infections from its induction to treatment. This current knowledge of immune responses to parasitic infections affecting humans, including interactions that occur during co-infections and how immune responses may be manipulated to develop therapeutic interventions against parasitic infection, has been thoroughly discussed. For easy reference, the most commonly studied parasites are examined in individual chapters written by investigators at the forefront of their fields. An overview of the immune system, as well as introduction to parasites, is included to guide background reading that will cover immunopathology of all types of parasitic infections i.e. protozoan, helminths and ectoparasites. Also included will be the discussions of drug development and associated challenges so far. An important feature of this book is the discussion of neglected parasitic infections and nanoparticles for antiparasitic drug delivery, a recently emerging topic. Neglected parasitic infections (NPIs) are a group of parasitic infections that are common in low-income populations in those who struggle to meet their daily basic needs and inaccessibility of primary health care services. The current status of neglected parasitic infections in and around the world and salient achievements resulting from the implementation of the proper control strategies recommended by World Health Organization has been covered in this book. As an emerging novel drug carrier, nanoparticles provide a promising way for the effective treatment of parasitic diseases by overcoming the shortcomings of low bioavailability, poor cellular permeability, nonspecific distribution and rapid elimination of antiparasitic drugs from the body. The progress of the enhanced antiparasitic effects of different nanoparticles payloads, the transport and disposition process in the body and their influencing factors has been elaborated. The challenges and prospects of nanoparticles for antiparasitic drug delivery is also discussed. This will help readers to understand the development trend of nanoparticles in the treatment of parasitic diseases and explore strategies in the development of more efficient nanocarriers to overcome the difficulty in the treatment of parasite infections in the future. The chapter "Microorganisms as drivers of host–parasite interactions" investigates the role of host- and parasite-associated microorganisms in host–parasite interactions at the individual, local and regional level leading to a holistic understanding of how the co-evolution of the different partners influences how the other ones respond, both ecologically and evolutionary. Some phenotypic alterations induced by parasites may also arise from conflicts of interests between the host or parasite and its associated microorganism. It is important and relevant to understand the proximate basis of parasite strategies, to predict their evolutionary dynamics and potentially to prevent therapeutic failures. The number of recently discovered molecular techniques for the study and diagnosis of parasite infection has completely re-structured our comprehension of how our defence system works to produce protection against infection/reinfection, or in some cases, how it becomes subverted by the offending pathogen to enable it to endure inside us for long periods of time. A plethora of molecular-based diagnostic tests have found their way into the routine of the parasitology diagnostic laboratory, improving the ease at which the offending pathogen can be rapidly identified. Newer drugs and vaccinations, many with less harmful side-effects, have come on the market that make controlling parasite populations at the community level possible without the risk of harming the very ones we wish to help. In parasitology, routine laboratory diagnosis involves conventional methods, such as optical microscopy, used for the morphological identification of parasites. Currently, molecular biology techniques are increasingly used to diagnose parasite structures in order to enhance the identification and characterization of parasites. The chapter "Molecular techniques for the study and diagnosis of parasite infection" reviews the current and new diagnostic techniques for the confirmation of parasite infections that affect people worldwide, helping to control parasitic disease mortality. The demand for novel antiparasitic drugs is extraordinarily high, but the pharmaceutical industry is not very enthusiastic about supporting development. Industrial partners are often only interested in compounds close to or already undergoing Phase-I trials. However, drug development costs are enormous, and focus more on the know-how of an industrial partner than on that of the scientist who developed the pharmacological model. Also, socio-economic conditions in endemic countries are often insufficient to allow the emergence of new weapons against parasitic diseases, and alternative strategies therefore need to be considered. Because vaccinations do not work in most instances and the parasites have sometimes become resistant to the available synthetic therapeutics, it is important to search for alternative sources of anti-parasitic drugs. Plants produce a high diversity of secondary metabolites with interesting biological activities, such as cytotoxic, anti-parasitic and anti-microbial properties. These drugs often interfere with central targets in parasites, such as DNA (intercalation, alkylation), membrane integrity, microtubules and neuronal signal transduction. Plant extracts and isolated secondary metabolites that can inhibit parasites have been discussed in detail. We also discuss the challenges and opportunities related to antiparasitic drug discovery, highlighting the progress that has been made in recent years.

In order to participate in the global initiative that is already under way and has everything to do with enhancing the fitness and survival of the vast majority of the human species, it is our intent that readers of this book will be adequately prepared with fundamental knowledge of parasites, immune responses, and therapeutics.

Acknowledgements

This book would not have been possible without the assistance of many people, especially the academic colleagues who worked alongside us along the way. It is the result of a collaborative effort, and it would not have been possible without the amazing people who supported us at every stage.

We are grateful to all the authors who choose our book as a way to help future generations traverse the complexity of parasitology. We hope that this edition proves useful for you and your students in the coming years. Thanks to all the experts (Luigi Milella, Ph.D.; Raffaele Pezzani, Ph.D.; Henrique Douglas Melo Coutinho, Ph.D.; Raja Loganathan, Ph.D.; Divya Kamaraju, Ph.D.; Ruma Jas, Ph.D.; Lubna Shamshad, Ph.D.; Amir Ali Khan, Ph.D.; Syed M. Hassan Shah, Ph.D. and Maria S. Atanassova, Ph.D.) who have given their time and expertise to review this book prior to its publication, and we thank them for the many corrections and suggestions they gave us.

We lack the space here to acknowledge all the other individuals whose special efforts went into this book. We offer instead our sincere thanks—and the finished book that they helped guide to completion. We, of course, assume full responsibility for errors of fact or emphasis.

We are grateful to the students and staff of our research groups, who helped us balance the competing demands on our time; to our colleagues in the Department of Biochemistry at the Hemvati Nandan Bahuguna Garhwal University, Uttarakhand, India and Department of Pharmacology, University of the Free State, Bloemfontein, Free State, South Africa, who helped us with advice and criticism. We hope our readers will continue to provide input for future editions.

Our hats are off to everyone at Wiley publishers. We enjoy collaborating with this enthusiastic, dedicated, and talented team of publishing professionals. Our thanks to the entire team.

Finally, we would like to thank our family and life partners for their unwavering support and patience with our book.

1

Introduction

Back to the Future – Solutions for Parasitic Problems

Rahime Şimşek[1], Aqsa Farooqui[2], Salah-Ud-Din Khan[3] and Shahanavaj Khan[4,5,]*

[1] Department of Pharmaceutical Chemistry, Faculty of Pharmacy, Hacettepe University, Ankara, Turkey
[2] Department of Nursing, Indian Institute of Health and Technology (IIHT), Deoband, Saharanpur, UP, India
[3] Department of Biochemistry, College of Medicine, Imam Mohammad Ibn Saud Islamic University (IMSIU), Riyadh, Saudi Arabia
[4] Department of Medical Lab Technology, Indian Institute of Health and Technology (IIHT), Deoband, Saharanpur, Uttar Pradesh, India
[5] Department of Health Sciences, Novel Global Community Educational Foundation, Australia
* Corresponding author

1.1 Introduction

Parasites are organisms that range widely in size. Parasites can be bacteria, protists, fungi, viruses, plants or animals. Parasites that live, feed, and harm the host in or on the host organism may be large enough to be visible. It is known that they infect millions of people around the world and cause thousands of deaths. In this respect, parasitic infections are an important health problem worldwide, especially in developing countries [1–3]. Today, parasitic diseases have ceased to be epidemics and have become the problem of the whole world. For example, according to the data of the World Health Organization, 1.5 billion people, in other words, 24% of the world's population, is infected with helminths transmitted from the soil (https://www.who.int/news-room/fact-sheets/detail/soil-transmitted-helminth-infections).

Diseases caused by parasites have not been adequately studied. The displacement of populations has led to the spread of regional parasitic diseases in the world and treatment and preventive health services are needed everywhere.

Parasitic Infections: Immune Responses and Therapeutics, First Edition.
Edited by Abhay Prakash Mishra and Manisha Nigam.

However, the pharmaceutical industry is still not as interested in developing drugs for the treatment of parasitic infections as in other drugs. Even antiparasitic drugs discovered at the beginning of the twentieth century are still used today. This is important in terms of showing how insufficient are the efforts to develop drugs for parasitic infections. In addition, the COVID-19 pandemic we have been through has also been a risk factor for the spread of parasitic infections. Both the disease itself and the drugs used in the treatment of COVID-19 infection have caused individuals with weakened immunity have become more susceptible to opportunistic parasitic infections.

The treatment of microbial infections or tumoral diseases caused by bacteria, protozoa, fungi, parasites and viruses with drugs is called chemotherapy. In this context, the treatment of infections caused by parasites is also called chemotherapy. Difficulties such as the development of resistance in the treatment of microorganisms other than parasites are also valid for the treatment of infections caused by parasites. Moreover, natural factors such as humidity and temperature also affect the incidence, spread and treatment of some parasitic infections. Mass movements such as travels to these affected regions, and the movement of immigrants also facilitate the spread of parasitic diseases [4–6].

Parasitic infections are most common in tropical climates due to poor hygiene conditions and are less common in developed countries. However, although they affect people who are disadvantaged in many respects, they spread all over the world with mass movements. Thus, parasitic infections affect millions of people around the world every year and can cause deaths. For these reasons, parasitic infections remain current and important as the subjects of drug research and development studies.

Today, there is a need for effective antiparasitic drugs with minimized side effects. In recent years, the factors related to the diseases caused by parasites have been clarified with mechanistic studies of parasitic infections. Thus, more data are available to derive selective and active compounds. With these data, it will be possible to reach new and active drugs having less side effects and increase the success of treatment in the near future.

1.2 Parasite and the Phenomenon of Parasitism

Parasite is a word of Greek origin, and it refers to microorganisms that have evolved to living in or on another organism and feeding by damaging it. Symbiosis is a word of greek origin and means “to live together”. It refers to a long-term and close relationship between two different biological organisms. Parasitism, which is a type of symbiosis, is an opportunistic relationship and expresses the relationship in which one species benefits while the other species is harmed [7–9].

1.3 Role of Parasites in Health and Diseases

The fight against parasites dates back to ancient times with the use of drugs and herbal methods being a success for humanity. According to studies, with the complete eradication of microorganisms such as viruses, bacteria and protozoa, human deaths will decrease and the quality of life will increase [10–13]. However, in a scenario where the parasites are completely destroyed, the ecological role of the parasites in the ecosystem can be better understood. The only parasite that humanity has managed to completely eradicate is variola, the smallpox parasite. Although only one parasite has been eradicated so far, intense efforts are also being made for the eradication of parasitic infections such as leprosy and Chagas disease, which are thought to be eradicated. In certain regions of the world, this eradication has been achieved [14].

In spite of the fact that they are harmful to human health, removing parasites from an ecosystem, especially from ecologically effective species, could be problematic in terms of disturbing the ecological balance. Fighting with parasites is an issue that is almost impossible to completely eradicate and needs to be evaluated in all its dimensions in terms of its consequences. The eradication of one parasite species can lead to the increase or extinction of other parasite species, as well as completely changing the balance of the ecosystem. Although it is thought that as a result of urbanization, healthier and more hygienic conditions will be created, this has not always been the case. In some cases, an increase in parasitic infections has been observed.

However, it is essential to fight parasites and to make efforts for the treatment of parasitic infections, but while doing this, they should not disturb the balance of nature, and strategies should be developed for all species in which life is possible without adversely affecting one another.

1.3.1 Ancient Information of Parasite Towards Health Problems

Medical struggles with microorganisms that cause disease in humans dates back to 3000 BCE. Parasites initially evolved long before the evolution of human beings, and early writings from Egypt mention some such as roundworms and tapeworms. Francesco Redi and Antonie van Leeuwenhoek in the seventeenth century began to research them and created an early version of the branch of parasitology. The pioneer experiments of various great scientists have helped us to understand the pathogenic mechanisms of the parasite.

Our information regarding infections of parasites extend into antiquity, and descriptions of parasites and parasitic infections are observed in the earliest writings and have been confirmed by the discovery of parasites in archaeological material. Human beings are hosts to approximately 300 different species of worms parasite and more than 70 species of various protozoans, several derived from our ancestors primate condition while others have been acquired from animals. Some parasites cause the most important diseases in the world; certainly these are parasites that have received the great attention. Since most of such parasite diseases occur in tropical regions, the parasitology field has tended to overlap with that of tropical medicine, and consequently the histories of these fields are intertwined. The organized study of parasites started with the rejection of the theory of spontaneous generation and the promulgation of the germ theory. After that, the history of human parasitology continued along two lines: the finding of a parasite and its subsequent connection with particular disease, and the identification of particular disease and the subsequent discovery that it was caused through the infection of the particular parasite.

It is known that many plants are used as chemotherapeutic treatments in Chinese medicine. Avicenna defined microbes and treated the diseases caused by microbes. Galenos tried to treat diseases by using the large plant flora of Anatolia. In the seventeenth and eighteenth centuries, plant extracts began to be used for the treatment of febrile illness. In this period, chemotherapy started with the use of *Cinchona* alkaloids in the treatment of malaria.

Parasites have been defined and characterized depending on a wide variety of factors, from parasite-related factors to host-related factors. Throughout history, humans have been confronted with a large number of parasites, including about 300 species of helminths, and 100 species of protozoa. These microorganisms have caused and continue to cause diseases in humans [10, 15].

1.3.2 Current Information on Parasites Causing Health Problems

Diagnostic methods of parasitic infections have been developed from the past to the present, and many drugs with various chemical compounds have been introduced for treatment. In addition, surgical operations are also performed for the treatment of some parasitic infections. Research on parasites and new drug development studies continue in line with the available data.

The pandemic period we are going through can also be associated with parasitism. COVID-19 treatment has been effective in the diagnosis, treatment and eradication of parasitic infections. For example, cough and dyspnea seen in COVID-19 are also symptoms of some parasitic infections, in which case misdiagnosis or treatment is possible. It is possible to confuse symptoms of parasitic infections, particularly those caused by protozoa, trematodes, and nematodes, with symptoms of COVID-19. In addition, an immune system weakened by parasitic infections can complicate the treatment of COVID-19 [16].

1.4 Control of Parasites

The control and proper treatment of parasites are essential requirements for the health of living beings. In recent years, the inflammatory response resulting from infections caused by parasites has been confirmed by numerous studies. In a study of dogs with visceral leishmaniasis caused by the *Leishmania donovani* complex, inflammatory lesions were observed in the brain, although no parasites were detected. The aim of the studies carried out in recent years were to treat only the infection caused by the parasite and to improve the clinical status. In recent years, mechanistic studies have increased, and all pathways in parasite infections, including the interaction between the host and the parasite, are being investigated. With all these studies, the host–parasite relationship, which involves complex interactions, is increasingly being understood. With the use of new and modern methods such as omic technologies, it will be possible to control parasitic infections in the near future [17, 18]. The control and treatment of parasites has been achieved using different therapeutics approaches.

1.4.1 Herbal Medicines

In the treatment of parasitic infections, herbal medicines as well as synthetic compounds can be the solution. For centuries, mankind has benefited from using plants to treat diseases. Today, some medicines have herbal active ingredients. It is known that many plants such as *Melissa officinalis*, *Citrus limon*, and *Thymus vulgaris* show antiparasitic activity. However, the use of herbal medicines in antiparasitic therapy is limited due to their cytotoxic effects and non-selective activity profiles.

In the current treatment of diseases caused by parasites, the inadequacy of vaccines and the resistance to the drugs used have initiated the search for new treatments. In this search, besides chemical molecules, plants have gained importance because of their secondary metabolites. Secondary metabolites can be effective in parasitism since these metabolites and parasites have common targets such as genetic material, membrane integrity, microtubules and neuronal signal transduction. It is also possible to reach active new derivatives by obtaining synthetic derivatives analogous to these compounds. Many plant products interact with the protein structures and microtubules, which are the main components of the parasite, and cause the death of the parasite. Synthetically derived drugs or herbal compounds must be active and not toxic to the host and compounds with a high selectivity index are needed. The cytotoxicity of herbal and synthetic drugs arises through interaction with genetic material, proteins and biomembranes. Alkylating agents cause genetic damage by alkylating DNA. This mechanism results in the death of the parasite.

Some secondary metabolites affect genetic mutations by chelating with DNA. In particular, some plants have alkaloids that cause the death of the parasite with this mechanism, such as topoisomerase enzyme inhibitors killing the parasite by disrupting DNA replication.

Destruction of the cell membrane results in impairment of the integrity of the cell. Cytotoxic effects are a result of cell damage, if the cell is a parasite cell, parasite death occurs. It is known that plants containing sesquiterpenes, phenylpropanoids and isothiocyanate structure show antiparasitic activity by this mechanism [4, 19].

1.4.2 Drugs and Drugs Analogs

Wide varieties of drugs and analogs are used for the treatment and management of parasites.

1.4.2.1 Antiprotozoal Drugs

Infections caused by *Plasmodium* species are still an important problem in the world and cause deaths. Alternative treatment methods and more importantly, new synthetic compounds are needed due to the development of resistance to existing treatment drugs. In this context, it is clear that parasite-selective and low-toxicity molecules will have therapeutic success [20].

The life cycle of *Plasmodium* parasites continues on two hosts, *Anopheles* mosquitoes and humans. During blood feeding of infected female *Anopheles*, sporocytes enter the host human and infect liver cells. Sporosides transform into tissue schizonts that cleave and release merosocytes. Merosides pass into the blood and form tarophosites. The trophocytes mature into blood schizonts. These schizonts break down and infect healthy red blood cells [21].

Quinine was the first known antimalarial drug for the treatment of malaria. It was first obtained from the bark of the *Cinchona* tree . The first synthetic antimalarial compound is quinacrine, a 9-aminoacridine derivative. Upon understanding the structure–activity relationships of quinine and its structural similarity with quinacrine, 4-aminoquinoline derivatives (chloroquine and hydroxychloroquine), 8-aminoquinoline derivatives (pamakine and primaquine) were synthesized. In the following years, the synthesis of compounds bearing the pyrimidine ring system continued and attempts were made to vaccinate against the disease.

Antimalarial drugs act by interacting with DNA, inhibiting dihydrofolate reductase, dihydroptereate synthesis, and protein synthesis.

According to their chemical structure, they can be classified as:

1) *Cinchona* alkaloids and analogues;
2) 4-aminoquinoline derivatives;

3) 8-aminoquinoline derivatives;
4) 9-aminoacridine derivative;
5) biguanidine derivatives;
6) diaminopyrimidine derivatives;
7) sesquiterpene lactones;
8) sulfonamides and sulfones; and
9) tetracyclines, clindamycin and azithromycin.

1.4.2.2 *Cinchona* Alkaloids and Analogues

It was known 400 years ago that *Cinchona* bark extract was used to reduce fever and treat malaria. A few centuries later the malaria parasite was discovered. *Cinchona* bark extract, first used by native South Americans, is the prototype for quinine and quinidine used to treat malaria.

For over a hundred years, studies have been carried out for the synthesis of quinine. Since it has four stereogenic centers, it has 16 different configurations, making the synthesis of pure quinine difficult. However, in 2001, the synthesis of pure' quinine was achieved [22].

Quinine ((*R*)-[(2*S*,4*S*,5*R*)-5-ethenyl-1-azabicyclo [2.2.2] octan-2-yl]-(6-methoxy-quinolin-4-yl)methanol), is mainly used in the treatment of infections caused by *P. falciparum*. In addition, due to its weak analgesic and antipyretic effect, it is used in muscle and joint pain, in the treatment of leg cramps, and rarely to induce labor because it causes uterine contraction. It causes hypotension by relaxing vascular smooth muscle through myocardial depression.

It is used as a sulfate salt in the treatment of malaria, in combination with pyrimethamine, sulfadiazine or tetracycline.

Quinidine ((*S*)-[(2*R*,4*S*,5*R*)-5-ethenyl-1-azabicyclo [2.2.2] octan-2-yl]-(6-methoxy-quinolin-4-yl)methanol), is the (+)-isomer of quinine and is obtained from the bark of the *Cinchona* tree. Although it has antiarrhythmic effects, its use in treatment is almost negligible due to its side effects.

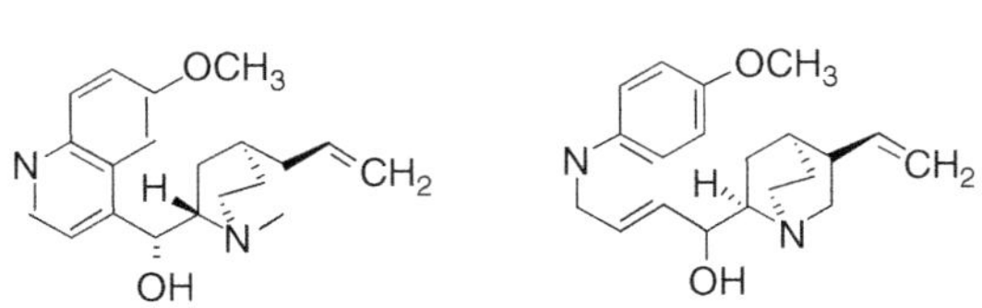

Quinine   Quinidine

1.4.2.3 4-aminoquinoline Derivatives

Chloroquine (4-*N*-(7-chloroquinolin-4-yl)-1-*N*,1-*N*-diethylpentane-1,4-diamine), hydroxychloroquine (2-[4-[(7-chloroquinolin-4-yl) amino] pentyl-ethylamino] ethanol), amodiaquine (4-[(7-chloroquinolin-4-yl) amino]-2-(diethylaminomethyl) phenol) and mefloquine ([2,8-bis(trifluoromethyl)quinolin-4-yl]-piperidin-2-yl-methanol) are 4-aminoquinoline derivatives available as therapies. These compounds are fast-acting, potent antiprotozoal drugs, especially against *P. falciparum*, acting on blood schizontocytes. However, resistance to these drugs is the most important problem in antiprotozoal therapy. Due to the high toxicity of chloroquine and the resistance of *Plasmodium* species to this drug, the -OH group was introduced, and the hydroxychloroquine synthesized. These drugs are used in the treatment of autoimmune diseases such as rheumatoid arthritis and systemic lupus erythematosus, and diseases such as anti-phospholipid syndrome, due to their immunomodulatory and antithrombotic properties. Because of their antithrombotic mechanism and their potential to inhibit proinflammatory mediators, these drugs have also been used clinically in the treatment of SARS-CoV-2 infections. It has been argued that these drugs can treat COVID-19 infections by inhibiting the binding of the virus to the host cell membrane and preventing the virus from entering the cell. They are also thought to inhibit the formation and spread of new viral particles. Another proposed antiviral mechanism for chloroquine is to inhibit the zinc ionophores that allow zinc to enter the cell, thereby reducing the intracellular zinc levels of RNA viruses and reducing their pathogenicity. It has also been predicted that chloroquine may inhibit virus-receptor binding by acting on the terminal glycosylation of the ACE-2 enzyme. Due to these features, chloroquine and its hydroxy derivative hydroxychloroquine have been tried in the treatment of COVID-19 infections. However, neither drug has demonstrated significant clinical success. In addition, the side effects of these drugs have led to the abandonment of the use of these compounds in the treatment of COVID-19 [23,24].

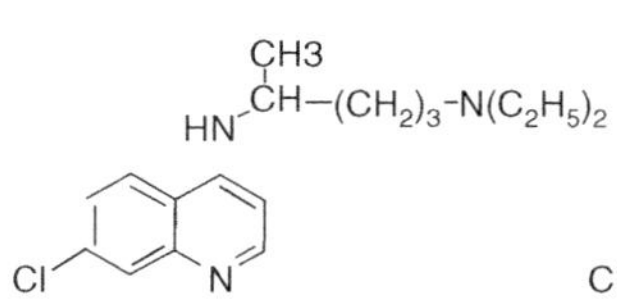

Chloroquine

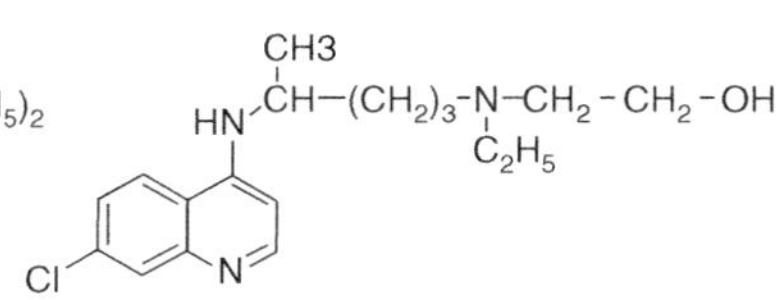

Hydroxychloroquine

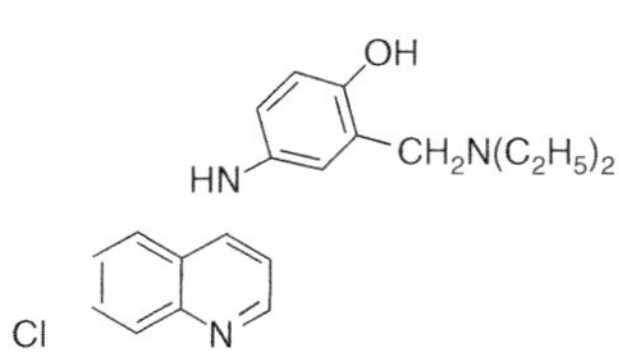

Amodiaquine

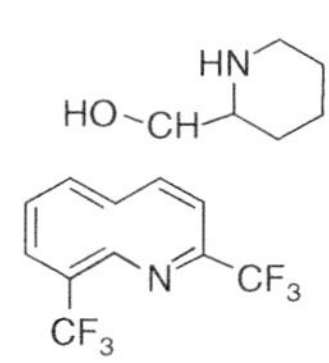

Mefloquine

8-aminoquinoline derivatives

1.4.2.4 8-aminoquinoline Derivatives

8-aminoquinoline derivatives exert their effects by killing hypnozoite forms. Although they have high antiprotozoal activities, their use is limited due to safety problems arising from interactions with glucose-6-phosphate dehydrogenase [25]. Approved by the FDA in 1952 and particularly effective against *Plasmodium vivax* and *Plasmodium ovale* species, primaquine is the prototype of 8-aminoquinoline derivative antimalarial compounds. It is used in combination with chloroquine or artemisinin in the treatment. It is effective against *Plasmodium falciparum* in a single dose and used as phosphate salt in treatment [26].

In addition to primaquine (4-*N*-(6-methoxyquinolin-8-yl)pentane-1,4-diamine), 8-aminoquinoline derivatives such as pamaquine (1-*N*,1-*N*-diethyl-4-*N*-(6-methoxyquinolin-8-yl)pentane-1,4-diamine), pentaquine (*N*-(6-methoxyquinolin-8-yl)-*N*′-propan-2-ylpentane-1,5-diamine) and isopentaquine (4-*N*-(6-methoxyquinolin-8-yl)-1-*N*-propan-2-ylpentane-1,4-diamine) are also used in the treatment.

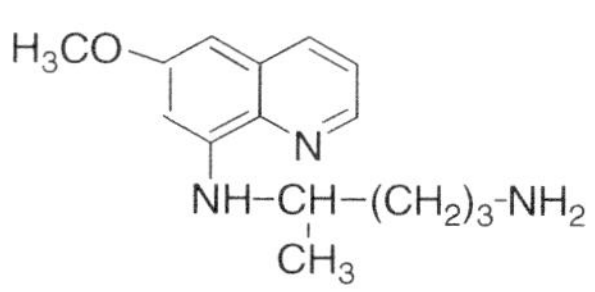

Primaquine

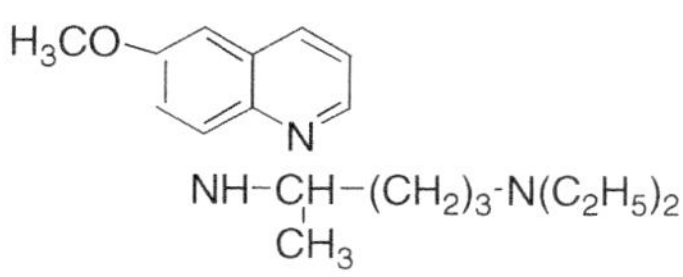

Pamaquine

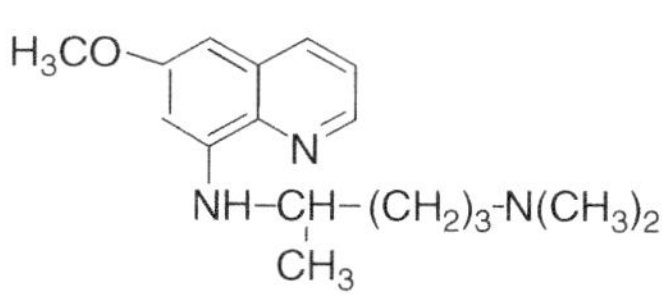

Pentaquine

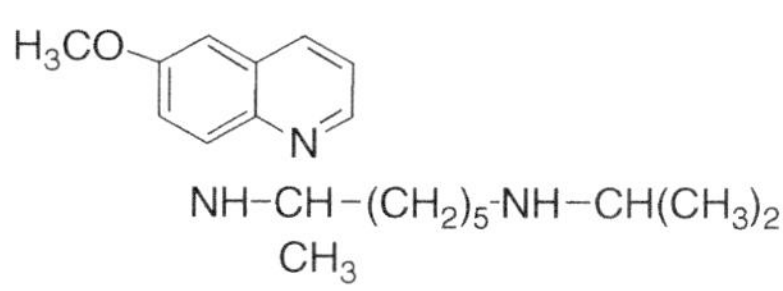

Isopentaquine

Tafenoquine (4-*N*-[2,6-dimethoxy-4-methyl-5-[3-(trifluoromethyl)phenoxy] quinolin-8-yl]pentane-1,4-diamine) is the latest approved 8-aminoquinoline derivative compound. It is a lipophilic derivative of primaquine, with a longer half-life and lower toxicity. It can also be used as a single dose due to its long half-life [27].

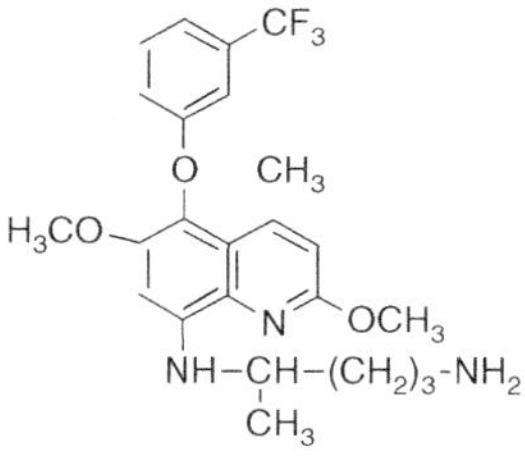

Tafenoquine

1.4.2.5 9-aminoacridine Derivative

Quinacrine (4-*N*-(6-chloro-2-methoxyacridin-9-yl)-1-*N*,1-*N*-diethylpentane-1,4-diamine) is a 9-aminoacridine derivative. It has a higher schizontoside effect than quinine. However, due to its high toxicity, it is not currently used in antiprotozoal therapy. It has toxic effects especially on the central nervous system.

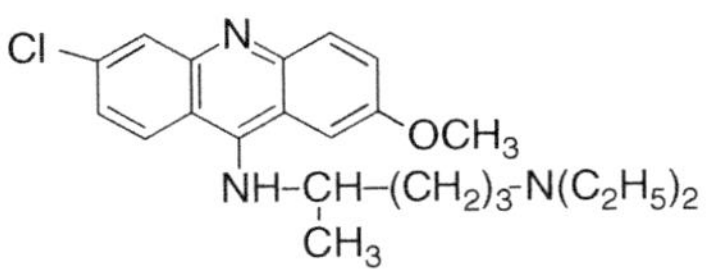

Quinacrine

1.4.2.6 Biguanidine Derivatives

Biguanide derivatives are antimalarial compounds that are frequently used for prophylactic purposes. Proguanil, chlorproguanil, and cycloguanil pamoate are drugs of this group.

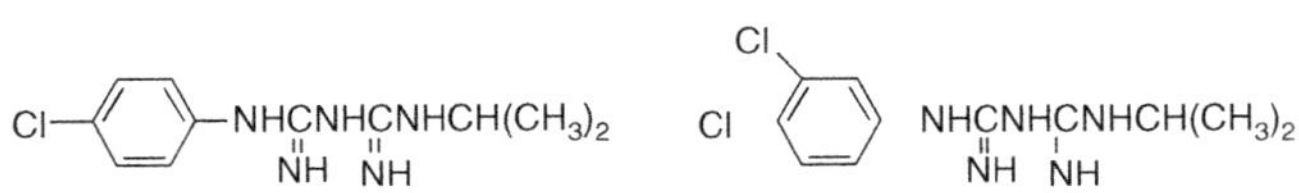

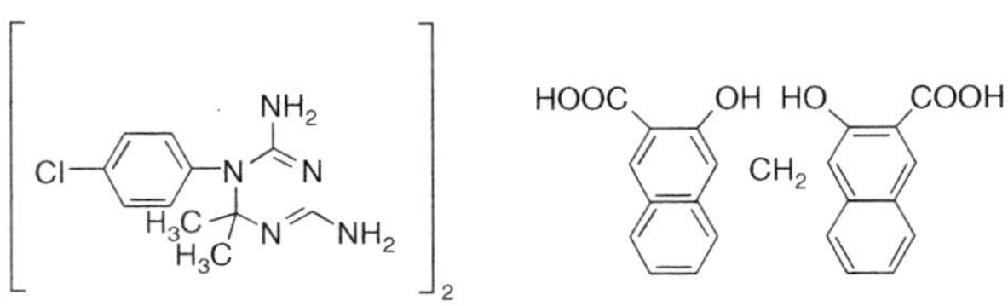

Cycloguanil pamoate

Proguanil (1-[amino-(4-chloroanilino) methylidene]-2-propan-2-ylguanidine) was found while investigating the antimalarial activity of pyrimidine derivatives. Compounds carrying nitrogen heterocycles such as pyrimidine, quinazoline and quinoline with basic side chain and arylamino or arylguanidino substituents have antimalarial activity. The noncyclic biguanide structure has a similar tautomeric system. The antimalarial activity of compounds with biguanide structure showed that the part responsible for the activity in heterocyclic rings was the side chain. Proguanil is a prodrug, its metabolite with dihydrotriazine cycloguanil structure has antimalarial activity.

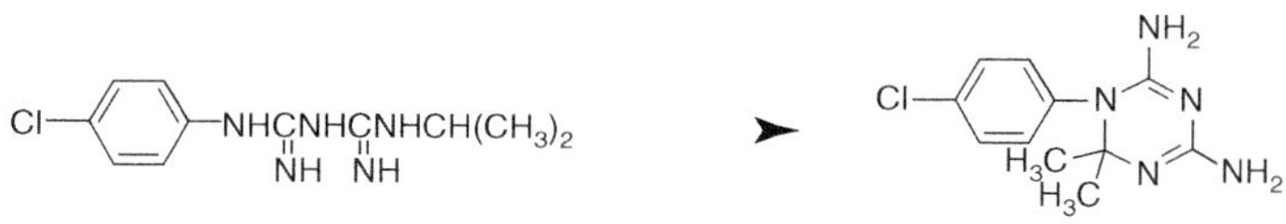

Proguanil Sikloguanil (active metabolite)

Another antimalarial drug in the biguanide structure is cycloguanil. Since cycloguanil is eliminated very quickly, the pamoate salt (cycloguanil pamoat: 1-(4-chlorophenyl)-6,6-dimethyl-1,3,5-triazine-2,4-diamine 4,4′-methylene-bis[3-hydroxy-2-naftaote] (2:1)) formed with pamoic acid is used as a depot formulation.

Chlorproguanil (1-[amino-(3,4-dichloroanilino) methylidene]-2-propan-2-yl-guanidine), a chlorinated derivative of proguanil, has a similar activity profile and is used in malaria tropica chemoprophylaxis [10, 28, 29].

1.4.2.7 Diaminopyrimidine Derivatives

Pyrimethamine (5-(4-chlorophenyl)-6-ethylpyrimidine-2,4-diamine) and trimethoprim (5-[(3,4,5-trimethoxyphenyl) methyl] pyrimidine-2,4-diamine), which are structurally similar to the pteridine part of dihydrofolic acid, are drugs that show dihydrofolate reductase enzyme inhibitory activity. The prototype of the group is pyrimethamine. Although pyrimethamine is safe at low doses, it causes megaloblastic anemia at high doses. Pyrimethamine inhibits the conversion of dihydrofolate to tetrahydrofolate by irreversibly binding to the enzyme dihydrofolate reductase. The compound shows higher affinity for the dihydrofolate reductase enzyme of *Plasmodium* than for the host dihydrofolate reductase enzyme. This feature makes the compound selective. Inhibition of the conversion of dihydrofolate to tetrahydrofolate reduces the folic acid synthesis required for nucleic acid synthesis in the parasite. Due to the development of pyrimethamine resistance, pyrimethamine-sulfadoxine combination is used in the treatment [10, 30, 31].

Trimethoprim is an antibacterial drug and its antimalarial activity was discovered later. Since it is less selective than pyrimethamine and has a shorter half-life, its use in therapy is limited. It is used in combination with other drugs [10].

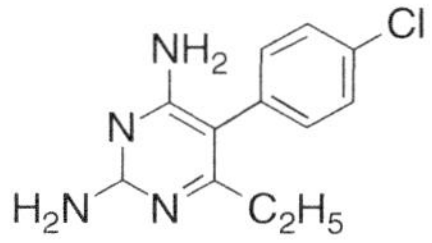

Pyrimethamine

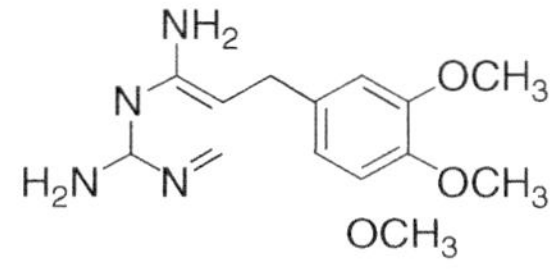

Trimethoprim

1.4.2.8 Sesquiterpene Lactones

Sesquiterpene lactones are secondary metabolites of more than 5000 compounds, most commonly found in the families Cactaceae, Solanaceae, Araceae, and Euphorbiaceae. Artemisinin ((1*R*,4*S*,5*R*,8*S*,9*R*,12*S*,13*R*)-1,5,9-trimethyl-11,14,15,16-tetraoxatetracyclo [10.3.1.04,13.08,13] hexadecan-10-one) and its derivatives are antimalarial compounds with sesquiterpene lactone structure. These compounds are used in combination with resistant synthetic derivatives. Thus, it is possible to use drug combinations with a longer half-life and to prevent the development of resistance [32].

The antimalarial activity of artemisinin was discovered in China in the mid-twentieth century. Artemisinin and its derivatives are antimalarial compounds with sesquiterpene trioxane lactone structure isolated from *Artemisia annua*. Artemisinin and its derivatives (artemether ((1*R*,4*S*,5*R*,8*S*,9*R*,10*S*,12*R*,13*R*)-10-methoxy-1,5,9-trimethyl-11,14,15,16-tetraoxatetracyclo [10.3.1.04,13.08,13] hexadecane),arteether((1*R*,4*S*,5*R*,8*S*,9*R*,10*S*,12*R*,13*R*)-10-ethoxy-1,5,9-trimethyl-11,14,15,16-tetraoxatetracyclo[10.3.1.04,13.08,13]hexadecane) and artesunate (4-oxo-4-[[(1*R*,4*S*,5*R*,8*S*,9*R*,10*S*,12*R*,13*R*)-1,5,9-trimethyl-11,14,15,16-tetraoxatetracyclo[10.3.1.04,13.08,13]hexadecan-10-yl]oxy]butanoic acid) have been included in antimalarial treatment protocols [33, 34].

These compounds can be administered in a variety of ways. Its metabolites have a longer half-life and therefore higher activity. The peroxide structure in the 1,2,4-trioxane ring system is essential for antimalarial activity. There is no activity in deoxy compounds that do not have a peroxide structure or in derivatives where one of the peroxide oxygens is replaced by carbon. One of the mechanisms of action predicted for artemisinins is that they show antimalarial activity by endoperoxide and carbon radical mechanism. According to this mechanism, artemisinins can be considered as prodrugs. In the infected erythrocyte, Fe^{2+} is present in the parasite's food vacuole, where hemoglobin is digested and free iron is detoxified by accumulating in hemozoin. Fe^{2+} reacts with endoperoxide to form free oxy radical and Fe^{3+}. The carbon radical formed by the oxy radical kills the *Plasmodium* in the erythrocyte. According to the second proposed mechanism, endoperoxide activation by the iron-dependent mechanism kills the parasite by targeting the parasite's endoplasmic reticulum [35].

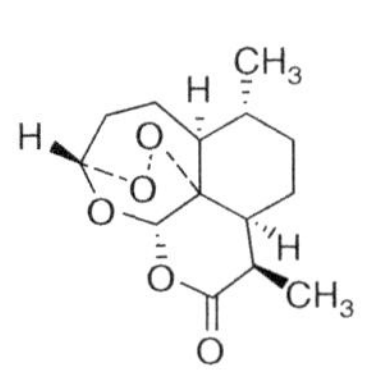

Artemisnin

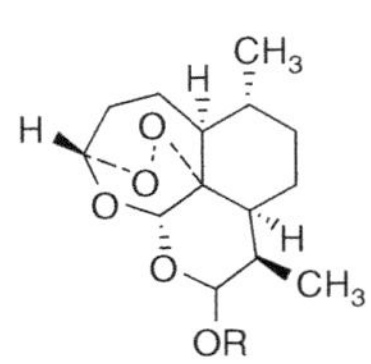

Arthemeter (R: CH_3)
Arteether (R: C_2H_5)

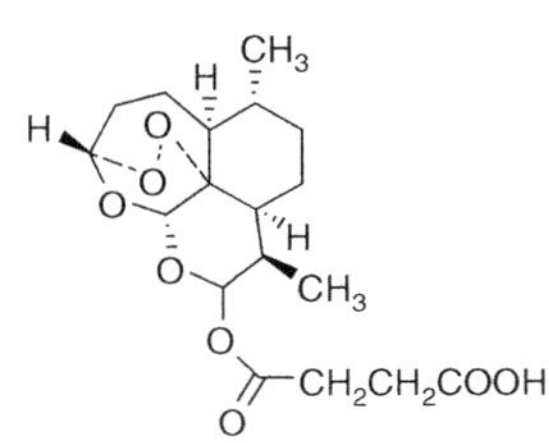

Artesunat

1.4.2.9 Sulfonamides and Sulfones

Sulfonamide and sulfone derivative compounds are used in the treatment of bacterial and fungal infections. These compounds target the folate metabolic pathway, where cofactors for amino acid and DNA synthesis are produced. This group of compounds has a wide variety of activities, including antimalarial activity. These compounds do not have strong antimalarial activity alone and are used in combination with pyrimethamine. While sulfadiazine, sulfadoxine, sulfamethoxypyridazine, sulphalen and sulfisoxazole are counted as antimalarial sulfonamide group compounds, dapsone and diformyldapsone are antimalarial sulfone group compounds.

1.4.2.10 Tetracyclines, Clindamycin and Azithromycin

Various antibiotics are used prophylactically alone or in combination with drugs such as quinine and artesunate in antimalarial therapy. Doxycillin, clindamycin, and azithromycin are examples of these antibiotics. These compounds act by inhibiting protein synthesis in the parasite. In addition, halofantrine (3-(dibutylamino)-1-[1,3-dichloro-6-(trifluoromethyl) phenanthren-9-yl] propan-1-ol), lumefantrine (2-(dibutylamino)-1-[(9*Z*)-2,7-dichloro-9-[(4-chlorophenyl) methylidene] fluoren-4-yl] ethanol) and atovaquone (3-[4-(4-chlorophenyl) cyclohexyl]-4-hydroxynaphthalene-1,2-dione) are also antimalarial compounds.

Halofantrine is a 9-phenanthrenemethanol compound. It is used in the treatment of malaria caused by chloroquine-sensitive and chloroquine-resistant *Plasmodium* species. It is taken orally. The most important side effect is cardiac toxicity.

Lumefantrine is an analogous compound to halofantrine. It is used in combination with artemether. It is an erythrocytic schizontocyte.

Atovaquone, a naphthoquinone derivative, is also used in combination with proquanil in the treatment of malaria, due to the side effects of existing drugs and resistance to these drugs. Compared to other drugs used in the treatment of malaria, it has fewer side effects [36].

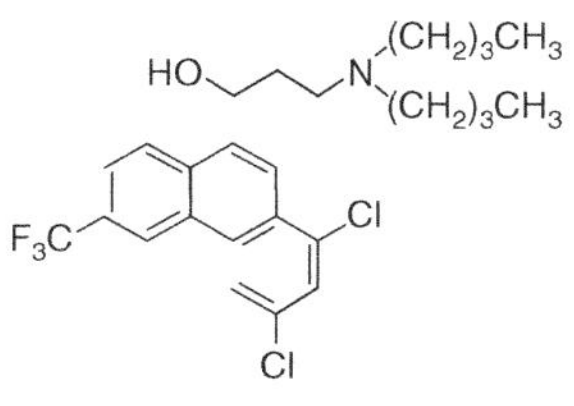

Halofantrine

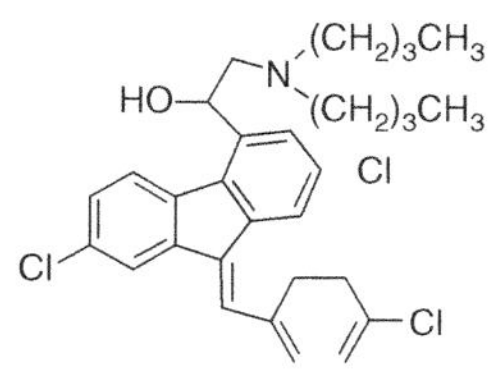

Lumefantrine

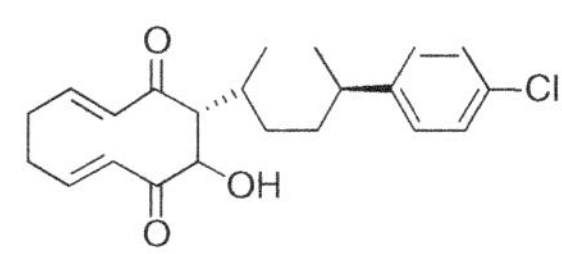

Atovaquone

Entamoeba species settle in the colon and *Entamoeba gingivalis* in the oral mucosa. The term amoebiasis refers to inflammation of the colon caused by *E. histolytica*. Amoebiasis is the most aggressive of the protozoal diseases and ranks second or third among the protozoal diseases that cause death. In infected individuals, amoebae in the colon lumen cause ulceration of the mucosa. The amoeba is excreted in the form of cysts in feces. The infection spreads through these cysts. Water and food contaminated with cysts cause the spread of the disease. Cysts ingested with water and food, pass through the small intestine and reproduce asexually to form trophocytes. The torophosites settle in the colon lumen and mature into amoebae. As a result of weakening of the host's resistance or virulence of the amoeba, the amoeba causes ulceration of the colonic mucosa and ultimately diarrhea in the patient.

In chronic cases, amoebae entering the mucous vessels can be transported to the liver by the portal circulation, causing hepatitis. In addition, they can be carried to the lungs, brain, genital area and form lesions.

Untreated *Entomoeba* infections can become chronic and fatal. Antiamoebic drugs used in the treatment should be effective on both intestinal and extraintestinal forms of the parasite. Extracts obtained from *Cephalis ipecacuanha* were first used in the treatment of the disease. The active ingredient of the extract is emetine, an ipecac alkaloid. Due to the cardiotoxic and gastrointestinal side effects of emetidine, alternative compounds have been discovered and entered into therapy. Later, carbarson and glycobiarsol were synthesized and the antiamoebic effect of chloroquine having antimalarial effect was determined. Following the development of haloquinoline derivatives, haloacetamide derivatives, paromomycin, azomycin, metronidazole, nimorazole, ornidazole and tinidazole have introduced treatment [10, 37, 38].
Antiamoebic drugs are divided into two groups according to their sites of action:

1) tissue amoebicides; and
2) luminal amoebicides.

Tissue amoebicides are drugs that kill amoebae that have settled in tissues such as intestines and liver. These compounds are ineffective against amoebae in the intestinal lumen.

Luminal amoebicides are compounds that act on amoebae in the intestinal lumen.

There are also compounds such as metronidazole that act on amoebae located in both tissue and lumen.

Antiamoebic compounds can be classified according to their chemical structure as follows:

1) 4-aminoquinoline derivatives;
2) antibiotics;

3) haloacetamides;
4) 8-hydroxyquinoline derivatives;
5) ipecac alkaloids;
6) 5-nitroimidazole derivatives;
7) organoarsenic compounds; and
8) other compounds.

1.4.2.11 4-aminoquinoline Derivatives

Chloroquine with 4-aminoquinoline structure, used in the treatment of malaria, also has antiamoebic activity. It has an equivalent activity to emetine in hepatic amoebiasis, since it accumulates in the hepatic parenchyma and reaches high concentrations. It is used orally [10].

Tetracyclines or paronomycin along with antiamoaebic drugs are used to treat invasive intestinal infection caused by amoebae. These drugs are ineffective in extracolon amobiasis. Tetracyclines have direct and indirect effects, respectively, by killing amoebae and the microorganisms with which amoebae live in symbiosis.

Paromomycin, on the other hand, has an antiamoebic effect by reaching high concentrations in the lumen of the large intestine. It acts by interacting with DNA [10].

1.4.2.12 Haloacetamides

Haloacetamide derivatives have the -N-$COCHCl_2$- structure. Diloxanide furoate ([4-[(2,2-dichloroacetyl)-methylamino] phenyl] furan-2-carboxylate), chlorbetamide (2,2-dichloro-*N*-[(2,4-dichlorophenyl) methyl]-*N*-(2-hydroxyethyl) acetamide), clefamide (2,2-dichloro-*N*-(2-hydroxyethyl)-*N*-[[4-(4-nitrophenoxy) phenyl] methyl] acetamide), etofamide (2,2-dichloro-*N*-(2-ethoxyethyl)-*N*-[[4-(4-nitrophenoxy)phenyl] methyl]acetamide), quinfamide ([1-(2,2-dichloroacetyl)-3,4-dihydro-2*H*-quinolin-6-yl] furan-2-carboxylate) and teclozan (2,2-dichloro-*N*-[[4-[[(2,2-dichloroacetyl)-(2-ethoxyethyl)amino]methyl]phenyl]methyl]-*N*-(2-ethoxyethyl) acetamide) are haloacetamide derivatives. Among the haloacetamide derivatives, the most used compound in therapy is diloxanide furoate.

Diloxanide furoate is used to treat amoebic colitis caused by *Entomoeba histolytica*. It shows activity in the intestinal lumen. Since it has low water solubility, formulations to increase its solubility should be developed. It is a luminal amoebicide, used alone in luminal cases, and in combination with a tissue amoebic drug for tissue amoeba. It acts by disrupting protein synthesis in ribosomes [39].

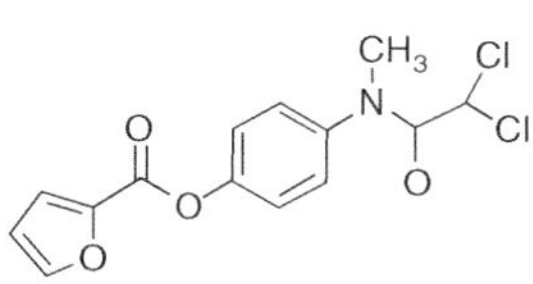

Diloxanide furoate

14.2.13 8-hydroxyquinoline Derivatives

8-hydroxyquinoline derivatives were used as popular oral antiparasitic agents from the 1950s to the 1970s. However, due to its toxic effects, its oral forms were withdrawn from the treatment in 1970, although the topical forms continued to be used. Although the antiamoebic activity of 8-hydroxyquinoline derivatives has been proven, these compounds are not widely used today due to their side effects.

However, although myelo-optic neuropathy was observed, especially in Japanese patients, with the use of chinofon (8-hydroxy-7-iodoquinoline-5-sulfonic acid), clioquinol (5-chloro-7-iodoquinolin-8-ol) was used safely during this period. Topical formulations continue to be used in some countries. Clioquinol has fewer side effects and little is known about resistance to this compound.

Currently, clioquinol is being investigated for new indications beyond its antiamoebic effects [40].

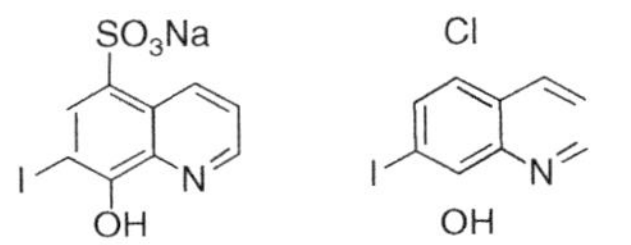

Chiniofon Clioquinol

1.4.2.14 Ipecac Alkaloids

Ipeca is a drug obtained from the plant *Cephalis ipecacuanhae.*

The *Cephaelis* plant was discovered in Brazil in the 1600s and was taken to Paris in the second half of the same century. Used to treat dysentery in France, it was discovered many years later to contain two alkaloids (emetine and cephaline). Emetine and cephaline are obtained from the dried roots and rhizomes of the plant *Cephalis ipecacuanhae.* In the nineteenth century, emetine was found to be effective in treating dysentery caused by amoebae.

These alkaloids, which have a reversible effect on DNA synthesis, cause cardiotoxicity and muscle weakness in addition to their side effects on the gastrointestinal tract [41, 42].

Emetine ((2*S*,3*R*,11*bS*)-2-[[(1*R*)-6,7-dimethoxy-1,2,3,4-tetrahydroisoquinolin-1-yl] methyl]-3-ethyl-9,10-dimethoxy-2,3,4,6,7,11*b*-hexahydro-1*H*-benzo[a] quinolizine) is a potent antiamoebic drug that kills trophosides. Although there are many stereoisomers of emetine with asymmetric centers in four positions, the most active isomer is the natural compound (-)-emetine. The active compounds were obtained by quaternizing the nitrogen atom in the emetine and replacing the substituents. Opening of the tetrahydroisoquinoline ring resulted in loss of activity.

The dehydro derivative obtained by saturating second and third positions is as active as emetine. Dehydroemetine ((11*bS*)-2-[[(1*R*)-6,7-dimethoxy-1,2,3,4-tetrahydro-isoquinolin-1-yl] methyl]-3-ethyl-9,10-dimethoxy-4,6,7,11*b*-tetrahydro-1*H*-benzo[a]quinolizine) is used when 5-nitroimidazole derivatives are ineffective or contraindicated. It has cardiovascular, gastrointestinal and neuromuscular side effects.

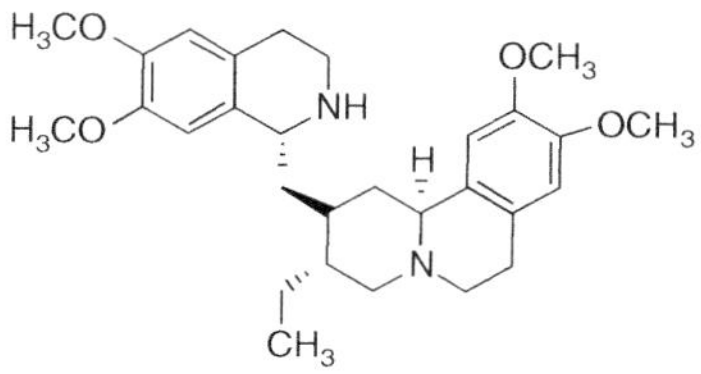

Emetine

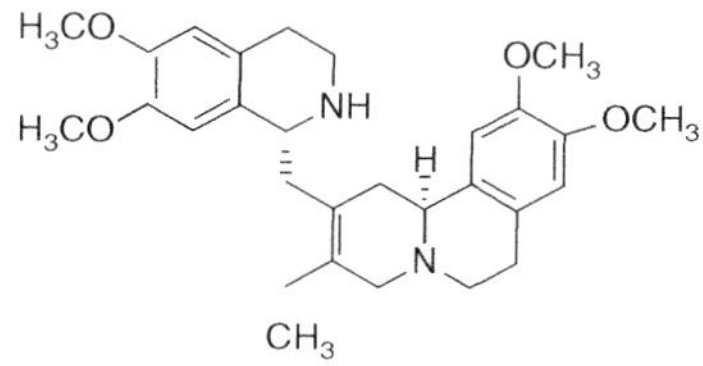

Dehydroemetine

1.4.2.15 5-nitroimidazole Derivatives

Nitroimidazole is an essential and unique ring system discovered in the 1950s. After its discovery, many drug molecules containing this ring system with various activities such as anticancer, antibacterial and antiviral have been synthesized. The most effective drugs used in the treatment of systemic and intestinal amoebiasis are nitroimidazole derivatives. The prototype of the group is metronidazole. This group of compounds shows activity by reductive bioactivation and formation of reactive intermediates. Nitroimidazole derivatives are prodrug and the nitro group is responsible for the activity of the drug. Activity occurs when the nitro group is converted to nitric oxide and reactive nitrogen species. These products inhibit growth by interacting with the cellular components of the protozoa.

Metronidazole (2-(2-methyl-5-nitroimidazol-1-yl) ethanol) was the first compound found for the treatment of trichomoniasis infections caused by *Trichomonas vaginalis*. Later, it was also used for the treatment of various diseases. Today it is used in the treatment of parasitic infections such as gastrointestinal infections, giardiasis (*G. duodenalis*) and amoebiasis (caused by *E. histolytica*). It is also effective against gram-positive and gram-negative bacteria. Metronidazole is a well tolerated drug and can also be used in pregnant women.

Ornidazole (1-chloro-3-(2-methyl-5-nitroimidazol-1-yl) propan-2-ol) is a 5-nitroimidazole derivative antiprotozoal drug that is successful in 87% of protozoal cases. Although it was first used to treat trichomoniasis, it was later discovered to be broad-spectrum antiprotozoal and antibacterial agent. Ornidazole is used clinically in raceme form.

Secnidazole (1-(2-methyl-5-nitroimidazol-1-yl) propan-2-ol) is a well-tolerated drug used in amoebiasis, giardiasis, trichomoniasis and urogenital infections. Its antiamoebic activity is due to the reduction of the nitro group on the imidazole ring.

Tinidazole (1-(2-ethylsulfonylethyl)-2-methyl-5-nitroimidazole) is effective on intestinal amoebiasis activity as well as some other protozoa species and some bacteria. Although its effect profile and side effects are similar to metronidazole, it reaches higher blood levels and has a longer duration of action compared to metronidazole.

Nimorazole (4-[2-(5-nitroimidazol-1-yl) ethyl] morpholine) is also known as nitrimidazine. Despite its antiinfective and antiamoebic activities, it is mainly used for trichomoniasis infections [10,43].

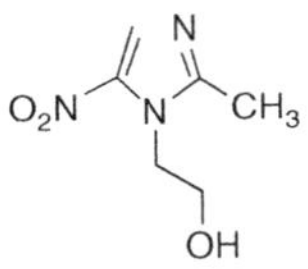

Metronidazole

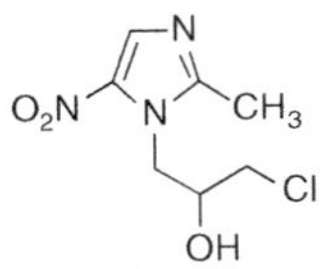

Ornidazole

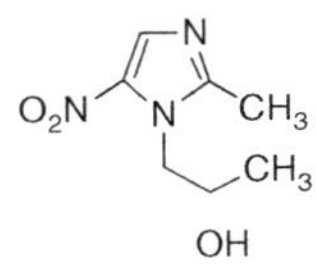

Secnidazole

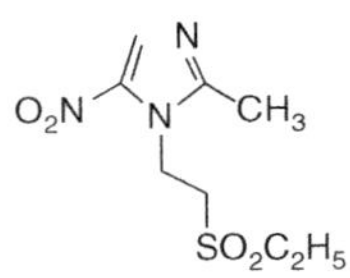

Tinidazole

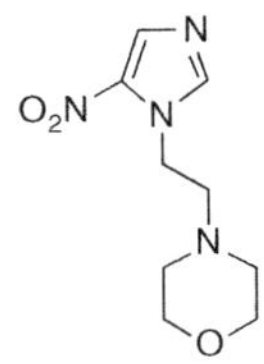

Nimorazole

1.4.2.16 Organoarsenic Compounds

Although carbarsone and glicobiarsol have been used as organoarsenic compounds in the past, today the use of organoarsenic compounds as antiamoebic agents is limited due to the availability of more active and less toxic compounds [10].

1.4.2.17 Other Compounds

Compounds with different chemical structures such as bialamicol and nitazoxanide are also used as antiamoebic. Bialamicol has the hydroxybiphenyl structure. It is used in the treatment of acute and chronic amoebiasis. Nitazoxanide is a thazolide compound, used in the treatment of invasive intestinal amoebiasis.

Leishmaniasis is the general name for infections caused by *Leishmania* species. The parasite is transmitted by the bite of humans by sandflies that feed on infected sources. Thus, infections caused by *Leishmania* occur and an immune response develops. Although cutaneous leishmaniasis is the most common form, visceral leishmaniasis is more serious and dangerous. *Leishmania donovani*, one of the *Leishmania* species that cause visceral leismaniasis or kala-azar and settles in the reticuloendothelial system, especially in the liver and spleen in humans. *L. tropica* causes cutaneous leishmaniasis and settles on the skin. In immunocompromized individuals, *Lesichmania* infections are seen as co-infection [2, 18, 44, 45].

Although many compounds have been synthesized subsequently, there is no definitive cure for the disease. Resistance to antileishmanial drugs reduces the treatment success of infections caused by leishmania by up to 33% [1]. Compounds used in the treatment of *Leishmania* can be classified as:

1) organic antimony compounds;
2) diamidine derivatives;

3) antibiotics; and
4) other compounds.

1.4.2.18 Organic Antimony Compounds

First, antimony potassium tartrate was found to be effective in leishmaniasis, then antimony compounds were studied and sodium stibogluconate (trisodium; (4*R*,5*S*,6*R*)-2-[[(4*R*,5*S*,6*R*)-4-carboxylato-6-[(1*R*)-1,2-dihydroxyethyl]-5-hydroxy-2-oxo-1,3,2λ^5-dioxastibinan-2-yl]oxy]-6-[(1*R*)-1,2-dihydroxyethyl]-5-hydroxy-2-oxo-1,3,2λ^5-dioxastibinane-4-carboxylate;nonahydrate) was synthesized. Among the antimony compounds discovered to date, the two compounds used today are sodium stibogluconate and meglumine antimonate (hydroxy(dioxo)-λ^5-stibane;(2*R*,3*R*,4*R*,5*S*)-6-(methylamino) hexane-1,2,3,4,5-pentol).

The structure of sodium stibogluconate was proven by chromatographic methods and meglumine antimonate structure was clarified by FAB-MS (fast-atom bombardment mass spectrometry) studies. Accordingly, two meglumine molecules are symmetrically coordinated with an Sb atom.

The action mechanism of antimony compounds is not fully known. One of the proposed models is the prodrug model, and antileishmanial activity occurs as a result of the conversion of the pentavalent antimony atom to the trivalent form. The second model explains antileishmanial activity of antimony compounds by the interference of the antimony atom in cellular mechanisms. The third hypothesis proposes that antimony compounds exert antileishmanial activity by activating the host's immune system [45].

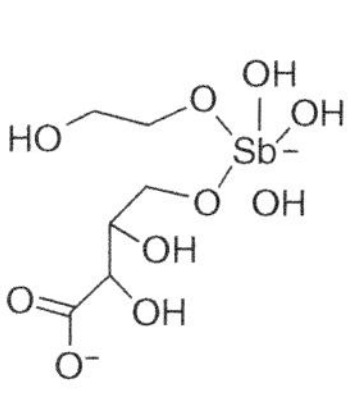

Stibogluconate

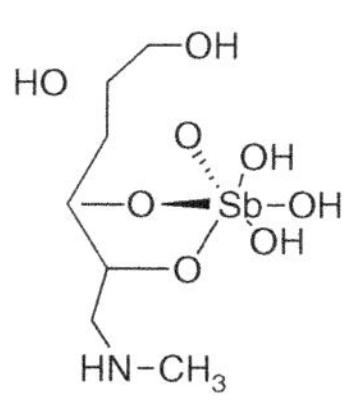

Meglumine antimonate

1.4.2.19 Diamidine Derivatives

Diamidines are a very large family of compounds with pharmacological and biological properties. From this group, hydroxystilbamidine isethionate (4-[(*E*)-2-(4-carbamimidoylphenyl) ethenyl]-3-hydroxybenzenecarboximidamide; 2-hydroxyethanesulfonic acid) and pentamidine isethionate [4-[5-(4-carbamimidoyl-phenoxy) pentoxy] benzenecarboximidamide; 2-hydroxyethanesulfonic acid] are compounds that show antiprotozoal activity as well as other pharmacological properties. Diamidine group compounds act by hydrogen bonding of amidine to DNA adenine.

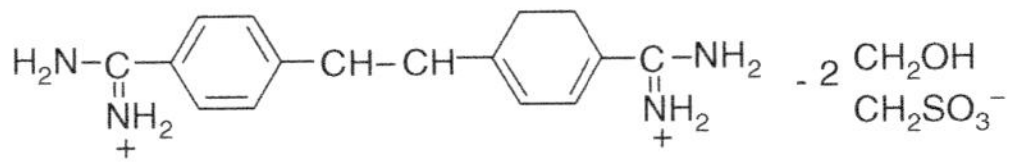

Hydroxstilbamidine isethionate

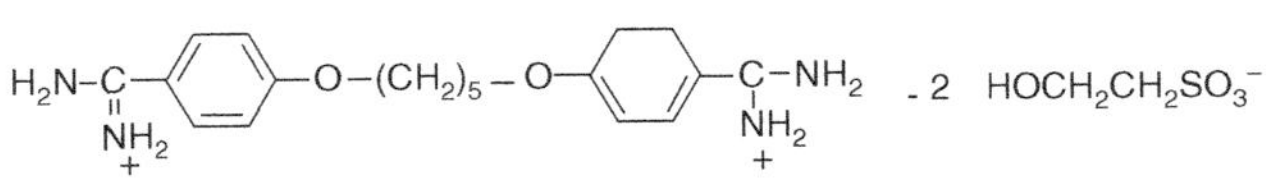

Pentamidine isethionate

1.4.2.20 Antibiotics

In cases where treatment with sodium stibogluconate was unsuccessful, a combination of amphotericin B and pentamidine isethionate is used. Additionally, there are some occasions in which Amphotericin B is used alone. In some cases of leishmaniasis, paronomycin with known antiprotozoal activity is used topically [10].

Other compounds include miltefosine (hexadecyl 2-(trimethylazaniumyl)ethyl phosphate), an alkyl phospholipid analogue developed as an antineoplastic agent, that is used orally for the treatment of visceral leishmaniasis. Due to the high toxicity of antimony compounds, miltefosine has gained importance. Since it does not affect the immune system of the host and is only effective on the parasite, it can be used safely in leishmaniasis infections seen in immunocompromized patients [10, 46].

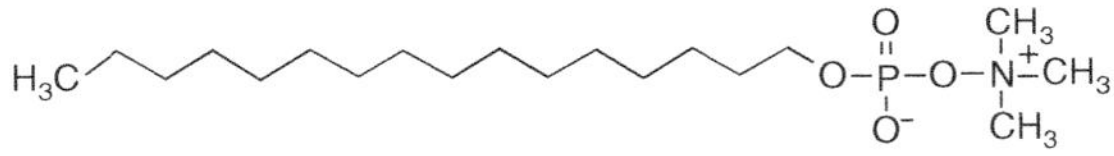

Miltefosine

Trichomoniasis is a non-viral sexually transmitted disease and one of the most common protozoal diseases worldwide. *Trichomonas* class protozoa produces intestinal and urogenital infections. Thychomoniasis caused by *T. vaginalis* is the most important of these infections and is seen in both men and women. Symptoms can range from asymptomatic to epithelial lesions characterized by foul odor and discharge in women. In addition, there is an increased risk of miscarriage and increased susceptibility to HIV-1 and cervical cancer in pregnant women. Although men can have urethritis, prostatitis, and discharge, the vast majority of infections are asymptomatic. *Trichomonas* infections are a serious problem for public health because of their symptoms, results, effects on both sexes and high incidence [47].

The drugs used in the treatment of trichomoniasis are classified as systemic compounds and topical compounds

1.4.2.21 Systemic Compounds

In the systemic treatment of trichominiasis, 5-nitroimidazole derivatives are used. Metronidazole is the most commonly used compound from this group. High treatment success can be achieved with metronidazole, which is used in both sexes. Tinidazole is more effective *in vitro* in trichomoniasis infections and has fewer side effects than metronidazole. Tinidazole is used when treatment with metronidazole is unsuccessful. In patients allergic to these compounds, intravaginal administration may be preferred over oral administration. In such cases, it should be noted that the success of the intravaginal application is lower. In addition to metronidazole and tinidazole, there are examples where other 5-nitroimidazole derivatives are used in the treatment of trichomoniasis [47].

1.4.2.22 Topical Compounds

In the treatment of trichomoniasis, there are drugs that are used topically in the form of vaginal tablets or suppositories. These compounds are useful in preventing the recurrence of vaginal infections and normalizing the vaginal flora. Povidone iodine is used in the form of a vaginal tablet or suppository, while iodoquinol is used as a vaginal tablet [10].

Drugs used in the treatment of infections caused by *Trypanosoma* group parasites are called antitrypanosomal drugs. The only alternative therapy is chemotherapy for the treatment of protozoal diseases caused by African trypanosomes (*Trypanosoma brucei*) and South American trypanosomes (*Trypanosoma cruzi*) species. However, like all available chemotherapeutics, these compounds have disadvantages such as low efficacy, high toxicity and development of resistance.

The disease is transmitted by the bite of *Triatoma* insects. The acute period begins following the bite of the insect, while the heart, esophagus, intestine and peripheral nervous system are affected in the chronic period. There are 15–20 years between these two periods. The infection caused by *T. cruzi* is known as Chagas disease because it was first discovered by Brazilian doctor Carlos Chagas. Chagas disease, also known as American trypanosomiasis, is more common in central and south America. The disease can be the cause of some serious conditions such as cardiomyopathy and death in adults. African trypanosomiasis infections caused by the bites of *T. gambiense* and *T. rhodesiense* species are also called "sleeping sickness". If the disease is not treated, it results in death. Benznidazole and nifurtimox are used in the clinic for the treatment of the disease. These compounds are effective in the acute phase of the disease, and treatment is almost impossible in the chronic phase [2].

Both compounds have nitroheterocyclic structure. Benznidazole is a nitroimidazole derivative and shows activity by inducing the oxidation of parasite nucleotides. Nifurtimox, a 5-nitrofuran derivative, creates a nitroanion radical and causes parasite DNA breaks. A change in the nitroreductase enzyme in the parasite causes to resistance to these compounds. Both compounds have limited use due to their serious side effects [1, 10, 48, 49].

Arsenic was first used for the treatment of sleeping sickness after its discovery in 1858. After arsenic was found to be effective against protozoa causing the disease, melarsoprol was synthesized. Later, pentamidine, nitrofurazone, nifurtimox and benznidazole were synthesized.

The drugs used in the treatment of trypanomosa infections can be classified as follows:

1) 5-nitrofuran derivatives;
2) urea derivatives;
3) organoarsenic compounds;
4) diamidine derivatives;
5) 2-nitroimidazole derivatives; and
6) amino acid derivatives.

1.4.2.23 5-nitrofuran Derivatives

Nifurtimox ((*E*)-*N*-(3-methyl-1,1-dioxo-1,4-thiazinan-4-yl)-1-(5-nitrofuran-2-yl) methanimine) is the prototype of 5-nitrofuran derivatives. It is the first choice in the treatment of Chagas disease. Although the mechanism of action of nifurtimox has not been fully elucidated, nifurtimox acts by a mechanism based on its metabolism through partial reduction to chemically reactive radicals leading to the production of toxic reduced oxygen species. The formed metabolic products can react with macromolecules such as DNA.

Myalgia, gastric side effects, muscle weakness, weight loss, memory loss, dermatitis and polyneuritis are side effects of nifurtimox [50,51].

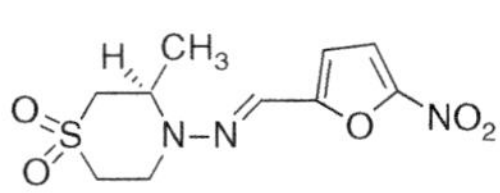

Nifurtimox

1.4.2.24 Urea Derivatives

Suramin sodium (hexasodium;8-[[4-methyl-3-[[3-[[3-[[2-methyl-5-[(4,6,8-trisulfona tonaphthalen-1-yl)carbamoyl]phenyl]carbamoyl]phenyl]carbamoylamino] benzoyl] amino]benzoyl]amino]naphthalene-1,3,5-trisulfonate) a urea derivative, has been used for 100 years in the treatment of African troponmyosis. After the discovery of the trypan blue dye, a series of potent compounds were synthesized. In the following period, with the synthesis and treatment of suramin, a new era began in African trypanosomiasis chemotherapy. It causes the death of the parasite by binding to the dehydrogenase and kinase enzymes of the protozoa, inhibiting the dihydrofolate reductase and thymidine kinase enzymes necessary for folate biotransformation, and also inhibiting the enzymes necessary for glycolysis.

Although suramin is used for the treatment of various parasitic infections such as *Plasmodium falciparum*, *Trypanosoma brusei*, *Trypanosoma evansi*, *Onchocerca volvulus* infections, ivermectin, with high oral bioavailability and lower toxicity, has now replaced suramin. Nephrotoxicity, dermatitis, anemia, peripheral neuropathy, bone marrow toxicity and hypersensitivity reactions are its side effects [52].

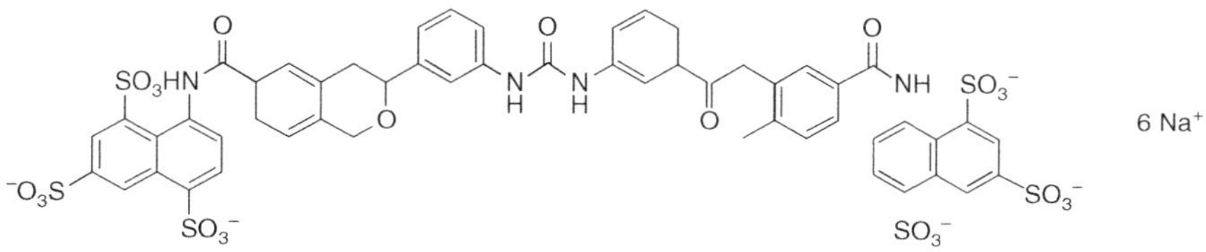

Suramin sodium

1.4.2.25 Organoarsenic Compounds

Organoarsenic compounds are selective because they are eliminated by mammalian cells by oxidation to pentavalent compounds, whereas the parasite does not have this mechanism, and this group of compounds has decreased penetration into mammalian cells.

Melarsoprol [2-[4-[(4,6-diamino-1,3,5-triazin-2-yl) amino] phenyl]-1,3,2-dithiarsolan-4-yl] methanol is a classical arsenic-derived compound that shows activity by inhibiting DNA synthesis. It is especially effective when the parasite affects the central nervous system. Melarsoprol shows its biological activity by turning into melarsen oxide when it enters the cell. Melarsoprol reacts with trypanothione to form a complex with the sulfhydryl groups of cysteine and inhibits the enzyme trypanothione reductase. This mechanism, is a proposed mechanism for the action of melapsoprol. In recent years, the clinical importance of melarsoprol has decreased due to the developments in the treatment of trypanosomiasis, especially the treatment success of the nifurtimox eflornithine combination, and the oral use of fexinidazole in tablet form. In order to reduce the toxic effect of melarsoprol, cyclodextrin complex was developed and approved in some countries and entered into treatment [53, 54].

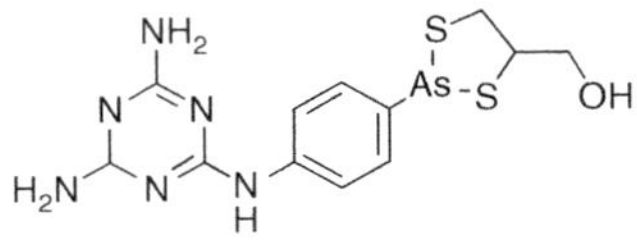

Melarsoprol

Tryparsamide ([4-[(2-amino-2-oxoethyl) amino] phenyl]-hydroxyarsinate) is an arsenic-containing compound used in the treatment of trypanosomiasis. This compound is ineffective in the early stages of the disease. Significant resistance to the compound has developed. It is used in the treatment of meningoencephalitis symptoms caused by the disease in the later stages of infection. It is used in combination with suramin. It has the side effects of organoarsenic derivative compounds [54].

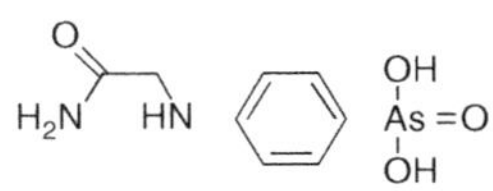

Tryparsamide

1.4.2.26 Diamidine Derivatives

Pentamidine isethionate, which is used in the treatment of leishmaniasis, is a compound that is effective against *Trypanosoma gambiense* and *Trypanosoma rhodesiense* in the early stages of the disease, and ineffective against *Trypanosoma cruzie*. It is a compound with a low therapeutic index and interferes with active transport in the parasite.

1.4.2.27 2-nitroimidazole Derivatives

Benznidazole (*N*-benzyl-2-(2-nitroimidazol-1-yl) acetamide) is the second compound approved after nifurtimox for the treatment of infections caused by *T. cruzi*. It is a 2-nitroimidazole derivative compound. The uptake mechanism is assumed to be by passive diffusion. Benznidazole is a prodrug and requires cellular activation to exert its effect. It is effective in acute and chronic phases of infection. Benznidazole activity emerges through electrophilic metabolites produced by the reduction of the nitro group to an amine group by free radicals and nitroreductases. It does not produce reactive oxygen species. The trypanocidal activity of benznidazole is hypothesized to arise by binding the reduced metabolites to the macromolecules of the parasite. It has also been determined that it causes oxidative stress in the parasite.

Fexinidazole (1-methyl-2-[(4-methylsulfanylphenoxy) methyl]-5-nitroimidazole), a 2-substituted 5-nitroimidazole derivative, was discovered as a result of new drug studies for the treatment of human African trypanomososiasis, due to the low efficacy, toxicity and administration problems of existing drugs. Its mechanism of action is similar to nifurtimox and benznidazole. It is used orally and acts in the acute and chronic stages of the disease [55–58].

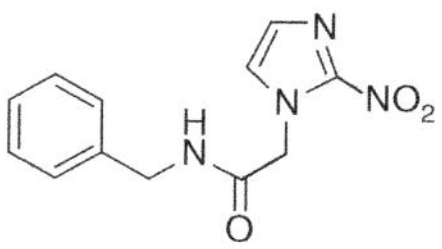

Benznidazole

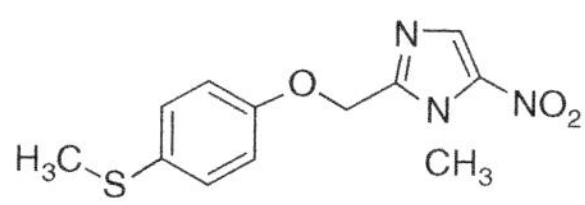

Fexinidazole

1.4.2.28 Amino Acid Derivatives

Eflornithine (2,5-diamino-2-(difluoromethyl)pentanoic acid) is an alpha-difluoroamino acid derivative compound, and is an inhibitor of the ornithine decarboxylase enzyme. Ornithine decarboxylase is the key enzyme in polyamine production. Polyamines are essential for DNA synthesis and cell proliferation in protozoa. In mammals, it is somewhat ineffective due to rapid turnover of the ornithine decarboxylase enzyme. It is used in the treatment of sleeping disease caused by *T. gambiense*. Its oral bioavailability is not good and high doses are required for treatment. It is also being investigated for cancer treatment as part of the drug repositioning strategy [54,56].

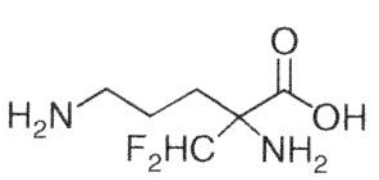

Eflornithine

Intestinal infection caused by *Giardia lamblia* is known as giardiasis or lambliasis. 4-aminoquinoline derivatives, nitroheterocyclic compounds, are used for the treatment of this disease. The most preferred compounds are metronidazole and quinacrine hydrochloride. Pneumocystosis, an infection known as interstitial plasma cell pneumonia, causes death in 50% of cases. Sulfomethoxazole-trimethoprim is used in the treatment. Although combinations of sulfadiazine and sulfapyrazine, tetracyclines, dapsone, pyrimethamine and spiramycin, which are sulfonamide derivatives, can be used in the treatment of toxoplasmosis caused by *Toxoplasma gondii*, pyrimethamine-sulfonamide combination and spiramycin is preferred in pregnant women. *Toxoplasma* infections are the most common parasitic diseases and one-third of all parasitic infections are caused by *T. gondii*. *Toxoplasma* infection in humans is transmitted by eating food contaminated with feces of an infected cat, by eating undercooked food contaminated with toxoplasma cysts, and by transmission from mother to fetus. Immunocompromized individuals are most susceptible to *Toxoplasma* infections [2,59].

1.4.2.29 Antihelmintic Drugs

Helminths are parasites that affect millions of people worldwide. The ineffectiveness of the drugs in their treatment, the possibility of developing resistance to these drugs, and the contraindication of existing drugs in some populations, limit the success of anthelmintic therapy. Anthelmintic drugs are drugs used against parasites called helminths that settle in the gastrointestinal tract lumen, blood and lymphatic vessels and some tissues. They show their effects by decomposing the helminth or expelling it from the infected patient's body. Zoologically, helminths are subdivided into helminths, nematodes, cestodes, trematodes, and filariae. The anthelmintic drugs available in the treatment act by paralyzing the parasite, by damaging the worm, causing it to be eliminated by the host immune system, or by altering the parasite's metabolism. Since the metabolic requirements of helminths differ from species to species, drugs that are effective on one type of helminth may be ineffective against other species. Anthelmintic treatment is carried out with the strategy of diagnosis and treatment of patients and administering a single dose of preventive treatment to individuals who are not diagnosed with the disease. Plant extracts were the first anthelmintic medicine, and *Aspidium* is still used. With the discovery of the anthelmintic activity of the piperazine ring, diethylcarbamazine citrate was developed. Later, pyrivinium pamoate, bephenium hydroxynaphthoate, niclosamide, metrifonate, thiabendazole, niridazole, pyrantel pamoate and pyraziquantel were developed and introduced as treatments.

Antihelminthic drugs used in treatment can be classified as follows according to their chemical structures:

1) chlorinated hydrocarbons;
2) phenols and derivatives;
3) cyanine dyes;
4) piperazine derivatives;
5) antimalarial compounds;
6) various heterocyclic compounds:
    i) benzimidazole derivatives;
    ii) imidazothiazole derivatives;
    iii) vinylpyrimidines;
    iv) halogenated salicylanilide derivatives;
    v) quaternary ammonium compounds;
    vi) quinoline derivatives;
    vii) heterocyclic compounds bearing nitro group; and,
    viii) organophosphate compounds;
7) antimony compounds; and
8) natural compounds.

1.4.2.30 Chlorinated Hydrocarbons

Tetrachloroethylene (1,1,2,2-tetrachloroethene), also known as perchlorethylene, is a chlorinated hydrocarbon compound with anthelmintic activity that acts by paralyzing the parasite. The helminth separated from the intestinal wall is excreted with feces. It is used in the treatment of helminthic infections caused by hookworms [10].

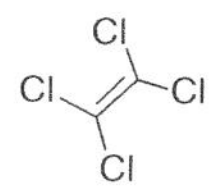

Tetrachloroethylene

1.4.2.31 Phenols and Derivatives

Dichlorophen (4-chloro-2-[(5-chloro-2-hydroxyphenyl) methyl] phenol), bithionol (2,4-dichloro-6-(3,5-dichloro-2-hydroxyphenyl) sulfanylphenol) and hexylresorcinol (4-hexylbenzene-1,3-diol) are anthelmintic compounds with a phenol structure.

Dichlorophen was used before praziquantel and niclosamide in the treatment of infections caused by *Taenia saginata*. In addition to its anthelmintic activity, it also has antifungal and antibacterial activity, and due to these properties, it is included in the structure of soaps and cosmetics. It shows its effect by proteolysis and digestion of the helminth.

Gastrointestinal side effects such as nausea, vomiting and diarrhea occur in up to one-third of patients following ingestion of a single oral dose. Side effects such as urticaria, contact dermatitis and photosensitivity may also occur.

Bithionol is a compound used as an anthelmintic in the treatment of infections caused by *Fasciola hepatica*. Its effectiveness in the treatment of helminthic infections seen as co-infection in HIV patients has also been investigated [60].

Hexylresorcinol is used in the treatment of hookworm infections. It also has an antibacterial effect [10].

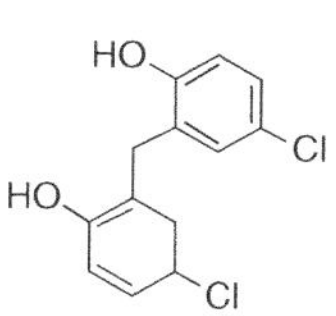

Dichlorophen

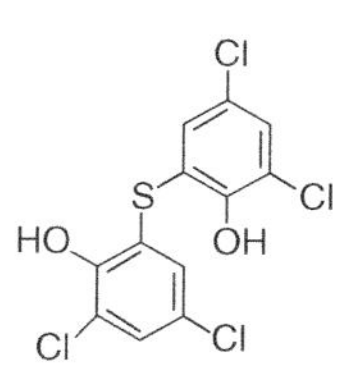

Bithionol

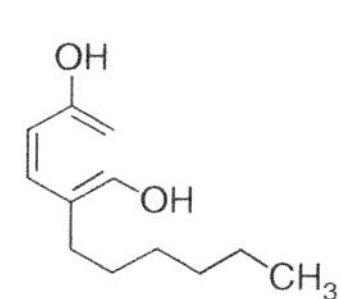

Hexylresocinol

1.4.2.32 Cyanine Dyes

Although pyrvinium pamoate (2-[(*E*)-2-(2,5-dimethyl-1-phenylpyrrol-3-yl) ethenyl]-*N*,*N*,1-trimethylquinolin-1-ium-6-amine) has been used for the treatment of pinworm infections, especially in children, its low bioavailability has made it the second choice after pyrantel pamoate and mebendazole due to poor absorption from the gastrointestinal tract [61].

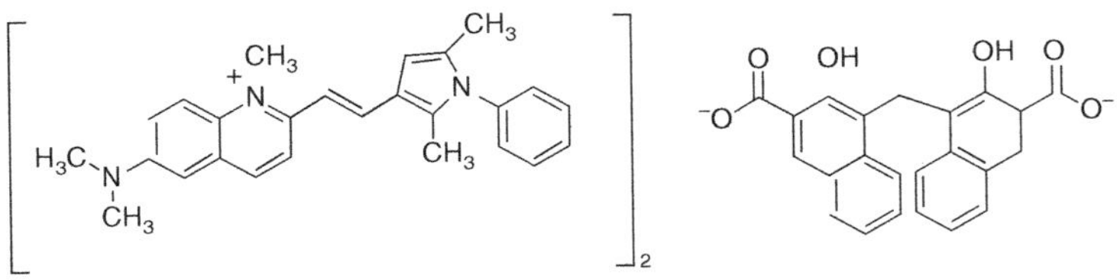

Pryvinium pamoat

1.4.2.33 Piperazine Derivatives

Piperazine, which is used in the treatment of *Ascaris lumbricoides* and *Enterobius vermicularis* infections, polarizes the muscle membrane by disrupting the ion channels at the neuromuscular junction, and reduces the contractile effect of acetylcholine by blocking the impulse transmission. As a result of this effect, the paralyzed helminth is expelled from the intestine. Although it is an alternative medicine in ascariasis, it has lost its importance due to other drugs used in enterobiasis [10].

The mechanism of action of diethylcarbamazine citrate (*N*,*N*-diethyl-4-methyl piperazine-1-carboxamide;2-hydroxypropane-1,2,3-tricarboxylic acid), which has been used as an anthelmintic for over 40 years, is not fully understood. It is an inactive drug *in vitro* and used in filaria infections. It is known that by reducing muscle activity, it provides immobilization of the helminth. Changes are also seen in the surface membrane of the helminth. It increases the adhesion of helminth granulocytes [62].

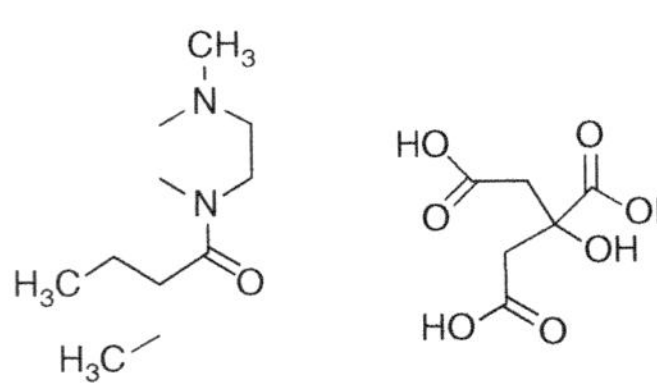

Diethylcarbamazine citrate

1.4.2.34 Antimalarial Compounds

Although mepacrin, which is used in the treatment of malaria, is used in the treatment of helminthic infections caused by *Taenia saginata* and *Taenia solium*, it has lost its importance in anthelmintic therapy with the introduction of niclozamide [10,63].

1.4.2.35 Various Heterocyclic Compounds

Different heterocyclic compounds are also used for the treatment and management of parasites. In the current topic we are discussing important heterocyclic compounds.

1.4.2.36 Benzimidazole Derivatives

Thiabendazole (4-(1*H*-benzimidazol-2-yl)-1,3-thiazole), mebendazole (methyl *N*-(6-benzoyl-1*H*-benzimidazol-2-yl) carbamate) and albendazole (methyl *N*-(6-propylsulfanyl-1*H*-benzimidazol-2-yl)carbamate) are benzimidazole derivative compounds used in the treatment of infections caused by gastrointestinal helminths. Initially, benzimidazole derivatives were thought to block several enzymes, including helminth fumarate reductase and malate dehydrogenase enzymes. It was later found that benzimidazoles bind to the beta-tubulin structure of the parasite with high affinity and inhibit microtubule polymerization. As a result, the cell structure deteriorates and the parasite dies. It has been observed that benzimidazoles used as anthelmintics selectively bind to the microtubules of the parasite and do not harm the host [10, 64].

Thiabendazole is first benzimidazole derivative and is relatively non-toxic. It entered treatment in 1961. It acts by inhibiting the parasite's fumarate reductase and malate dehydrogenase enzymes. As a result, the release of acetylcholinesterase is blocked. It is a broad-spectrum compound and is ineffective only against filaria type nematodes [10, 64].

Mebendazole, a benzimidazole derivative, is a safe and broad-spectrum anthelmintic that has been used for many years for nematode infections. Its mechanism of action is that it binds to nematode tubulin and inhibits intestinal microtubule polymerization. Inhibition of microtubule synthesis prevents glucose absorption in the parasite. This mechanism results in the elimination of the parasite from the body. The dissociation rate from the microtubule of the parasite is lower than that of the human microtubule [65].

Albendazole, like other benzimidazole derivative anthelmintic compounds, is an oral drug for the treatment of intra- and extraluminal parasitic infections. Although its absorption from the gastrointestinal tract is weak, absorption can be increased with fatty foods. The compound, which is also used in the treatment of hydatid cyst, is also used in *Strongyloides stercoralis* and tapeworm infections [10, 66].

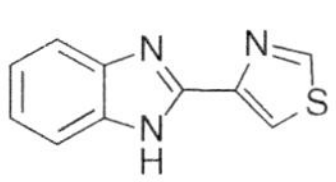

Thiabendazole

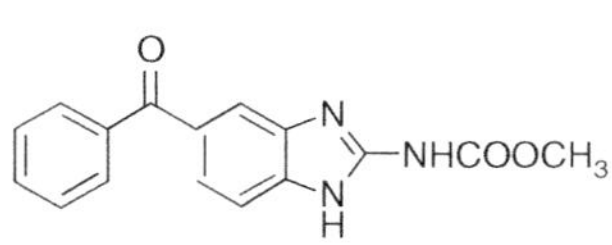

Mebendazole

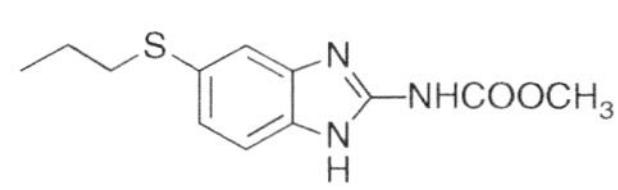

Albendazole

1.4.2.37 Imidazothiazole Derivatives

Imidazothiazoles, which are nicotinic acetylcholine receptor antagonists, are compounds that allow the elimination of the parasite by paralyzing it. The first anthelmintic compound discovered from this group was tetramisole. It has also been used in the treatment of rheumatoid arthritis.

Tetramisole (6-phenyl-2,3,5,6-tetrahydroimidazo[2,1-b][1,3]thiazole) is a compound in raceme form. The L isomer of the compound is levamisole ((6*S*)-6-phenyl-2,3,5,6-tetrahydroimidazo[2,1-b][1,3]thiazole). The activity of the L isomer is higher than the D isomer, and the L isomer is less toxic. It is a broad spectrum anthelmintic. It has been used successfully in the treatment of ascarid infections. It has also been used in the treatment of rheumatoid arthritis [67–70].

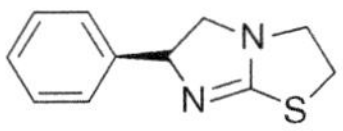

Levamisole

1.4.2.38 Vinylpyrimidines

Pyrantel pamoate is a broad spectrum antinematodal anthelmintic compound with a vinyl pyrimidine structure. It causes paralysis by depolarizing at the neuromuscular junction of the nematode. These effects are the opposite of the effects of piperazine and these two anthelmintic drugs should not be used together. It is especially used in pinworm infections such as *Enterobius vermicularis* and *Ascaris lumbricoides*. It has low hepatotoxicity due to poor absorption [71].

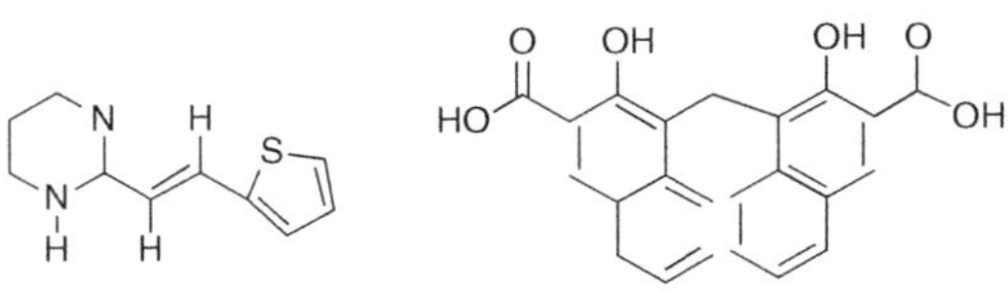

Pyrantel pamoate

1.4.2.39 Halogenated Salicylanilide Derivatives

Niclosamide (5-chloro-*N*-(2-chloro-4-nitrophenyl)-2-hydroxybenzamide) was the first commercialized compound to act as an anticestode. The mechanism of action of all salicylanilide derivatives and niclosamide is also because they are hydrogen ionophores, displacing their protons across the inner membrane of the mitochondria. Thus, it inhibits ATP production, causing oxidative phosphorylation to be separated from electron transport. This process is observed in isolated mitochondria of the helminth and the host. It is used in parasitic infections caused by taenia species, *Dipylidium caninum* and *Joyeuxiella pasqualei*. It is also used as molluscite [72].

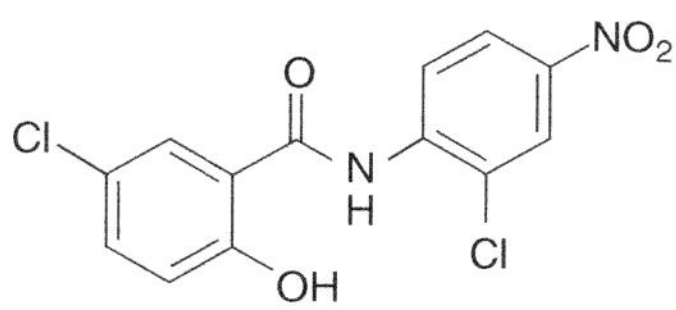

Niclosamide

1.4.2.40 Quaternary Ammonium Compounds

Bephenium hydroxynaphtoate (benzyl-dimethyl-(2-phenoxyethyl)azanium;3-carboxynaphthalen-2-olate) is used in the treatment of infections caused by *Ascaris lumbricoides*, *Ancylostoma duodenale* and *Necator americanus*. Since the salt form is less soluble in water, it is not absorbed from the gastrointestinal tract and passes into the intestine. Nowadays, its use has been superseded due to the introduction of new compounds with better pharmacokinetic properties [10].

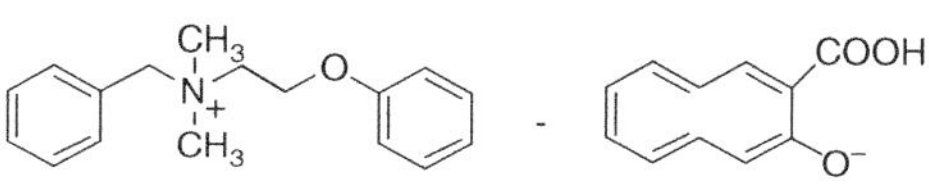

Bephenium hydroxynaphtoate

1.4.2.41 Quinoline Derivatives

Praziquantel (2-(cyclohexanecarbonyl)-3,6,7,11*b*-tetrahydro-1*H*-pyrazino[2,1-a] isoquinolin-4-one) is a medicine used to treat infections caused by *Schistosoma* species. Although the exact mechanism of action is not known, it is known to antagonize voltage-gated calcium channels, although it was thought to impair calcium homeostasis in the past. The uncontrolled flow of calcium ions as a result of antagonism of these channels causes uncontrolled muscle contraction and paralysis of the parasite. In addition, experimental studies have shown that myosin regulatory light chains and uptake of adenosine are important in the drug's mechanism of action. The R isomer has been shown to be anthelmintic, therefore, it is the R isomer that is used in therapy [73].

Oxamniquine ([7-nitro-2-[(propan-2-ylamino) methyl]-1,2,3,4-tetrahydroquinolin-6-yl] methanol), the hydroxylated metabolite of 2-aminotetrahydroquinoline, is used in the treatment of infections caused by *Schistosoma mansoni*. It has been superseded by the introduction of praziquantel into treatment. Oxamniquine is a prodrug that is sulfonated by endogenous sulfotransferases in the presence of 3′-phosphoadenosine-5′-phosphosulfate (PAPS). The resulting sulfate ester is unstable and alkylates macromolecules by forming an electrophilic intermediate. Thus, the cellular functions of the parasite are disrupted and this mechanism results in parasite death [74].

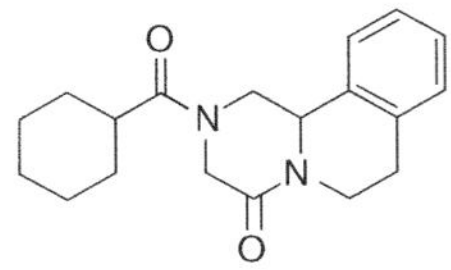

Praziquantel

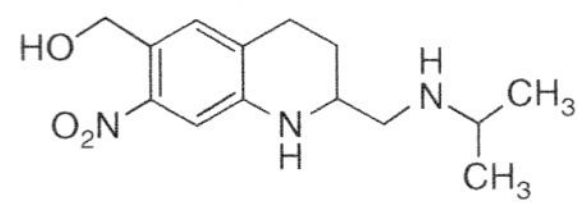

Oxamniquine

1.4.2.42 Heterocyclic Compounds Bearing Nitro Group

Niridazole (1-(5-nitro-1,3-thiazol-2-yl) imidazolidin-2-one) is used in the treatment of infections caused by the *Schistosoma* species *S. haematobium*, *S. mansoni*, and *S. japonicum*, and *Dracunculus medinensis*. Niridazole shows its effect by binding to the macromolecules of the parasite with the reactive intermediates formed by the nitroreductase. The 5-nitro group is essential for anthelmintic activity [75].

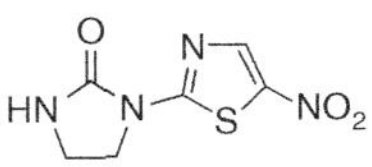

Niridazole

1.4.2.43 Organophosphate Compounds

Metrifonate (2,2,2-trichloro-1-dimethoxyphosphorylethanol) is an organophosphorus cholinesterase inhibitor. It has been used to treat human schistosomiasis. It has no place in treatment protocols today due to its low therapeutic efficacy and the fact that it requires repeated doses. It is still used in veterinary medicine. It is being researched for the treatment of Alzheimer's disease [76, 77].

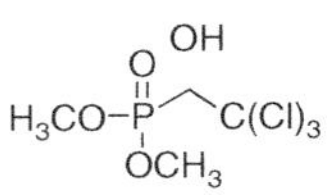

Metrifonate

1.4.2.44 Antimony Compounds

One of the oldest known anti-schistosome drugs is antimony potassium tartrate (dipotassium; antimony(3+);2,3-dioxidobutanedioate). This compound is active against *S. mansoni* and *S. haematobium* species. Antimony potassium tartrate shows its effect by paralyzing the parasite. It also prevents the parasite from laying eggs. Due to its high toxicity, side effects and application difficulties, it is not currently used in the treatment of schistosomiasis [77].

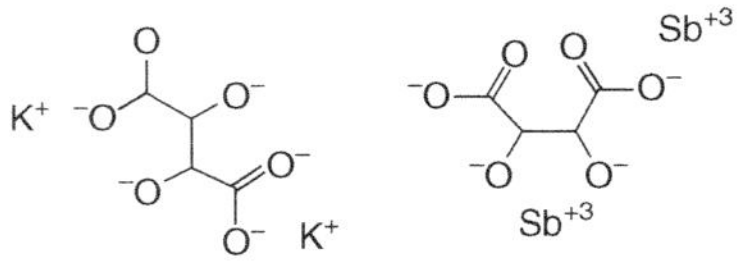

Antimony potassium tartrate

Stibophen (pentasodium;2-(2-oxido-3,5-disulfonatophenoxy)-1,3,2-benzodioxastibole-4,6-disulfonate) shows activity by affecting the phosphofructokinase activity of the parasite. Stibophen has a much higher affinity for parasite enzymes than for host liver enzymes. Despite its high toxicity and especially cardiovascular side effects, it is included in schistosomiasis treatment protocols [78].

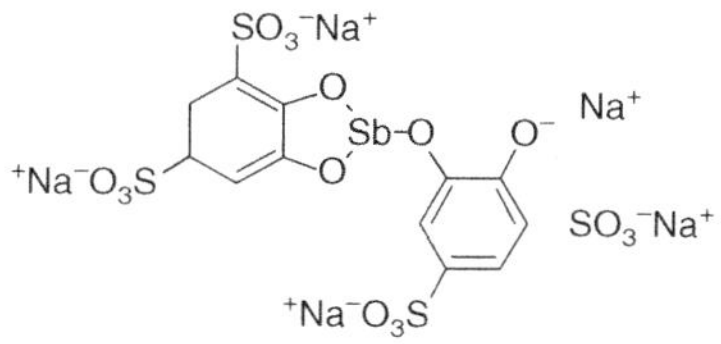

Stibophen

1.4.2.45 Natural Compounds

Discovered in Japan in the 1970s, ivermectin ((1*R*,4*S*,5'*S*,6*R*,6'*R*,8*R*,10*E*,12*S*,13*S*,14*E*,16*E*,20*R*,21*R*,24*S*)-6'-[(2*S*)-butan-2-yl]-21,24-dihydroxy-12-[(2*R*,4*S*,5*S*,6*S*)-5-[(2*S*,4*S*,5*S*,6*S*)-5-hydroxy-4-methoxy-6-methyloxan-2-yl]oxy-4-methoxy-6-methyloxan-2-yl]oxy-5',11,13,22-tetramethylspiro[3,7,19-trioxatetracyclo[$15.6.1.1^{4,8}.0^{20,24}$]pentacosa-10,14,16,22-tetraene-6,2'-oxane]-2-one), is a new antiparasitic drug. Ivermectin is a 16-membered macrocyclic lactone derived from its precursor, avermectin, from *Streptomyces avermitilis*. Ivermectin is obtained by saturating the double bond between 22–23 in avermectin. It has a broad spectrum and is effective against gastrointestinal roundworms, lungworms, mites, lice and hornflies. It is used for the oral treatment of various internal nematode infections including onchocerciasis, strongyloidiasis, ascariasis, cutaneous larval migrans, filariasis, gnathostomiasis and trichuriasis, as well as ectoparasitic infections such as lice and scabies. Since it is an effective drug against *Onchocerca volvulus*, it has replaced diethylcarbamazine in the treatment of that parasite. In the treatment of onchocerciasis, a single dose is used orally for six months or once a year. Initially, it was believed that ivermectin blocked neurotransmitters acting on GABA-gated Cl- channels. GABA is considered the primary inhibitory neurotransmitter in the somatic neuromuscular system of nematodes. Later, ivermectin was found to target glutamate-gated Cl- channels. By this mechanism, while ivermectin affects nematodes, it cannot reach the central nervous system where GABA receptors are localized, as it cannot cross the blood–brain barrier and thus does not affect mammals [10,79,80]. In the current COVID-19 pandemic, a large number of drugs for treatment have been tried in the clinic as part of the drug repositioning strategy. One of these drugs is ivermectin. The therapeutic effect on COVID-19 infection is still under investigation. The general opinion is that ivermectin does not reduce either the risk of mortality or the need for mechanical ventilation in severe COVID-19 infection [24, 81–86].

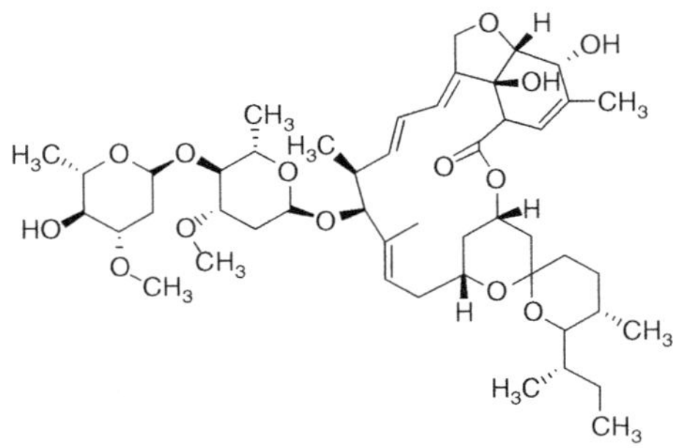

Ivermectin

Paromomycin ((2*S*,3*S*,4*R*,5*R*,6*R*)-5-amino-2-(aminomethyl)-6-[(2*R*,3*S*,4*R*,5*S*)-5-[(1*R*,2*R*,3*S*,5*R*,6*S*)-3,5-diamino-2-[(2*S*,3*R*,4*R*,5*S*,6*R*)-3-amino-4,5-dihydroxy-6-(hydroxymethyl)oxan-2-yl]oxy-6-hydroxycyclohexyl]oxy-4-hydroxy-2-(hydroxymethyl)oxolan-3-yl]oxyoxane-3,4-diol) is a macrolide antibiotic used in the treatment of giardiasis, amoebaiasis and cryptosporidiosis. It acts by interacting with ribosomal subunits [87].

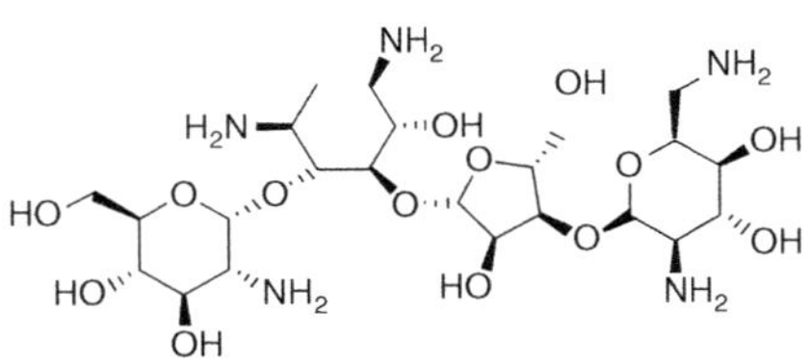

Paromomycin

1.4.2.46 Drugs Used against Ectoparasites

Ectoparasites are insects that settle on the skin or in the stratum corneum of the skin due to inadequate personal care. They cause dermatitis with itching in the area where they are located. Infections are caused mainly by scabies (*Sarcoptes scabiei*), the common bed bug (*Cimex lectularius*), fleas, and lice, including the body louse (*Pediculus humanis*), pubic louse (*Phthirius pubis*), and head louse (*Pediculus humanus capitis*). Ectoparasite infections are more common in underdeveloped countries. Sulfur preparations were used for the treatment of ectoparasitic infections until modern drugs became available. Then, compounds such as benzyl benzoate and chlorophenotan were synthesized and their insecticidal effects were investigated. Active compounds isolated from *Pyrethrum* led to the synthesis of pyrethrin analogs.

Drugs used against ectoparasites can be classified as:

chlorinated hydrocarbons;
pyrethrins and prethyroids;

sulfur compounds; and
other compounds.

1.4.2.47 Chlorinated Hydrocarbons

Having a hexachlorocyclohexane structure, lindane (1,2,3,4,5,6-hexachloro-cyclohexane) is one of the first-generation anthelmintic compounds that inhibits insect and GABA A receptors. Thus, it inhibits the flux of chlorine channels. Lindane is used as a model compound for insecticide activity determination. Shampoo and lotion lindane formulations are used to treat scabies and lice. It is toxic to the central nervous system, reproductive system, cardiovascular system, kidney and liver. It is not a first line therapy drug due to its toxicity [88, 89]. Chlordane (1,3,4,7,8,9,10,10-octachlorotricyclo [5.2.1.$0^{2,6}$]) dec-8-ene, like other chlorinated hydrocarbons, is a highly toxic compound. It is a broad-spectrum insecticide. It disrupts the hormone balance by causing endocrine damage [90, 91].

Clofenotane, which has 1-chloro-4-[2,2,2-trichloro-1-(4-chlorophenyl) ethyl] benzene structure, has been used in insect control since the 1940s. Due to its widespread use, it has caused significant environmental pollution. Its use is banned in most countries today. It acts by blocking voltage-gated sodium channels [92].

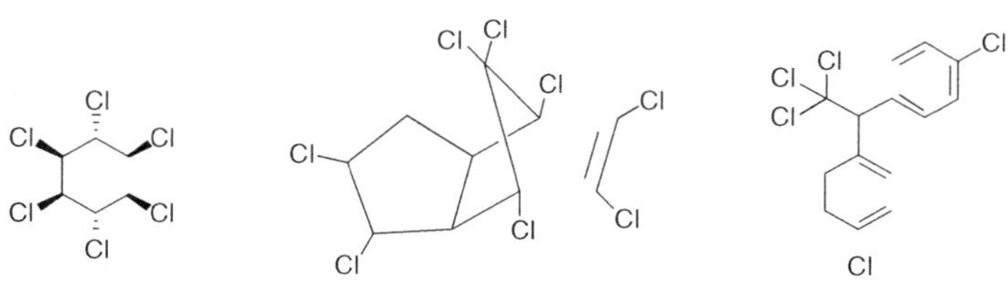

Lindane Chlordane Clofenotane

1.4.2.48 Pyrethrins and Prethyroids

Pyretrum is a natural insecticide obtained from the plant *Chrysanthemum cinerariaefolium*. It shows structural similarity with pyrethroid insecticides and pyrethrin I, one of the active components of pyrethrum. Chemically, this group of compounds is called pyrethrins because they are a mixture of esters (pyrethrin I ([(1*S*)-2-methyl-4-oxo-3-[(2*Z*)-penta-2,4-dienyl]cyclopent-2-en-1-yl] (1*R*,3*R*)-2,2-dimethyl-3-(2-methylprop-1-enyl)cyclopropane-1-carboxylate)) and II ([(1*S*)-2-methyl-4-oxo-3-[(2*Z*)-penta-2,4-dienyl]cyclopent-2-en-1-yl](1*R*,3*R*)-3-[(*E*)-3-methoxy-2-methyl-3-oxoprop-1-enyl]-2,2-dimethylcyclopropane-1-carboxylate) and (cinerin I ([(1*S*)-3-[(*Z*)-but-2-enyl]-2-methyl-4-oxocyclopent-2-en-1-yl] (1*R*,3*R*)-2,2-dimethyl-3-(2-methylprop-1-enyl)cyclopropane-1-carboxylate)) and II([(1*S*)-3-[(*Z*)-but-2-enyl]-2-methyl-4-oxocyclopent-2-en-1-yl](1*R*,3*R*)-3-[(*E*)-3-methoxy-2-methyl-3-oxoprop-1-enyl]-2,2-dimethylcyclopropane-1-carboxylate) of *Chrysanthemum cinerariaefolium*, the plant from which they are obtained. Synthetic pyrethroids are more stable than natural pyrethroids. Synthetic derivatives are quite diverse and toxic to the insect species. The high fat solubility of pyrethrins makes them selective in the insect. The first synthetic insecticides in the prethyroid structure were developed in the 1950s. Pyrethroids affect the insect's nervous system by interacting with voltage-gated chloride, calcium and sodium channels. Since the effect of pyrethrins starts quickly and lasts for a short time, it is combined with piperonyl butoxide to create a synergistic effect. In this combination, piperonyl butoxide delays the oxidative inactivation of pyrethrins by inhibiting cytochrome P450 monooxygenases that metabolize pyrethrins in the insect [93–95].

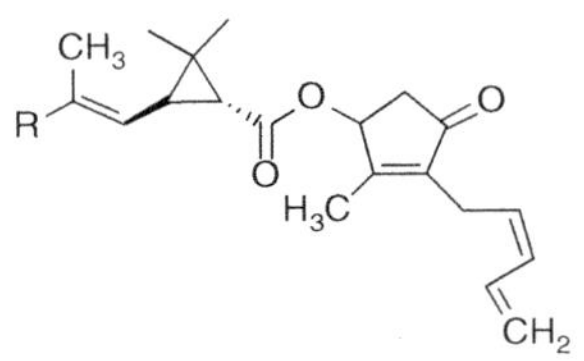

Permethrin I:R = CH_3
Permethrin II:R = $COOCH_3$

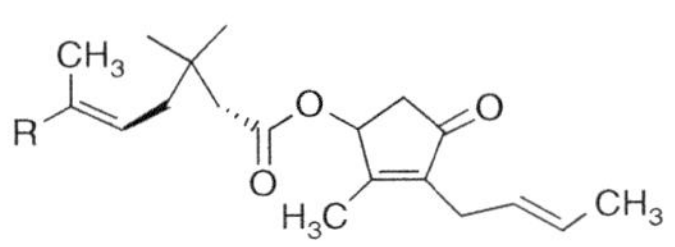

Cinerin I:R = CH_3
Cinerin II:R $-COOCH_3$

Permethrin ((3-phenoxyphenyl) methyl 3-(2,2-dichloroethenyl)-2,2-dimethylcyclopropane-1-carboxylate) is an FDA-approved compound for the treatment of lice and scabies, as a pet shampoo, to treat wooden furniture, and on the surface of clothing. Since it is a synthetic compound with high stability, it cannot be used together with a synergist. Its insecticidal effect is as much as natural compounds, its toxicity to mammals is low [10, 84, 85, 96–98].

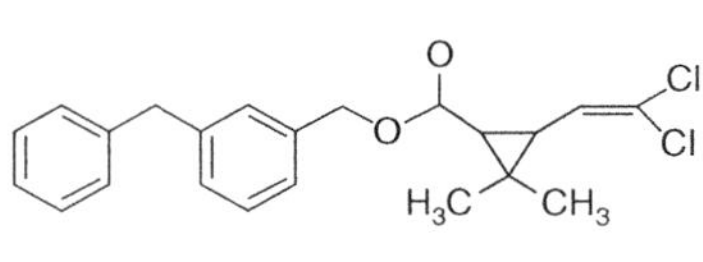

Permethrin

1.4.2.49 Sulfur Compounds

Sulfur is the oldest compound used in the treatment of scabies. It is used in ointment form, its effect is lower than other drugs, and it can be used on the whole body. It has low toxicity [99, 100].

1.4.2.50 Other Compounds

Benzyl benzoate is one of the oldest drugs used as a scabicide. It kills the parasite by acting on its nervous system. It has the advantages of being odorless and not irritating to the skin. Benzyl benzoate is also used in permethrin resistance infections [10, 101, 102].

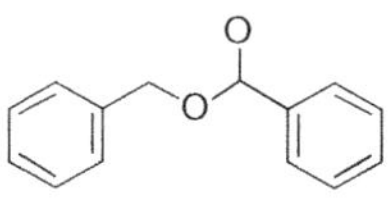

Benzyl benzoate

Malathion (diethyl 2-dimethoxyphosphinothioylsulfanylbutanedioate) is an organophosphate insecticide and shows its effect by converting its active metabolite malaoxone. It irreversibly inhibits acetylcholine esterase. It is safe because it does not show serious systemic adverse reactions in topical use. On topical application, it is converted to inactive rather than active metabolites. It is used in the treatment of head and pubic lice [84].

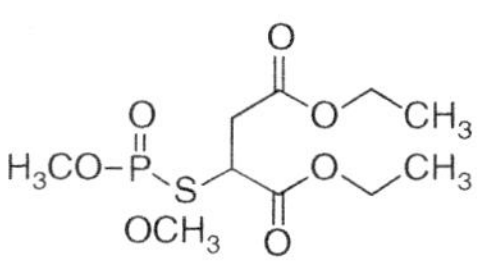

Malathion

Crotamiton ((*E*)-*N*-ethyl-*N*-(2-methylphenyl) but-2-enamide) has been used topically for many years to treat lice and relieve itching caused by lice. It can also be used in pediatric patients due to its low toxicity, but it must be applied multiple times for treatment success [85,103].

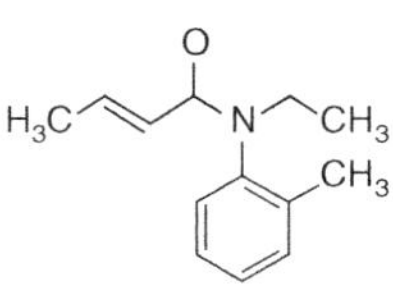

Crotamiton

Spinosad is the chemical compound formed by the fermentation of *Saccharopolyspora spinosa*, a naturally occurring soil actinobacteria. It is a macrolide compound consisting of a mixture of spinosyn A and D. In spinosad, Spinosyn A is the major component and Spinosyn D is the minor component. it is approved by the FDA for the topical treatment of head lice. It acts by excitation in the insect's nervous system and disrupting the function of nicotinic and GABA ion channels [10, 104, 105].

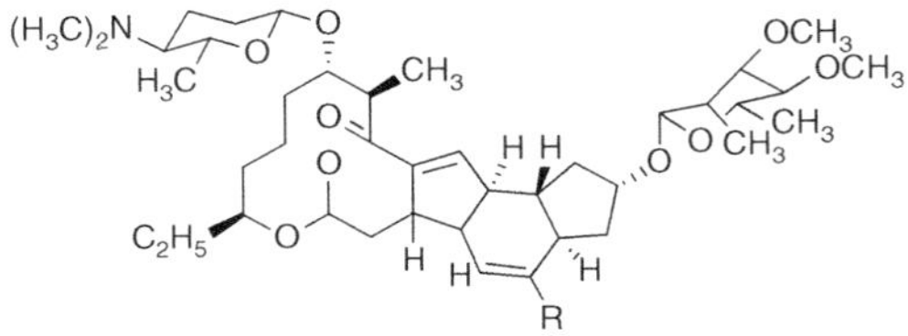

Spinosyn A : R = H
Spinosyn D : R = CH_3

Abametapir is a pediculicide drug with 5-methyl-2-(5-methylpyridin-2-yl) pyridine structure. It shows its effect by chelating with metals such as iron, copper and zinc. These metals are essential for the vital functions of the insect. It is used in cases where other resistant drugs such as permethrin are not successful in treatment [106, 107].

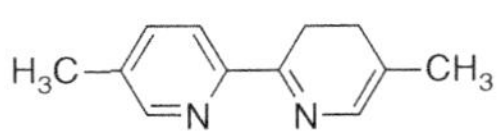

Abametapir

1.5 Conclusions

Parasites are unicellular or multicellular organisms that have the potential to cause disease in the host and have been present throughout the existence of humanity. Chronic illnesses, which lower the quality of life for the host, and even death can result from parasitic infections. In the past, parasite-related infections have been treated with herbal remedies. Later, medications that had active ingredients that were chemically manufactured entered the therapy. Research on parasites is now being forward in two primary directions: medication development and understanding how parasites interact with their hosts to cause disease. As our understanding of parasites and their relationship to disease advances this will enable the development of medications with more logical approaches and to control parasitic infections.

1.5.1 Future Prospects

Chronic diseases and infections are phenomena that threaten human life and cause economic and physical losses all over the world. Comprehensive studies have not been conducted on the development of antiparasitic drugs, and the number of drugs offered for treatment is limited, as is research on existing drugs. Some of the drug research and development studies carried out aim to discover new molecules as drug candidates for the treatment of these diseases. Due to the resistance to the drugs used in the treatment of infections, the difficulties experienced in getting the drug to some parts of the world and the developing technology, the knowledge gained at the molecular level has increased the development of new drug candidate molecules. Drugs for parasitic infections also have an important place in these studies. In the future, it can be said that there will be important developments in the treatment of parasitic infections and the prevention of diseases caused by parasites. In this context, studies on compounds with optimized pharmacokinetic properties, which are more effective, and whose side effects are minimized, are continuing. In addition, due to the similarity between cancer and parasitic diseases, antiparasitic drugs are thought to be a new starting point in cancer treatment within the scope of drug repositioning strategy. In this context, trials are underway for tumor treatment of some antiparasitic drugs (eg macrolides, benzimidazole derivatives and artemisinin). The drug repositioning strategy, which is a popular approach today, is also used to develop antiparasitic drugs. According to this strategy, compounds with various mechanisms of action, such as kinase inhibitors or antirheumatic drugs, are being tested for their antiparasitic activities in vitro.

References

1 Muhaj, F.F., George, S.J., Nguyen, C.D., and Tyring, S.K. (2022). Antimicrobials and resistance part II: antifungals, antivirals, and antiparasitics. *Journal of the American Academy of Dermatology* 86 (6): 1207–1226.

2 Carnero, P.R., Mateo, P.H., Martin-Garre, S. et al. (2017). Unexpected hosts: imaging parasitic diseases. *Insights into Imaging* 8: 101–125.

3 Elsheikha, H.M. (2014). The future of parasitology: challenges and opportunities. *Frontiers in Veterinary Science* 1: 1–3.

4 Dziduch, K., Greniuk, D., and Wujec, M. (2022). The current directions of searching for antiparasitic drugs. *Molecules* 27 (5): 1534–1558.

5 Kolodziej-Sobocinska, M. (2019). Factors affecting the spread of parasites in populations of wild European terrestrial mammals. *Mammal Research* 64: 301–318.

6 Khater, H.F. (2017). *Natural Remedies in the Fight against Parasites*. IntechOpen.

7 Dimijian, G.G. (2000). Evolving together: the biology of symbiosis, part 1. *Proceedings (Baylor University Medical Center)* 13 (3): 217–226.

8 Solomon, N.U., James, I.M., Alphonsus, N.O.O., and Nkiruka, R.U. (2015). A review of host-parasite relationships. *Annual Research & Review in Biology* 5 (5): 372–384.

9 Hahn, M.A. and Dheilly, N.M. (2016). Experimental models to study the role of microbes in host-parasite interactions. *Frontiers in Microbiology* 7: 1300.

10 Akgün, H., Aytemir, M, Balkan, A. et al. (2021). *Farmasötik ve Medisinal Kimya*. Hacettepe Üniversitesi Yayınları.

11 Allen, J.A. and Sutherland, T.E. (2014). Host protective roles of type 2 immunity: parasite killing and tissue repair, flip sides of the coin. *Seminars in Immunology* 26 (4): 329–340.

12 Hotez, P.J., Brindley, P.J., Bethony, J.M. et al. (2008). Helminth infections: the great neglected tropical diseases. *Journal of Clinical Investigation* 118 (4): 1311–1321.

13 McNair, C.M. (2015). Ectoparasites of medical and veterinary importance: drug resistance and the need for alternative control methods. *Journal of Pharmacy and Pharmacology* 67 (3): 351–363.

14 Wood, C.L. and Johnson, P.T.J. (2015). A world without parasites: exploring the hidden ecology of infection. *Frontiers in Ecology and the Environment* 13 (8): 425–434.

15 Cox, F.E.G. (2002). History of human parasitology. *Clinical Microbiology Reviews* 15 (4): 595–612.

16 Whitehead, B., Christiansen, S., Ostergaard, L., and Nejsum, P. (2022). Helminths and COVID-9 susceptibility, disease progresion, and vaccination efficacy. *Trends in Parasitology* 38 (4): 277–279.

17 Antushevich, H. (2020). Interplays between inflammasomes and viruses, bacteria (pathogenic and probiotic), yeasts and parasites. *Immunology Letters* 228: 1–14.

18 Melo, G.D., Silva, J.E.S., Grano, F.G. et al. (2015). *Leishmania* infection and neuroinflammation: specific chemokine profile and absence of parasites in the brain of naturally-infected dogs. *Journal of Neuroimmun* 289: 21–29.

19 Wink, M. (2012). Medicinal plants: a source of anti-parasitic secondary metabolites. *Molecules* 17 (11): 12771–12791.

20 Faheem, D.S., Johri, S., Abirami, M. et al. (2022). Search for structurally diverse heterocyclic analogs as dual-acting antimalarial and antileishmanial agents: an overview. *European Journal of Medicinal Chemistry Reports* 4: 100031.

21 Parhizgar, A.R. and Tahghighi, A. (2017). Introducing new antimalarial analogues of chloroquine and amodiaquine: a narrative review. *Iranian Journal of Medical Sciences* 42 (2): 115–128.

22 Sullivan, D.J. (2011). Cinchona alkaloids: quinine and quinidine. In: *Treatment and Prevention of Malaria* (eds H.M. Staines and S. Krishna). Springer.

23 Unlu, B., Simsek, R., Kose, S.B.E. et al. (2022). Neurological effects of SARS-CoV-2 and neurotoxicity of antiviral drugs against COVID-19. *Mini-Reviews in Medicinal Chemistry* 22 (2): 213–231.

24 Gluchowska, K., Dzieciatkowski, T., Sedzikowska, A. et al. (2021). The new status of parasitic diseases in the COVID-19 pandemic-risk factors or protective agents? *Journal of Clinical Medicine* 10 (11): 2533–2546.

25 Ashley, E.A., Recht, J., and White, N.J. (2014). Primaquine: the risks and the benefits. *Malaria Journal* 13: 418–424.

26 Chairat, K., Jittamala, P., Hanboonkunupakarn, B. et al. (2018). Enantiospecific pharmacokinetics and drug-drug interactions of primaquine and blood-stage antimalarial drugs. *Journal of Antimicrobial Chemotherapy* 73 (11): 3102–3113.

27 Duparc, S., Chalon, S., Miller, S. et al. (2020). Neurological and psychiatric safety of tafenoquine in plasmodium vivax relapse prevention: a review. *Malaria Journal* 19: 111.

28 White, N.J. (2014). *Manson's Tropical Infectious Diseases,* 23e (eds J. Farrar et al.). Elsevier.

29 Aronson, J.K. (2016). *Meyler's Side Effects of Drugs*, 16e (ed. J.K. Aronson). Elsevier.

30 Enna, S.J. and Bylund, D.B. (2009). *xPharm: The Comprehensive Pharmacology Reference* (eds S.J. Enna and D.B. Bylund). Elsevier.

31 Amusengeri, A., Tata, R.B., and Bishop, O.T. (2020). Understanding the pyrimethamine drug resistance mechanism via combined molecular dynamics and dynamic residue network analysis. *Molecules* 25: 904–927.

32 Chadwick, M., Trewin, H., Gawthrop, F., and Wagstaff, C. (2013). Sesquiterpenoids lactones: benefits to plants and people. *International Journal of Molecular Sciences* 14 (6): 12780–12805.

33 Krishna, S., Bustamante, L., Haynes, R.K., and Staines, H.M. (2008). Artemisinins: their growing importance in medicine. *Trends in Pharmacological Sciences* 29 (10): 520–527.

34 Ho, W.E., Peh, H.Y., Chan, T.K., and Wong, W.S.F. (2014). Artemisinins: pharmacological actions beyond anti-malarial. *Pharmacology&Therapeutics* 142 (1): 126–139.

35 Krishna, S., Uhlemann, A.C., and Haynes, R.K. (2004). Artemisinins: mechanisms of action and potential for resistance. *Drug Resistance Updates* 7 (4–5): 233–244.

36 Baggish, A.L. and Hill, D.R. (2002). Antiparasitic agent atovaquone. *Antimicrob. Agents Chemother.* 46 (5): 1163–1173.

37 Salles, J.M., Moraes, L.A., and Salles, M.C. (2003). Hepatic amoebiasis. *The Brazilian Journal of Infectious Diseases* 7 (2): 96–110.

38 Espinosa-Cantellano, M. and Martinez-Palomo, A. (2000). Pathogenesis of intestinal amoebiasis: from molecules to disease. *Clinical Microbiology Reviews* 13 (2): 318–331.

39 Aloisio, C. and Longhi, M. (2018). Diloxanide furoate binary complexes with -, methyl–, and hydroxypropyl–cyclodextrins: inclusion mode, characterization in solution and in solid state and in vitro dissolution studies. *Pharmaceutical Development and Technology* 23 (7): 723–731.

40 Pippi, B., Reginatto, P., Machado, G.M. et al. (2017). Evaluation of 8-hydroxyquinoline derivatives as hits for antifungal drug design. *Medical Mycology* 55 (7): 763–773.

41 Lee, M.R. (2008). Ipecacuanha: the South American vomiting root. *Journal of the Royal College of Physicians of Edinburgh* 38 (4): 355–360.

42 Jain, S., Chaudhary, P.N., and Gawade, V.B. (2012). Late stages in the biosynthesis of ipecac alkoloids-cephaeline, emetine and psychotrine in Alangium lamarckii Thw. (Alangiaceae). *Scientific World* 10 (10): 24–28.

43 Gupta, R., Sharma, S., Singh, R. et al. (2022). Functionalized nitroimidazole scaffold construction and their pharmaceutical applications: a 1950–2021 comprehensive overview. *Pharmaceuticals* 15: 561.

44 Lee, N., Bertholet, S., Debrabant, A. et al. (2002). Programmed cell death in the unicellular protozoan parasite Leishmania. *Cell Death & Differentiation* 9: 53–64.

45 Haldar, A.K., Sen, P., and Roy, S. (2011). Use of antimony in the treatment of leishmaniasis: current status and future directions. *Molecular Biology International* 2011: 571242.

46 Tunalı, V., Harman, M., Çavuş, İ. et al. (2022). Leishmaniasis'te Miltefosin Etkinliğinin *in vitro* Olarak Araştırılması. *Türkiye Parazitoloji Dergisi* 46 (2): 97–101.

47 Secor, W.E. (2012). Trichomonas vaginalis: treatment questions and challenges. *Expert Review of Anti-Infective Therapy* 10 (2): 107–109.

48 Thomas, J.A., Baker, N., Hutchinson, S. et al. (2018). Insights into antitrypanosomal drug mode-of-action from cytology-based profiling. *PLOS Neglected Tropical Diseases* 12 (11): e.0006980.

49 Alsford, S., Eckert, S., Baker, N. et al. (2012). High-throughtput decoding of antitrypanosomal drug efficacy and resistance. *Nature* 482: 232–236.

50 Falk, N., Berenstein, A.J., Moscatelli, G. et al. (2022). Effectiveness of nifurtimox in the treatment of Chagas disease: a long-term retrospective cohort study in children and adults. *Antimicrobial Chemotherapy* 66 (5): 1–10.

51 Boiani, M., Piacenza, L., Hernandez, P. et al. (2010). Mode of action of nifurtimox and N-oxide-containing heterocycles against Tyrpanosoma cruzi: is oxidative stress involved? *Biochem. Pharmacol* 79 (12): 1736–1745.

52 Wiedemar, N., Hauser, D.A., and Maser, P. (2020). 100 years of suramin. *Antimicrobial Agents and Chemotherapy* 64 (3): e01168–19.

53 Larson, S., Carter, K., and Hovel-Miner, G. (2021). Effects of trypanocidal drugs on DNA synthesis: new insights into melarsoprol growth inhibition. *Parasitology* 148 (10): 1143–1150.

54 De Koning, H.P. (2020). The drugs of sleeping sickness: their mechanisms of action and resistance, and a brief history. *Tropical Medicine and Infectious Disease* 5 (1): 14–36.

55 Torrele, E., Trunz, B.T., Tweats, D. et al. (2010). Fexinidazole-A new oral nitroimidazole drug candidate entering clinical development for the treatment of sleeping sickness. *PLOS Neglected Tropical Diseases* 4 (12): e923.

56 Franco, J., Marcelo, L.S., and Comini, M.A. (2018). Drugs and drug reistance in African and American trypanosomiasis. *Annual Reports in Medicinal Chemistry* 51: 97–133.

57 Kaiser, M., Bray, M.A., Cal, M. et al. (2011). Antitrypanosomal activity of fexinidazole, a new oral nitroimidazole drug candidate for treatment of sleeping sickness. *Antimicrobial Agents and Chemotherapy* 55 (12): 5602–5608.

58 Rajao, M.A., Furtado, C., Alves, C.L. et al. (2014). Unveiling benznidazole'e mechanism of action through overexpression of DNA repair proteins in *Trypanosoma cruzi*. *Environmental and Molecular Mutagenesis* 55: 309–321.

59 Silva, L.A., Fernandes, M.D., Machado, A.S. et al. (2019). Efficacy of sulfadiazine and pyrimetamine for treatment of experimental toxoplasmosis with strains obtained from human cases of congenital disease in Brazil. *Experimental Parasitology* 202: 7–14.

60 Walson, J.L. and John-Stewart, G. (2007). Treatment of helminth co-infection in individuals with HIV-1: a systematic review of the literature. *PLoS Neglected Tropical Diseases* 1 (3): e102.

61 Gunderson, E.L., Bryant, C., Bulman, C.A. et al. (2022). Pyrvinium pamoate and structural analogs are early macrofilaricide leads. *Pharmaceuticals* 15 (2): 189–205.

62 Howard, B. (2007). Diethylcarbamazine. In: *xPharm: The Comprehensive Pharmacology Reference* (eds S.J. Enna and D.B. Bylund). Elsevier.

63 Agube, A.A., Ajaghaku, D.L., and Uzochukwu, I.C. (2020). In silico and in vitro investigations of antihelmintic activities of selected approved drugs. *Archives of Community Medicine and Public Health* 6 (2): 261–269.

64 Abongwa, M., Martin, R.J., and Robertson, A.P. (2017). A brief review on the mode of action of antinematodal drugs. *Acta Veterinariat (Beogr)* 67 (2): 137–152.

65 Larsen, A.R., Bai, R., Chung, J.H. et al. (2015). Repurposing the antihelmintic mebendazole as a hedgehog inhibitor. *Molecular Cancer Therapeutics* 14 (1): 3–13.

66 Pawluk, S.A., Roels, C.A., Wilby, K.J., and Ensom, M.H.H. (2015). A review of pharmacokinetics drug-drug interactions with the anthelmintic medications albendazole and mebendazole. *Clinical Pharmacokinetics* 54: 371–383.

67 Shankerrao, S., Bodke, Y.D., and Santoshkumar, S. (2017). Synthesis and antimicrobial activity of some imidazothiazole derivatives of benzofuran. *The Arabian Journal of Chemistry* 10 (1): 554–558.

68 Lee, K.C., Ladizinski, B., and Federman, D.G. (2012). Complications associated with use of levamisole-contaminated cocaine: an emerging public health challenge. *Mayo Clinic Proceedings* 87 (6): 581–586.

69 Soudkolaei, A.S., Kalidari, G.A., and Borji, H. (2021). Anthelmentic efficacy of fenbendazole and levamisole in native fowl in northern Iran. *Parasites & Vectors* 14: 104–111.

70 Clarkson, M.J. and Beg, M.K. (1971). Critical tests of levamisole as an anthelmintic in the horse. *Annals of Tropical Medicine & Parasitology* 65 (1): 87–91.

71 Horton, J. (2003). Human gastrointestinal helminth infections: are they now neglected diseases? *Trends in Parasitology* 19 (1): 527–531.

72 Page, S.W. (2008). Antiparasitic drugs. In: *Small Animal Clinical Pharmacology*, 2e (eds J.E. Maddison, S.W. Page and D.B. Church). 198–260.

73 Thomas, C.M. and Timson, D.J. (2020). The mechanism of action of praziquantel: can new drugs exploit similar mechanisms? *Current Medicinal Chemistry* 27 (5): 676–696.

74 Taylor, A.B., Pica-Mattocica, L., Polcaro, C.M. et al. (2015). Structural and functional characterization of the enantiomers of the antischistosomal drug oxamniquine. *PLoS Neglected Tropical Diseases* 9 (10): 1–13.

75 Huang, H., Rigouin, C., and Williams, D.L. (2012). The redox biology of schistosome parasites and applications for drug development. *Current Pharmaceutical Design* 18 (24): 3595–3611.

76 Davis, A. and Bailey, D.R. (1969). Metrifonate in urinary schistosomiasis. *The Bulletin of the World Health Organization* 41: 209–241.

77 Trainor-Moss, S. and Mutapi, F. (2016). Schistosomiasis therapeutics: what's in the pipeline? *Expert Review of Clinical Pharmacology* 9 (2): 157–160.

78 Cours, J.R. and Conceicao, M.J. (2010). Specific schistosomiasis treatment as astrategy for disease control. *Memórias do Instituto Oswaldo Cruz* 105 (4): 598–603.

79 Crump, A. and Ivermectin, O.S. (2011). "Wonder drug" from Japan: the human use perspective. *Proceedings of the Japan Academy Ser B Physical and Biological Sciences* 87 (2): 13–28.

80 Reis, G., Silva, E.A.S.M., Silva, D.C.M. et al. (2022). Effect of early treatment with ivermectin among patients with covid-19. *The New England Journal of Medicine* 386: 1721–1731.

81 Lim, S.C.L., Hor, C.P., Tay, K.H. et al. (2022). Efficacy of ivermectin treatment on disease progression among adults with mild to moderate COVID-19 and comorbidities the I-TECH randomized clinical trial. *JAMA Internal Medicine* 182 (4): 426–435.

82 Marcolino, M.S., Meira, K.C., Guimaraes, N.S. et al. (2022). Systematic review and meta-analysis of ivermectin for treatment of COVID-19: evidence beyond the hype. *BMC Infectious Diseases* 22: 639–663.

83 Okumus, N., Demirturk, N., Cetinkaya, R.A. et al. (2021). Evaluation of the effectiveness and safety of adding ivermectin to treatment in severe COVID-19 patients. *BMC Infectious Diseases* 21 (1): 411–421.

84 Idriss, S. and Levitt, J. (2009). Malathion for head lice and scabies: treatment and safety considerations. *Journal of Drugs in Dermatology* 8 (8): 715–720.

85 Chandler, D.J. and Fuller, L.C. (2019). A review of scabies: an infestation more than skin deep. *Dermatology* 235: 79–90.

86 Wolday, D., Gebrecherkos, T., Arefaine, Z.G. et al. (2021). Effect of co-infection with intestinal parasites on COVID-19 severity: a prospective observational cohort study. *EClinical Medicine* 39: 101054.

87 Chawla, B., Jhingran, A., Panigrahi, A. et al. (2011). Paromomycin affects translation and vesicle-mediated trafficking as revealed by proteomics of paromomycin-susceptible-resistant *Leishmania donovani*. *PLoS One* 6 (10): e26660.

88 Islam, R. and Lynch, J.W. (2012). Mechanism of action of the insecticides, lindane and fipronil on glycine receptor chloride channels. *British Journal of Pharmacology* 165 (8): 2707–2720.

89 Piskac-Collier, A.L. and Smith, M.A. (2009). Lindane-induced generation of reactive oxygen species and depletion of glutathione do not result in necrosis in renal distal tubule cells. *Journal of Toxicology and Environmental Health, Part A* 72 (9): 1160–1167.

90 Pridmore, R.D., Thrush, S.F., Wilcock, R.J. et al. (1991). Effect of the organochlorine pesticide technical chlordane on the population structure of suspension and deposit feeding bivalves. *The Marine Ecology Progress Series* 76: 261–271.

91 Mendes, V., Ribeiro, C., Delgada, I. et al. (2021). The association between environmental exposures to chlordanes, adiposity and diabetes-related features: a systematic review and meta-analysis. *Scientific Reports* 11: 14546.

92 Pietrantoni, I., Rossi, M., Scarselli, M. et al. (2018). Dichlorodiphenyltrichloroethane, an old pesticide with a new mechanism of toxicity. *Current Topics in Pharmacology* 22: 69–76.

93 Tiwari, B., Khaewar, S., and Tiwari, D.N. (2018). Pesticides and rice agriculture. In: *Cyanobacteria* (eds A.K. Mishra, D.N. Tiwari and A.N. Rai), 303–325. Elsevier.

94 Gleave, K., Lissenden, N., Chaplin, M. et al. (2021). Piperonyl butoxide (PBO) combined with pyrethroids in insecticide-treated nets to prevent malaria in Africa. *The Cochrane Database of Systematic Reviews* 5 (5): CD012776.

95 Syme, T., Gbegbo, M., Obuobi, D. et al. (2022). Pyrethroid-piperonyl butoxide (PBO) nets reduce the efficacy of indoor residual spraying with pirimiphos-methyl against pyrethroid-resistant malaria vectors. *Scientific Reports* 12: 6857.

96 Sharma, R. and Singal, A. (2011). Topical permethrin and oral ivermectin in the management of scabies: a prospective, randomized, double blind, controlled study. *Indian J Dermatol Venereol Leprol* 77: 581–586.

97 Omotoso, G., Oloyede, O., Lawal, S. et al. (2002). Permethrin exposure affects neurobehavior and cellular characterization in rats' brain. *Environmental Analysis Health and Toxicology* 35 (4): e2020022.

98 Yamada, T., Lake, B.G., and Cohen, S.M. (2022). Evaluation of the human hazard of the liver and lung tumors in mice treated with permethrin based on mode of action. *Critical Reviews in Toxicology* 52 (1): 1–31.

99 Sharquie, K.E., Al-Rawi, J.R., Noaimi, A.A., and Al-Hassany, H.M. (2012). Treatment of scabies using 8% and 10% topical sulfur ointment in different regimens of application. *Journal of Drugs in Dermatology* 11 (3): 357–364.

100 Mila-Kierzenkowska, C., Wozniak, A., Krzyzynska-Malinowska, E. et al. (2017). Comparative efficacy of topical pertmehrin, crotamiton and sulfur ointment in treatment of scabies. 11(3):357-364. *Journal of Arthropod-Borne Diseases* 11 (1): 1–9.

101 Sharma, G., Dhankar, G., Thakur, K. et al. (2016). Benzyl benzoate-loaded microemulsion for topical applications: enhanced dermatokinetic profile and better delivery promises. *AAPS PharmSciTech* 17: 1221–1231.

102 Karthikeyan, K. (2005). Treatment of scabies: newer perspectives. *Postgraduate Medical Journal* 81: 7–11.

103 Choi, D., Ji, Y., Jang, Y. et al. (2020). Crotamiton, an anti-scabies agent, suppresses histamine- and chloroquine-induced itch pathways in sensory neurons and alleviates scratching in mice. *Biomolecules & Therapeutics (Seoul)* 28 (6): 569–575.

104 Seiler, J.C, Keech, R.C., Aker, J.L. et al. (2022). Spinosad at 0.9% in the treatment of scabies: efficacy results from 2 multicenter, randomized, double-blind, vehicle-controlled studies. *Journal of the American Academy of Dermatology* 86 (1): 97–103.

105 Kirst, H.A (2010). The spinosyn family of insecticides: realizing the potential of natural products research. *The Journal of Antibiotics* 63: 101–111.

106 Woods, A.D, Porter, C.L, and Feldman, S.R (2021). Abametapir for the treatment of head lice: a drug review. *Annals of Pharmacotherapy* 56 (3): 352–357.

107 Aditya, S. and Abametapir:, R.A (2021). A novel pediculicide with a unique mechanism. *Journal of Dr. NTR University of Health Sciences* 10 (2): 69–71.

2

Induction of Immune Responses and Inflammation to Parasitic Infections

Gurdeep Singh[1,], Abhishek Tiwari[2,*], Varsha Tiwari[2], and Mukesh Kr Singh[3]*

[1] *School of Pharmaceutical Sciences, Lovely Professional University, Phagwara, Punjab, India*
[2] *Pharmacy Academy, IFTM University, Moradabad, Uttar Pradesh, India*
[3] *School of Pharmaceutical Sciences, IFTM University, Moradabad, Uttar Pradesh, India*
* *Corresponding authors*

2.1 Introduction

Senescent multimorbidity, which is the co-existence of numerous chronic pathologies including the principal diseases of advanced age, is mostly caused by aging. Slowing down the aging process can simultaneously avoid the onset of a number of age-related disorders [1]. Worm therapy is a form of immunotherapy that treats immunological problems and autoimmune diseases by purposefully contaminating a patient's body with parasites in any of their developmental stages, including eggs, larvae, or adult worms. Studies showing that individuals who routinely contract parasitic worms have low rates of autoimmune illnesses, so all parasitic worms cannot be damaging to the body [2]. In affluent nations, the prevalence of autoimmune inflammatory illnesses has dramatically increased over the past three decades. Just a few examples include type 1diabetes (T1D), multiple sclerosis (MS), rheumatoid arthritis (RA), and Crohn's disease. An immune-mediated attack on a target organ that causes the immune system to no longer identify itself is the hallmark of autoimmune disease. Both antibody and cell-mediated elements can contribute to autoimmune disease [3].

Various age-related diseases, such as cardiovascular disease, dementia, cancer, chronic obstructive pulmonary disease (COPD), osteoporosis, and age-related macular degeneration, are made worse by this inflammation [4]. Gut dysbiosis, an

Parasitic Infections: Immune Responses and Therapeutics, First Edition.
Edited by Abhay Prakash Mishra and Manisha Nigam.

imbalance in the composition of the intestinal microbiome, is one factor contributing to inflammatory aging. Such an imbalance, which is aggravated by pro-dysbiotic components of contemporary life like antibiotic use and the so-called Western diet, can show up as a decrease in immunomodulatory microbial species [5].

When considered collectively, the studies reviewed here propose multiple methods for triggering an immune response and inflammation in response to parasite infections.

2.1.1 Induction of Immunity to Parasite Infections

Immune responses must be induced by the immune system's innate components, which react quickly and relatively non-specifically, and other specialized components, which detect specific epitopes on an antigen. Dendritic cells (DCs), which operate as "professional antigen-presenting cells", are a crucial component of this relationship. In combination with major histocompatibility complex (MHC) molecules, DCs endocytose and process the antigen to create the peptide that is displayed on the cell surface. The issue with many infectious diseases is that the immune system only responds after the organism has been invaded, at which point disease symptoms have already shown. Vaccination with a harmless form of the disease-producing agent, which is nevertheless capable of imitating the immune system, circumvents this issue and benefits the immune system. This method produces immunological responses that shield the body from further pathogen attack. After crucial interactions between the two main components of the immune system – innate and specific – immune defenses are formed. Natural killer (NK1) cells, granulocytes, macrophages (Ms1), and dendritic cells are the cellular elements of innate defenses (DCs1) [6]. Specific immune responses take longer to develop, but they are more lasting and specific. These particular defenses must interact with intrinsic components to become effective. The critical interaction between the innate and specific parts of the immune system involves the role played by antigen-presenting cells (APCs1), which include monocytes, MS, endothelial cells, fibroblasts, fibrocytes, and DCs [7]. B-cells can also serve as APCs from the specific immunological component, but not for primary immune responses that require activating naive lymphocytes. DCs are referred to as "professional APCs" since they are at the center of antigen presentation for both primary and recall immune responses. Despite the fact that DCs are particularly significant APCs, not all monocytes exhibit APC characteristics. Due to their "scavenger" phagocyte activity they are frequently more active in effector immune responses and they can even block the emergence of antigen-specific responses [8]. The aim of the chapter deals with the immunity response and inflammation to parasite infections with special emphasis on the different treatment regimens for parasite infections.

Many parasites such as *Blastomyces dermatitis* and *Staphylococcus aureus* elicit a ferocious immunological response from the host; unfortunately, this response may be ineffectual or may interfere with a normal immune response due to generations of contact and genetic modification. Furthermore, parasites can migrate away from an inflamed location to avoid being trapped, some have tough body walls, and most are relatively large. Due to these distinctions, parasite infection differs from the majority of viral and bacterial infections, in which the immune system typically provides protection. One author stated that there is now "no coherent explanation of immunity in any parasite disease" due to the intricacy of the host response to parasitic infection [9].

Recent research has shown that basophils, a rare form of white blood cell, are important participants in type 2 inflammation, although it is still unclear how basophils work in this setting [10]. A potential possibility for regulating basophil responses after parasite infection is the Notch signaling pathway, a molecular lock-and-key that may quickly transmit messages of inflammation to a range of cell types [11].

2.2 Defense Systems

2.2.1 Innate Immune Defenses

Innate immune defenses are composed of numerous and variable components. The skin, mucosal secretions, stomach, and intestinal pH are physical/chemical barriers that form the first line of innate defense. Phagocytic cells (neutrophils and macrophages), the complement system, natural killer cells, antimicrobial peptides, and sentinel cell production of innate defense cytokines are among the main innate defense systems. Different stressors have the tendency to decrease the effectiveness of innate defense mechanisms, increasing infection vulnerability. The development of adaptive (or acquired) immunity (antibody and T cell-mediated immunity) to a particular pathogen by the animal increases the activity of all of these innate defense mechanisms. The only aspect of innate immunity to be covered in detail will be the function of sentinel cells in the early detection of microbial invasion, activation of innate defense mechanisms, and induction of acquired immunity.

2.2.2 Production of Pro-Inflammatory Cytokines by Sentinel Cells

Sentinel cells play key roles in the early detection of microbial invasion and the stimulation of the immune response including mast cells, dendritic cells, and macrophages. These cells have receptors known as toll-like receptors (TLRs), which can bind chemicals that are specific to infectious pathogens. They are concentrated at or beneath epithelial surfaces. These substances include bacterial DNA (which is rich in CpG motifs), flagellin, bacterial lipoproteins, and bacterial lipopolysaccharide (endotoxin). These microbial compounds are seen as danger signals, and the sentinel cells respond quickly by secreting pro-inflammatory cytokines such as tumor necrosis factor-alpha (TNF), interleukin-1 (IL-1), and IL-6. These substances cause a localized inflammatory response at the site of their creation and stimulate the neutrophils, macrophages, natural killer cells, and endothelial cells to generate more of the proteins needed to effectively clear a local infection. The pro-inflammatory cytokines will start to circulate in the bloodstream if they are secreted in sufficient quantities. When they get to the hypothalamus, they cause fatigue, appetite loss, and fever. They cause the liver to create a lot of acute-phase proteins and stimulate neutrophilia in the bone marrow. These act to improve the body's natural defensive mechanisms' capacity to contain an infection. The animal interacts with other members of its species less due to lethargy and loss of hunger, which lessens the spread of infection. The replication of some viruses is slowed down by fever, and some inherent defenses that work to destroy viruses are strengthened.

2.2.3 Role of Dendritic Cells, Macrophages, and B Cells in Antigen Processing and Presentation to T-lymphocytes

Only peptide antigens displayed on major histocompatibility class II (MHC II) molecules are recognized by T-helper (CD4+) cells. These peptides originate from endocytosis proteins that are subsequently broken down into brief peptides by enzymes and attached to MHC II molecules. These molecules are then transported to the cell surface and presented to T-helper cells. Only peptide antigens shown on MHC I molecules are recognized by T-cytotoxic (CD8+) cells. These peptides are made from proteins that are produced in the antigen-presenting cell's cytoplasm. A proteasome in the cytoplasm processes a sample of all proteins made within a cell before being transferred into the endoplasmic reticulum, where brief peptide fragments are linked to MHC I molecules. Because the healthy animal is tolerant of normal self-proteins, there are no cytotoxic T cells that can detect them. Cytotoxic T cells will be exposed to peptides from foreign proteins if a cell contracts a virus or mutates and starts generating such proteins in its cytoplasm. The cell that is generating an alien protein will be killed in response by the cytotoxic T-cell. After initial exposure, it takes the cytotoxic T cells 7 to 10 days to become proficient at controlling viral infection. On a second encounter with the virus, they react very quickly.

Both killed and live vaccination antigens can be endocytosed, digested, and presented to T-helper cells on MHC II molecules. However, the creation of pathogen proteins in the cytoplasm of infected cells can only be induced by live vaccinations. As a result, live vaccines are often thought to be more successful at stimulating the production of cytotoxic T cells after vaccination. Researchers have created techniques to transfer vaccination antigens into the cytoplasm without utilizing conventional modified live vaccines now that this endogenous channel of antigen presentation has been identified. To transfer protein antigens into the cytoplasm for processing and presentation on MHC I molecules to cytotoxic T cells, one technique is to use adjuvants, such as ISCOMS (immune stimulating complexes), which fuse with the cell membrane. Utilizing recombinant virus vectors, which have one or more pathogen-specific genes inserted into their genetic material, is an additional strategy. When the vector infects a cell, this genetic material triggers the production of pathogen proteins in the cytoplasm, which is then processed and displayed on MHC I molecules. Instead of utilizing a standard modified live virus to deliver the protein, a viral vector may be more cost-effective or less dangerous to use overall than the MLV vaccination. It might also be less likely to be blocked by maternal antibodies (Figure 2.1).

2.2.4 Stimulation of Specific Immune Responses

2.2.4.1 T Lymphocytes

The bulk of specific immune responses, including antigen-activated B lymphocyte antibody generation and antigen-specific cytotoxic T cell (Tc) effector activity, depend on CD4+Th lymphocytes for effective execution. The fact that not all Th lymphocytes detect the same peptide and that only unique Th cell clones can recognize specific

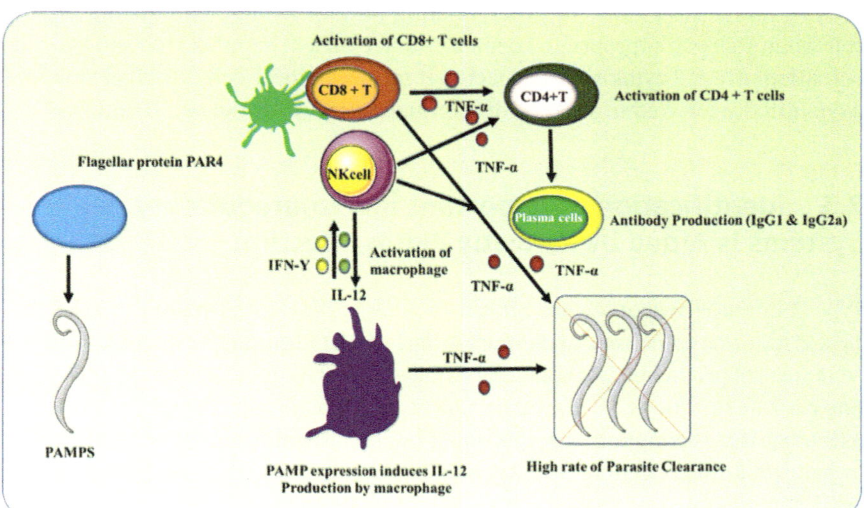

Figure 2.1 Role of dendritic cells, macrophages, and B-cells in antigen processing.

peptide sequences is a crucial factor to consider. This specificity derives from the APCs' digestion of the antigen, which yields a number of peptides that can stimulate the target Th cells. The conversion of antigen-stimulated B and Tc lymphocytes into effector and antibody-producing cells, respectively, is mediated by cytokines. Of course, proper cognate contacts are necessary for such cytokine signaling.

2.2.4.2 B Lymphocytes

Continued stimulation of the surface cytokine receptors by Th cell-produced substances as well as the membrane immunoglobulin-like receptors on B lymphocytes is necessary for B lymphocyte activation. A physical interaction between the B and Th cells is additionally necessary. As an illustration, CD40 and its ligand are essential for enabling selective differentiation. Similar to T lymphocyte responses, B lymphocyte responses are likewise limited to specific clones that can detect one of the antigen's epitopes, also known as the B cell epitopes. Although it is not necessary for APCs to process and present antigens in order to excite B lymphocytes, it appears that they do so in order to "transport" antigens to the B lymphocyte regions of lymphoid follicles. Because the epitopes recognized by B lymphocytes are frequently discontinuous and depend on the tertiary and quaternary (conformational) structure of the antigenic determinants on the antigen, this antigen must be "given" somewhat intact. A B-lymphocyte is prompted to cap its B cell receptor (BCR) in preparation for division and differentiation after it connects with the epitope for which it is specified via the BCR. The interaction with the Th lymphocytes at this time, particularly the contribution of the T cell cytokines, is crucial. The B lymphocyte returns to its resting stage, as it did before the engagement with the antigen, in the absence of this interaction. The B lymphocyte starts the process of differentiating into an antibody-producing plasma cell when there is appropriate cognate contact with T lymphocytes and the participation of T cell cytokines. The fact that several of the T cell cytokines involved were initially referred to as B cell differentiation factors is the result of this.

2.3 Identification of Important Immunoregulatory Systems Is Aided by Parasitic Worm Infection

Type 2 immune responses are related to allergic illness and helminth infection. Type 2 immune responses, in contrast to Th1 or Th17 immune responses, have the advantage of producing less collateral damage, which is often necessary for controlling parasite infections. On the other hand, type 2 reactions in allergies can result in pathology that is crippling or even lethal. The crucial significance of these responses in parasite killing/ejection, healing, metabolic alterations, and allergy disease has been demonstrated in models employing animals lacking in key components of type 2 immune pathways. Th2 cells that produce IL-4, IL-5, and IL-13 and antigen-specific IgE immunoglobulins are characteristics of type 2 immunity. Eosinophils are activated in response to antigen identification by Th2 cells, while primed mast cells are degranulated in response to IgE cross-linking. Allergy-related illnesses such as atopic dermatitis, asthma, and food allergies have clinical symptoms that are brought on by these effector immune cells working in concert.

It's interesting to note that in both people with chronic helminth infections and people who have repeatedly been exposed to allergens in the environment or in the therapeutic setting [12], high antigen-specific IgG4 levels have been observed, and regulatory T cells (Treg) and regulatory B cells that secrete IL-10 and TGF are more prevalent in circulation. [13, 14]. Thus, the type 2 adaptive immune response is tolerant to changes in the innate immune system that are necessary for their activation, external substances acting on adaptive immunity cells, or the cells' inherent exhaustion.

a) A population of innate immune cells called dendritic cells (DCs) is crucial for the creation of the best effector. Immune system responses in cells, including Th2 responses [15, 16]. As they are closely linked to barrier locations like the lungs, DCs can pick up antigens both inside and outside the epithelial barrier. Upon the discovery of an environmental allergy or helminth infection [17], as they get activated and go to the draining lymph nodes, DCs can potentially trigger a Th2 immune response by presenting antigens to T cells. Although it is obvious that DCs play a crucial role in Th2 formation, our understanding of the precise signals that prime a DC to produce Th2 is still lacking. Thymic stromal lymphopoietin (TSLP), interleukin-25 (IL-25), and IL-33 are examples of epithelial-derived cytokines that are important in the treatment of allergies and parasite infection [18]. These cytokines have the ability to directly activate type 2 innate lymphoid cells (ILC2s), causing a rapid innate type 2 response, as well as directly operate on DCs, skewing subsequent responses to Th2. Lacking antigen-specific receptors, ILC2s are innate lymphocytes that make a lot of the type 2 cytokines IL-5, IL-13, and IL-9 as well as the pro-resolving proteins amphiregulin [19] and IL-10 [20].

Additionally, class II MHC expression, peptide antigen presentation, and IL-4R signals for type 2 response initiation are all possible with activated ILC2s [21].

2.4 The Role of Regulatory Cells, Cytokine Inhibitors, and Immunoglobulins in the Modulation of Allergic Illness by Helminths

2.4.1 Immunoglobulins

Functional allergy as evaluated by skin prick test (SPT) reactivity highly corresponds with allergen-specific Ig E in high-income countries. However, this link frequently fails in helminth-endemic areas, particularly rural areas with poor socioeconomic status [22, 23]. Wheeze and/or atopy have also been linked positively by epidemiological research to antihelminth Ig E (ascariasis, schistosomiasis, filariasis). The cross-reactivity of antihelminth Ig E to specific allergens may be one of the causes of differences between Ig E reactivity and allergy in high-income versus helminth-endemic areas. For instance, Ig E against *Onchocerca volvulus* tropomyosin reacts with Derp10 tropomyosin from house dust mites, increasing allergic reactions to HDM. Despite the fact that IgE against carbohydrates on *Schistosoma* egg glycoproteins can react with cross-reactive carbohydrate determinants (CCDs) on peanut antigens, this Ig E has a low affinity, so cross-linking and degranulation of IgE-coated mast cells that are specific for carbohydrates do not take place. Consequently, these cross-reactive reactions could prevent clinical responses to allergens like peanuts [24]. Further research is required to determine if elevated levels of circulating CRP or carbohydrate-specific IgE play a role in causing or avoiding allergic reactions [22]. Following allergen-specific immunotherapy, the immunoglobulin isotype IgG4 is frequently linked to a tolerized allergic response; many people with helminth infections also produce more of this antibody [25,26]. IgG4 has been linked to a number of systemic disorders, although it is not yet obvious how this pathogenic role of IgG4 differs from active tolerance development to allergens or during helminth infection. Despite the growing knowledge of potential negative effects, IgG4 is still not completely understood. Despite the new reports of illnesses connected to IgG4, IgG4 antibodies are thought to be the least inflammatory of all isotypes since they do not activate IgG4 complement and, unlike IgE, do not result in mast cell degranulation. IgG4 is characterized as functionally monovalent because of its exceptional capacity to switch antigen-binding arms, which prevents immunological complex formation. Because of this, its primary role in this situation seems to be a blocking one, which may be helpful in preventing Ig E-mediated inflammation. While in a *S. mansoni* endemic area, higher levels of both Ig E and IgG4 were found in infected individuals, a higher ratio of Ig E to IgG4 predicted clinical allergic symptoms, just as in allergen-specific immunotherapy, high levels of anti-*Ascaris* IgG4 have been negatively associated with allergen SPT positivity [24]. Due to the homology of many helminth products to common allergens, IgG4 responses generated in response to helminth products may also bind and block IgE epitopes on allergens, decreasing responses to allergens and directly reducing SPT responses [27].

2.4.2 Regulatory Cells

In animal models, regulatory B cells (Bregs) and regulatory T cells (Tregs) play key roles in the regulation of allergic airway inflammation and type 2 immune responses. Levels of Tregs and Bregs are elevated and necessary for the maintenance of tolerance in those who have become tolerant to allergens through high-dose environmental exposure or allergen-specific immunotherapy [28]. The immune-suppressive cytokines IL-10 and TGF- may be produced by both Tregs and Bregs, which can reduce harmful inflammation [29]. The immunosuppressive effects of a variety of helminths, such as *Onchocerca*, *Ascaris*, *Trichuris*, or *Toxocara* species, are also mostly mediated by IL-10 and TGF-β [30, 31, 32] and seem to play a key role in the inhibition of allergic reactions. In fact, IL-10 was associated with a decreased likelihood of allergic skin reactions in Gabonese students who had schistosome infections [33]. In *Schistosoma haematobium*, higher concentrations of circulating FOXP3+ CD25+ Treg cells have been found [34] and in filaria-infected people [35]. Antihelmintic therapy for *S. haematobium* or geohelminth infected people results in a normalization of circulating FOXP3 Treg or PD-1 and CTLA-4-expressing CD4+ cells116 and/or later elevated in vitro cytokine responses to both helminth and bystander antigens [34]. Similar to MS patients who have helminth infections, more Breg cells have been seen in Gabonese people who have *S. haematobium* infections [36].

2.4.3 Myeloid Cells

The development of effector versus regulatory T-cells is determined by dendritic cells (DC), the vital link between innate and adaptive immunity, depending on their ontogeny, tissue location, and/or the presence of environmental cues. Conventional type 1 (cDC1) and type 2 dendritic cells (cDC2), two distinct myeloid DC subsets, can be separated by a number of surface expression markers that have recently been discovered using an impartial methodology across tissues and species [37, 38].

cDC2 can enhance both Th17 or Th2 cells depending on the environment and exhibits improved allergen absorption compared to cDC1, whilst cDC1 can create large levels of IL-12p70 and activate cytotoxic CD8 T-cell and antitumor responses. Intriguingly, cDC1 can also play a tolerogenic role in allergy models; they do this by inducing Treg cells through retinoic acid and peroxisome proliferator-activated receptor gamma (PPAR) and reducing inflammation in schistosome infections as well as in an HDM and ovalbumin model of allergic airway inflammation. More DCs are seen in the blood, induced sputum, and bronchoalveolar lavage when an allergen is introduced to asthma patients, but only cDC2 migrated into the bronchial tissue [39]. The Th2 polarization molecule OX-40L and the Fc RI are expressed more by the DCs. However, other myeloid cell types, such as monocyte-derived dendritic cells, may also be crucial to support Th2 cell formation in either allergy models or helminth infection, as type 2 responses in allergy models are substantially disrupted in the absence of DCs [40].

2.4.4 Innate Lymphoid and Epithelial Cells

It has only lately been clear how crucial early, innate, epithelial cell-derived cytokines are for the onset of type 2 responses. At barrier locations, the epithelial cell cytokines IL-25, IL-33, and TSLP activate ILC2s, which release a lot of IL-5,

IL-13, and IL-9. ILC2s have recently been demonstrated to be capable of activation to release IL-10, establishing an immunoregulatory pathway (similar to that of T cells) that may be accessible to parasite immunomodulation [20]. Therefore, proximal epithelial cell responses are a prime candidate for intervention since inhibiting these cytokines could reduce subsequent T-cell, dendritic cell, and ILC responses. The study of the repression of these pathways, however, is still in its early stages because they have been identified only recently. The IL-33 pathway in *H. polygyrus* infection is one instance, in which numerous parasite immunomodulatory components have been found. Excretory/secretory products of *Heligmosomoides polygyrus* (HES) duplicate the suppressive effect of parasite infection in reducing airway allergic inflammation [41]. The initial ILC2 reactions to an allergen preparation from *Alternaria alternata* have been eliminated [42], a stimulation with clinical relevance. HES administration blocks the IL-33 pathway by inducing IL-1 [43], RNA-containing HES extracellular vesicles that reduce IL-33 receptor expression, and HpARI, a protein in HES that directly binds IL-33 and prevents its release. *Alternaria* allergen administration is a remarkably potent stimulus for IL-33 release [5]

2.5 Conclusion

Parasites can suppress, alter, or even trigger some immune activation pathways, which effectively but subtly control the host immune system. Our interpretations of the immune response to parasites as they were understood just a few years ago now appear to be overly simplistic, and many of the facts that have been seen have had to be reinterpreted in light of the new findings. These recent discoveries have made it possible for us to comprehend many of the observed events much more clearly. It is simplest to demonstrate these ideas using a few specific parasitic diseases and to interpret them in terms of human disorders.

It is still unclear which immunological pathways are necessary for the development, growth, and maintenance of defenses against parasite infections. As we learn more about induction of immune responses and inflammation to parasitic infections, we will also be revealing new immune pathways that are modulated by parasitic infections and looking at new options to use therapeutics to treat it.

References

1 Austad, S.N. (2016). The geroscience hypothesis: is it possible to change the rate of aging? In *Advances in Geroscience* (eds F. Sierra and R. Kohanski), 1–36. Springer, Cham.

2 Zibaei, M. (2018). Helminths and approach to treatment of immune-mediated diseases. *International Journal of Enteric Pathogens* 6 (2): 31–32.

3 Zaccone, P., Fehervari, Z., Phillips, J.M. et al. (2006). Parasitic worms and inflammatory diseases. *Parasite Immunology* 28 (10): 515–523.
4 Xia, S., Zhang, X., Zheng, S. et al. (2016). An update on inflamm-aging: mechanisms, prevention, and treatment. *Journal of Immunology Research* 2016.
5 Burbank, A.J., Sood, A.K., Kesic, M.J. et al. (2017). Environmental determinants of allergy and asthma in early life. *Journal of Allergy and Clinical Immunology* 140 (1): 1–2.
6 Marshall, J.S., Warrington, R., Watson, W., and Kim, H.L. (2018). An introduction to immunology and immunopathology. *Allergy, Asthma & Clinical Immunology* 14 (2): 1–0.
7 Gaudino, S.J. and Kumar, P. (2019). Cross-talk between antigen presenting cells and T cells impacts intestinal homeostasis, bacterial infections, and tumorigenesis. *Frontiers in Immunology* 6 (10): 360.
8 Hirayama, D., Iida, T., and Nakase, H. (2017). The phagocytic function of macrophage-enforcing innate immunity and tissue homeostasis. *International Journal of Molecular Sciences* 19 (1): 92.
9 Zander, R.A. and Butler, S.N. (2013). Dysfunctional adaptive immunity during parasitic infections. *Current Immunology Reviews* 9 (3): 179–189.
10 Lumb, F.E., Doonan, J., Bell, K.S. et al. (2017). Dendritic cells provide a therapeutic target for synthetic small molecule analogues of the parasitic worm product, ES-62. *Scientific Reports* 7 (1): 1704.
11 Webb, L.M., Oyesola, O.O., Früh, S.P. et al. (2019). The notch signaling pathway promotes basophil responses during helminth-induced type 2 inflammation. *Journal of Experimental Medicine* 216 (6): 1268–1279.
12 Van de Veen, W., Wirz, O.F., Globinska, A., and Akdis, M. (2017). Novel mechanisms in immune tolerance to allergens during natural allergen exposure and allergen-specific immunotherapy. *Current Opinion in Immunology* 1 (48): 74–81.
13 Allen, J.E. and Maizels, R.M. (2011). Diversity and dialogue in immunity to helminths. *Nature Reviews Immunology* 11 (6): 375–388.
14 Nutman, T.B. (2015). Looking beyond the induction of Th2 responses to explain immunomodulation by helminths. *Parasite Immunology* 37 (6): 304–313.
15 Phythian-Adams, A.T., Cook, P.C., Lundie, R.J. et al. (2010). CD11c depletion severely disrupts Th2 induction and development in vivo. *Journal of Experimental Medicine* 207 (10): 2089–2096.
16 Van Rijt, L.S., Jung, S., KleinJan, A. et al. (2005). In vivo depletion of lung CD11c+ dendritic cells during allergen challenge abrogates the characteristic features of asthma. *The Journal of Experimental Medicine* 201 (6): 981–991.
17 Hammad, H. and Lambrecht, B.N. (2015). Barrier epithelial cells and the control of type 2 immunity. *Immunity* 43 (1): 29–40.

18 Han, H., Roan, F., and Ziegler, S.F. (2017). The atopic march: current insights into skin barrier dysfunction and epithelial cell-derived cytokines. *Immunological Reviews* 278 (1): 116–130.

19 Monticelli, L.A., Osborne, L.C., Noti, M. et al. (2015). Artis D. IL-33 promotes an innate immune pathway of intestinal tissue protection dependent on amphiregulin–EGFR interactions. *Proceedings of the National Academy of Sciences* 112 (34): 10762–10767.

20 Seehus, C.R., Kadavallore, A., Torre, B.D. et al. (2017). Alternative activation generates IL-10 producing type 2 innate lymphoid cells. *Nature Communications* 8 (1): 1900.

21 Oliphant, C.J., Hwang, Y.Y., Walker, J.A. et al. (2014). MHCII-mediated dialog between group 2 innate lymphoid cells and CD4+ T cells potentiates type 2 immunity and promotes parasitic helminth expulsion. *Immunity* 41 (2): 283–295.

22 Amoah, A.S., Boakye, D.A., Yazdanbakhsh, M., and van Ree, R. (2017). Influence of parasitic worm infections on allergy diagnosis in sub-Saharan Africa. *Current Allergy and Asthma Reports* 17: 1–9.

23 Obeng, B.B., Amoah, A.S., Larbi, I.A. et al. (2014). Schistosome infection is negatively associated with mite atopy, but not wheeze and asthma in Ghanaian s school children. *Clinical & Experimental Allergy* 44 (7): 965–975.

24 McSorley, H.J., Chayé, M.A., and Smits, H.H. (2019). Worms: pernicious parasites or allies against allergies? *Parasite Immunology* 41 (6): e12574.

25 Aalberse, R.C. and Schuurman, J. (2002). IgG4 breaking the rules. *Immunology* 105 (1): 9–19.

26 Aalberse, R.C., Stapel, S.O., Schuurman, J., and Rispens, T. (2009). Immunoglobulin G4: an odd antibody. *Clinical & Experimental Allergy* 39 (4): 469–477.

27 Doenhoff, M.J., El-Faham, M., Liddell, S. et al. (2016). Cross-reactivity between *Schistosoma mansoni* antigens and the latex allergen Hev b 7: putative implication of cross-reactive carbohydrate determinants (CCDs). *PLoS One* 11 (7): e0159542.

28 Palomares, O., Akdis, M., Martín-Fontecha, M., and Akdis, C.A. (2017). Mechanisms of immune regulation in allergic diseases: the role of regulatory T and B cells. *Immunological Reviews* 278 (1): 219–239.

29 Hussaarts, L., van der Vlugt, L.E., Yazdanbakhsh, M., and Smits, H.H. (2011). Regulatory B-cell induction by helminths: implications for allergic disease. *Journal of Allergy and Clinical Immunology* 128 (4): 733–739.

30 Alcântara-Neves, N.M., de Sg Britto, G., Veiga, R.V. et al. (2014). Effects of helminth co-infections on atopy, asthma and cytokine production in children living in a poor urban area in Latin America. *BMC Research Notes* 7: 1-1.

31 Doetze, A., Satoguina, J., Burchard, G. et al. (2000). Antigen-specific cellular hyporesponsiveness in a chronic human helminth infection is mediated by Th3/

Tr1-type cytokines IL-10 and transforming growth factor-β but not by a Th1 to Th2 shift. *International Immunology* 12 (5): 623–630.

32 Satoguina, J., Mempel, M., Larbi, J. et al. (2002). Antigen-specific T regulatory-1 cells are associated with immunosuppression in a chronic helminth infection (onchocerciasis). *Microbes and Infection* 4 (13): 1291–1300.

33 Van den Biggelaar, A.H., van Ree, R., Rodrigues, L.C. et al. (2000). Decreased atopy in children infected with Schistosoma haematobium: a role for parasite-induced interleukin-10. *The Lancet* 356 (9243): 1723–1727.

34 Schmiedel, Y., Mombo-Ngoma, G., Labuda, L.A. et al. (2015). CD4+ CD25hiFOXP3+ regulatory T cells and cytokine responses in human schistosomiasis before and after treatment with praziquantel. *PLoS Neglected Tropical Diseases* 9 (8): e0003995.

35 Wammes, L.J., Hamid, F., Wiria, A.E. et al. (2012). Regulatory T cells in human lymphatic filariasis: stronger functional activity in microfilaremics. *PLoS Neglected Tropical Diseases* 6 (5): e1655.

36 Van der Vlugt, L.E., Zinsou, J.F., Ozir-Fazalalikhan, A. et al. (2014). Interleukin 10 (IL-10)–producing CD1dhi regulatory B cells from *Schistosoma haematobium*–infected individuals induce IL-10–positive T cells and suppress effector T-cell cytokines. *The Journal of Infectious Diseases* 210 (8): 1207–1216.

37 Alcántara-Hernández, M., Leylek, R., Wagar, L.E. et al. (2017). High-dimensional phenotypic mapping of human dendritic cells reveals interindividual variation and tissue specialization. *Immunity* 47 (6): 1037–1050.

38 Guilliams, M., Dutertre, C.A., Scott, C.L. et al. (2016). Unsupervised high-dimensional analysis aligns dendritic cells across tissues and species. *Immunity* 45 (3): 669–684.

39 El-Gammal, A., Oliveria, J.P., Howie, K. et al. (2016). Allergen-induced changes in bone marrow and airway dendritic cells in subjects with asthma. *American Journal of Respiratory and Critical Care Medicine* 194 (2): 169–177.

40 Williams, J.W., Tjota, M.Y., Clay, B.S. et al. (2013). Transcription factor IRF4 drives dendritic cells to promote Th2 differentiation. *Nature Communications* 4 (1): 2990.

41 McSorley, H.J., O'Gorman, M.T., Blair, N. et al. (2012). Suppression of type 2 immunity and allergic airway inflammation by secreted products of the helminth H eligmosomoides polygyrus. *European Journal of Immunology* 42 (10): 2667–2682.

42 McSorley, H.J., Blair, N.F., Smith, K.A. et al. (2014). Blockade of IL-33 release and suppression of type 2 innate lymphoid cell responses by helminth secreted products in airway allergy. *Mucosal Immunology* 7 (5): 1068–1078.

43 Zaiss, M.M., Maslowski, K.M., Mosconi, I. et al. (2013). IL-1β suppresses innate IL-25 and IL-33 production and maintains helminth chronicity. *PLoS Pathogens* 9 (8): e1003531.

3

Animal Parasites

Insight into Natural Resistance

Nasib Zaman[1], Muhammad Rizwan[1], Kishawar Sultana[2], Abdur Rauf[3,], Yahya S. Al-Awthan[4,5], and Omar Bahattab[4]*

[1] *Centre for Biotechnology and Microbiology, University of Swat, KP, Pakistan*
[2] *Centre of Biotechnology and Microbiology, University of Peshawar, KP, Pakistan*
[3] *Department of Chemistry, University of Swabi, Swabi, Anbar, Khyber Pakhtunkhwa (KP), Pakistan*
[4] *Department of Biology, Faculty of Science, University of Tabuk, Tabuk, Saudi Arabia*
[5] *Department of Biology, Faculty of Science, Ibb University, Ibb*
* *Corresponding author*

3.1 Introduction

Food animals (cattle, poultry, and pigs) have been widely raised throughout the world, not only as a source of food but also as a source of revenue [1]. Helminths are parasitic worms that are mainly transmitted through food or feces; infection has been documented all over the world [2]. Parasite infection prevention in animals is important not only for animal welfare and the human–animal interaction, but it also minimizes the potential risk of human infection [3]. Once an infectious agent has been recognized, hosts have some options for reducing the pathogen's influence on their health. As part of their total array of defences, animals may employ three ways to protect themselves against the detrimental consequences of parasite infestations [4, 5]. Firstly, they can use infection-minimizing or infection-preventing mechanisms (qualitative resistance), such as behavioral avoidance of surroundings, conspecifics, or foods connected to diseases, or "avoidance" [5]. Once an infection or parasite has been established, the second line of defence

Parasitic Infections: Immune Responses and Therapeutics, First Edition.
Edited by Abhay Prakash Mishra and Manisha Nigam.

consists of mechanisms that reduce the amount of growth that occurs, with both specific and nonspecific immune responses being important (referred to resistance). Third, by limiting the damage caused by parasites and repairing it, hosts can lessen the negative effects of infection known as "tolerance", allowing for infection without parasite prevention [4, 5].They can, for example, fight the expanding pathogen population directly in order to eliminate it. However, hosts can try to reduce the amount of harm produced by a given number of infections by increasing tissue healing and detoxifying pathogen by-products [6]. Animal adopt some mechanisms to resistance and remove animal pathogenic parasites. Anthelmintic resistance is on the rise in livestock in countries all around the world, including the United Kingdom, according to some reports. Control of helminth parasites is a major concern in both animal and human medicine. Individual susceptibility to various parasitic diseases varies significantly (due to genetic factors). The identification of genes that contribute to resistance variation leads to a better understanding of resistance mechanisms [7]. Some studies have found that some breeds of domestic ruminants are more resistant to parasite illnesses such as coccidian protozoa, worms, ticks, and flies than others. The phenomena of trypanotolerance demonstrated by West African humpless cattle (N'dama) and West African shorthorn cattle, which continue to exist in trypanosome-challenged environments, is perhaps the most obvious illustration of this [8]. Toxoplasmic encephalitis (TE) is mostly prevented by interferon (IFN)-dependent cell-mediated immunity. *Toxoplasma gondii* is controlled by the humoral immune system in the brain. Combining T and B cells with non-T cells that produce IFN, microglia, astrocytes, and dendritic cells results in resistance [9]. Some of the exceptional host specificity of the numerous species of *Eimeria* is one example of how many parasites do not grow at all in hosts other than their natural ones. For example, some sheep-borne larvae of the bovine parasite *Ostertagia ostertagi* undergo development in sheep but very few of them reach the adult stage [8]. As they mature, many animals have a greater resistance to developing initial infections with certain parasites. For instance, animals who are a few months old are more prone to acquire a patency in their ascarid infection. In the same way, very young ruminants and horses are most commonly observed to have patent *Strongyloides* infections. When hosts are infected later in life, the parasites stop growing or fail to stop at the larval stage in the tissues. *Nematodirus battus* is not a very dangerous parasite for sheep older than three months [8]. Among the fly-repelling tactics used by ungulates are tail flipping, head tossing, muzzle flicking, and muscle twitching. When the number of biting flies is large, the cattle repel the flies more actively, and the animals that do the best repelling have the fewest biting flies around them [10, 11]. Mast cells have been identified in abundance in fish parasitized by helminths, showing that this cell type is linked to the response to these parasites. When parasites enter the host's tissues, alarmins, also known as damage-associated molecular patterns

(DAMPs), are generated by harmed or dead cells and start inflammatory processes. Alarmins trigger the production of pro-inflammatory cytokines and chemokines by local mast cells, dendritic cells, and macrophages, causing leukocyte recruitment into the infected area [12]. Nonspecific immune response mechanisms to also help the animal to resist against some nematodes [13]. The availability of resources may have an impact on an animal's ability to resist parasites and tolerate them. For many vertebrate host species, a better food supply lowers the amount of parasites that feed on host tissue [14].

3.1.1 Host Resistance against Animal Parasites

Once an infectious agent has been recognized, hosts have two options for reducing the pathogen's influence on their health. They can, for example, fight the expanding pathogen population directly in order to eliminate it. However, hosts can try to reduce the amount of harm produced by a given number of infections by increasing tissue healing and detoxifying pathogen by-products: "tolerance" and "resistance" [6]. Animals adopt the following mechanisms to resistance and remove animal pathogenic parasites:

1) genetic resistance of ruminants to gastrointestinal nematodes;
2) host resistance to *Toxoplasma gondii* in the brain;
3) species and breed resistance in animals against parasitic diseases;
4) resistance due to age;
5) removal strategies for biting flies;
6) avoidance and eradication strategies for ticks;
7) avoidance and removal strategies for fleas;
8) medicinal herbs used by animals for the treatment of parasites and pathogens
9) mechanism of the immune response of teleost fish to helminth parasite infection;
10) nonspecific immune response mechanisms to *Haemonchosis contortus*;
11) effector immune mechanisms against helminths;
12) the function of eosinophils against nematode infections; and
13) host resistance of gut nematoses depending on resource availability

3.1.1.1 Genetic resistance of ruminants to gastrointestinal nematodes

Susceptibility to numerous parasite infections varies significantly across individuals. Genetic factors account for some of this diversity in vulnerability. The current issue is identifying the most effective means of utilizing the variety to enhance our comprehension of parasite infection and lessen the damage caused by parasitic disease. Understanding of resistance mechanisms is improved by locating the genes that affect resistance variance. However, further research is required to establish if these genes, individually or collectively, account for a significant enough fraction of the diversity in resistance to support marker-assisted selection. Comparing reactions between susceptible and resistant stock offers a valuable method for separating pathological, irrelevant, and protective responses [7]. Breed differences should make understanding the mechanisms underlying resistance to such infections easier, which should help improve treatments and vaccinations or identification of resistant sheep [7] or other animals The fact that some varieties of domestic ruminants are more resistant to certain parasite infections, such as coccidian protozoa, nematodes, ticks, and flies, than other breeds has aroused the curiosity of some researchers. The phenomena of trypanotolerance exhibited by West African humpless cattle, such as the N'dama, are likely the best illustration that survives in areas of heavy trypanosome challenge. Although it is believed that immune reactions may be involved, how these cattle control their parasitaemias is still unknown. In studies on helminth infections, Red Masai sheep from East Africa are more resistant to *Haemonchus contortus* disease than other imported breeds. In contrast, South African Merinos have been less susceptible to trichostrongylosis than some other breeds [8]. Compared to Merino or European breeds, the St. Croix, Barbados Blackbelly, and Florida Native (Gulf Coast) sheep breeds. are significantly more resistant to *H. contortus*. Merino, Scottish Blackface, and Finn Dorset sheep, all homozygous for hemoglobin A, develop smaller worm burdens after infection than their hemoglobin B homozygous or heterozygous counterparts, indicating that variations in hemoglobin genotypes within breeds reflect differences in susceptibility to *H. contortus* infection. However, under severe stress, these genotypic variations in susceptibility frequently falter. According to research conducted within a single breed in Australia, individual Merino lambs can be classified as responders or non-responders based on how they reacted to *Trichostrongylus colubriformis* infection and their genetic differences are passed down to the next generation. It has been demonstrated that genetics plays a role in tick resistance, particularly *Rhipicephalus* (Boophilus) resistance, which is higher in the humped (*Bos indicus*) Zebu breeds and lower in the European (*Bos taurus*) varieties. However, where cattle have a genetic makeup that is 50% or more Zebu, a high level of resistance is still feasible, permitting a restricted use of acaricides [8].

3.1.1.2 Host resistance against *Toxoplasma gondii* in the brain

To prevent the onset of toxoplasmic encephalitis, interferon (IFN)-dependent cell-mediated immunity is crucial (TE). *Toxoplasma gondii* in the brain is likewise controlled by the humoral immune system. T and B cells, non-T cells that make IFN, microglia, astrocytes, and dendritic cells work together to generate resistance. Interactions between these cells and the activation of effector cells that prevent intracellular parasite reproduction are mediated by other cytokines. In mice, the Ld gene gives resistance to the development of TE. The major histocompatibility complex contains these genes, which governs immunological responses. These genes seem necessary for determining the host's resistance to this infection. *T. gondii* strains also influence TE development [9].

3.1.1.3 Species resistance in animals against parasitic diseases

Animal parasites like *Cryptosporidium parvum, Fasciola hepatica, Trichinella spiralis*, and the asexual stages of *Toxoplasma* have a vast host range. Some of the exceptional host specificity of the numerous species of *Eimeria* is one example of how many parasites do not grow at all in hosts other than their natural ones. For instance, some *Ostertagia* larvae grow in sheep but relatively few of them mature into adults. However, with these unnatural hosts, particularly parasites that migrate through tissue, there can occasionally be serious effects, especially if the migratory path becomes unpredictable. An example of visceral larva migrans in children caused by *Toxocara canis* is accompanied by hepatomegaly and sporadically involves the brain and eyes [8].

3.1.1.4 Age resistance

As they mature, many animals have a greater resistance to developing initial infections with certain parasites. For instance, animals who are a few months old are more prone to acquire a patency in their ascarid infection. The majority of patent *Strongyloides* infections in horses and ruminants are found in very young animals. The parasites either stop growing or stop at larval stages in the tissues if hosts become infected at an older age. *Nematodirus battus* is not a very dangerous parasite for sheep older than three months [8]. Similarly, dogs gradually become resistant to *Ancylostoma* infection during their first year of life. On the other hand, most parasite organisms appear to have evolved a potent counter-mechanism if age resistance is present. *Toxocara vitulorum, Toxocara canis* and *Strongyloides* spp. are affected; they remain in the host's tissues as larval stages and only become active in the late stages of pregnancy to infect the unborn child in utero or through the transmammary pathway. The key hatching conditions for the egg in the case of *Nematodirus battus*, which enabling the parasites to survive as a lamb-to-lamb infection from one season to the next. These conditions include a prolonged cold followed by a temperature above 10°C [8].

3.1.1.5 Removal strategies for biting flies

Many studies have shown the high expenses incurred by flying parasites like Tabanid flies, which are widespread in Asia, East Africa, and the US. According to a study, a horse can be bitten by as many as 4000 tabanid flies in a single day, resulting in a blood loss of up to 0.5 liters; these insects also can transmit several diseases. Among the fly-repelling tactics used by ungulates include tail flipping, head tossing, muzzle flicking, and muscle twitching. When the severity of the biting flies is high, cattle engage in more fly repelling activity, The lowest number of biting flies are around those who actively participate in fly repelling [10, 11]. Elephants have thick, fly-resistant skin that covers much of their body and is impervious to gigantic tabanids. Asian elephants kept in captivity in Nepal had 43% less flies on their flanks when a branch was present to serve as a fly switch than when one was not [15]. In other experiments, when branches were excessively long or bushy, elephants changed them into switch-size branches [16].

3.1.1.6 Avoidance and eradication strategies for ticks

Due to the fact that ungulates graze on grasslands, ticks prey on them. It is possible to lose a lot of blood after ingesting pregnant female ticks. A single engorging tick, for instance, has been found in studies on growing cattle to reduce a growing calf's annualized weight gain by 3 kg [17, 18].

Before making their way to the head, neck, or hindquarters, where oral grooming is ineffective, ticks climb up legs and over shoulders. The first line of defence against ticks for the majority of ungulates is to swipe the tongue or lower incisors over the shoulders and trunk. Grooming in eastern and southern African antelope has been thoroughly studied. Antelope sweep their lateral incisors over their shoulders, bellies, and flanks while grooming, likely catching the majority of flying ticks [19]. These lower incisors form a tooth comb that makes it simpler to remove mature ticks [20].

3.1.1.7 Avoidance and removal strategies for fleas

Fleas are the most common ectoparasite on cats. Fleas are difficult to get rid of and bounce around the body, in contrast to ticks which move slowly. Cats have tongues with papillae that have cornified. The papillae aid in both flea catching and maintaining the health of the cat's fine fur coat [21].

3.1.1.8 Medicinal herbs used by animals for the treatment of parasites and pathogens

Herbal medicine and modern pharmaceuticals have their roots in ancient herbal therapies for parasite and pathogen eradication, as well as preventing serious diseases in humans and animals [22]. The observation of a sick chimp extracting and consuming the bitter pith of a *Vernonia amygdalina* plant, which is known to have antibacterial qualities, provided the most persuasive proof of the use of herbal therapy for ailments in animals [23]. A special feature of animal herbal treatment is the use of anthelminthics from a "pharmacy in the woods" to remove intestinal parasites from an animal's digestive tract. The regular ingestion of full leaves of indigenous plants (*Manniophyton fulvum*) by chimps [24] and bonobos [25] resulted in the purging of intestinal parasites, according to studies. Chimpanzees in the wild eat plant leaves that pass through their intestines. In other instances, the plant material increases intestinal motility, removing nematodes from the intestine.

3.1.1.9 Mechanism of the immune response of teleost fish to helminth parasite infection

Environmental, hormonal, dietary, and life stage variables can all have an impact on the immune response in teleost fish [26]. One of the most important defence mechanisms of fish intestines against parasites is hyperplasia of mucus-producing cells [27]. When helminth infections induce tissue damage, innate immune cells, such as ILC2s (type 2 innate lymphoid cells), respond quickly by producing cytokines and chemokines, which deploy neutrophils, basophils, and eosinophils [28]. Depending on the type of damage a parasite does and how deeply it infiltrates a cell, different responses result from infection. Granular leukocytes called rodlet cells are only seen in teleost fish. They are typically called in when hazardous chemicals or helminth infections are present in tissues [29]. These cells support the inflammatory response when helminths are merely clinging to the intestinal epithelium; however, when penetration into the intestinal wall deepens, the presence of granulocytes and macrophages increases [30]. Mast cells have been found in significant numbers in fish parasitized by helminths, demonstrating that this cell type is connected to the defence against these parasites. Damage-associated molecular patterns (DAMPs), also known as alarmins, are released by damaged/dead cells when parasites infiltrate the host's tissues and encourage inflammatory responses. Alarmins stimulate leukocyte recruitment into the infected area by inducing the production of pro-inflammatory cytokines and chemokines by local mast cells, dendritic cells, and macrophages [12]. The local release of cytokines and chemokines is stimulated by tumor necrosis factor (TNF), which in turn enhances the presentation of antigens, phagocytosis, and co-stimulation of T cells. The transition from the inflammatory phase, which is dominated by neutrophils, to the phase, which is dominated by macrophages, appears to be mostly mediated by interleukin-6 [31].

3.1.1.10 Nonspecific immune response mechanisms to *Haemonchosis contortus*

The *H. contortus* larvae need a gastrointestinal niche that promotes development and expansion while shielding them from host barriers such as mechanical (peristaltic movement) and chemical (abomasum mucus). The larvae's motility and parasite load both contribute to parasite colonization in the host abomasum. After being sensitized by past infections, certain hosts can adjust the microenvironmental properties of their niche to expel the parasite [13]. Complement fixation is one of the earliest intrinsic reactions to *H. contortus* infection. Numerous studies have discovered that helminths bind certain molecules (opsonins) on their surface and activate the alternative complement pathway [32]. Vasoactive and chemotactic peptides (C3a and C5a) are produced after complement activation in larvae, and they mobilize eosinophils to the infection site without relying on particular pathways (CD4+ and IL-5). Additionally, *H. contortus* releases chemo-attractants for neutrophils and eosinophils, which intensifies the inflammatory response [33]. The thymus-independent rise in tissue eosinophils, which enhances complement activation and eosinophil cytotoxicity against larvae in the early stages of infection, is a crucial innate response in the absence of particular antibodies. Inflammation caused by the alternative complement pathway and driven by mast cells and eosinophils is linked to the rapid clearance of parasites in rats following the initial infection [32].

3.1.1.11 Effector Immune Mechanisms against Helminths

Infections with helminths are accompanied by hypereosinophilia, IgE production, mucosal mastocytosis, and goblet cell hyperplasia [34]. Depending on where the helminth is located, these immunological characteristics play a significant role in various effector pathways. There have been described a number of defences against parasites that live on tissues. These parasites are frequently the larval stages of trematodes (trematodes that move through tissue, such as *Schistosoma* species, *Fasciola* species, or nematodes). Eosinophils, neutrophils, macrophages, or platelets are examples of effector cells in antibody-dependent cellular cytotoxicity (ADCC), and IgE, IgG, or IgA are the antibodies [35–38].

3.1.1.12 The role of eosinophils in nematode infection protection

Parasite infections, especially those caused by nematodes that invade tissues, trigger a potent Th2-type immune response that raises the levels of immunoglobulin E and eosinophil in the blood and tissues. Eosinophils are effective against nematode larvae, although not all helminth infections can be treated with them. The activation of acquired immunity results in a significant accumulation and activation of eosinophils, which affects the worm larvae, when a host is infected by a nematode species that has previously been encountered. It has been demonstrated that eosinophils play a part in host defence, inflammation, and immunomodulation. Eosinophil generation and activation are mainly regulated by the innate lymphoid cells of group 2 and Th2 cells, interleukin (IL)-5.

Non-antigen-specific immune activation that defends the host by causing group 2 innate lymphoid cells and Th2 cells to produce the cytokines interleukin (IL)-5 and IL-13 in response to IL-33 stimulation. Migration of larvae from a species distinct from the one previously encountered causes this immunological response.

Eosinophils are essential in the destruction of migratory larvae. So, both antigen-specific and non-antigen-specific mechanisms by which eosinophils support host defence [39].

3.1.1.13 Host resistance of gut nematodes depends on resource availability

Dynamics between hosts and parasites can be significantly changed by resource availability. In addition to giving hosts more resources to fight infections, an abundance of food can make hosts more tolerant of illnesses by lowering the competition for food between hosts and parasites [40]. The availability of resources may affect an animal's capacity to resist and tolerate parasites, regardless of whether the parasites consume the resources the host requires or host tissue. For many vertebrate host species, a better food supply lowers the amount of parasites that feed on host tissue [14], suggesting that these hosts are devoting additional energy to fend off their parasites and the tissue damage that goes along with them. Contrarily, greater availability of food for other animal hosts, like *Daphnia*, encourages increasing tolerance of the parasites that they share food with, may be because less competition exists when there is more food available to both the host and the parasite [41]. Contrary to what researchers predicted, a low-resource diet improved host resistance to worm formation in the stomach [42]. It was also suggested that by starving the parasite, a low-resource diet may improve host immunity and resistance to infection. For instance, the increasing consumption of glucose by domesticated animal hosts can reduce their immune system's defences against gastrointestinal nematodes.

3.2 Future Prospects

In the context of animal ecology and evolution, the study of host tolerance is still in its infancy and is being put to the test in various settings. Over the course of the infection, the life of the host, and in response to environmental cues, tolerance might change. Diversity in tolerance is a pervasive feature, just as hosts differ in resistance. Given that some theoretical models suggest that tolerance will be driven to fixation in a population, the existence of diversity in tolerance is intriguing. The idea of tolerance is proving to be valuable, but as others have pointed out before us, different research applies it in different ways. The parasite's qualities and the host's traits work together to determine the host's fitness. In fact, separating the two and determining correlation can be challenging or perhaps impossible. However, research that emphasizes tolerance has the potential to increase both general immunology and our understanding of the ecology of host–parasite relationships.

3.3 Conclusion

This chapter discussed the importance of understanding natural resistance of prospective hosts toward animal parasites, particularly their relationships. Such interactions may have a considerable impact on host–parasite evolution, which may help us predict the outcomes of future parasite–host interactions. Numerous studies have demonstrated the genetic interaction between hosts and parasites as well as the wide genetic heterogeneity in host responses to parasites. The main factor affecting a host's resistance to parasite infections is its immunological makeup. In addition to age, other factors such as resource accessibility, host avoidance behaviors, or others may enhance host resistance against animal parasite control.

References

1 Ilea, R.C. (2009). Intensive livestock farming: global trends, increased environmental concerns, and ethical solutions. *Journal of Agricultural and Environmental Ethics* 22 (2): 153–167.
2 Otero-Abad, B., Rüegg SR, Hegglin D., et al. (2017). Mathematical modelling of *Echinococcus multilocularis* abundance in foxes in Zurich, Switzerland. *Parasites & Vectors* 10 (1): 1–12.
3 Deplazes, P., Eckert, J., Mathis, A., et al. (2016). *Parasitology in Veterinary Medicine*. Wageningen Academic Publishers.
4 Boots, M., Best, A, Miller, M.R. and White, A. (2009). The role of ecological feedbacks in the evolution of host defence: what does theory tell us? *Philosophical Transactions of the Royal Society B: Biological Sciences* 364 (1513): 27–36.
5 De Roode, J.C. and Lefèvre, T. (2012). Behavioral immunity in insects. *Insects* 3 (3): 789–820.
6 Cobb, N.A. (1894). *Contributions to an Economic Knowledge of Australian Rusts (Uredineae): Improving Wheat by Selection*. C. Potter, Government printer.
7 Stear, M. and Murray, M. (1994). Genetic resistance to parasitic disease: particularly of resistance in ruminants to gastrointestinal nematodes. *Veterinary Parasitology* 54 (1–3): 161–176.
8 Taylor, M.A., Coop, R.L., and Wall, R.L. (2015). *Veterinary Parasitology*. John Wiley & Sons.
9 Suzuki, Y. (2002). Host resistance in the brain against toxoplasma gondii. *The Journal of Infectious Diseases* 185 (Supplement_1): S58–S65.
10 Okumura, T. (1977). Relationship of attacking fly abundance to behavioral responses of grazing cattle. *Japanese Journal of Applied Entomology and Zoology* 21: 119–122. 21

11 Harris, J.A., Hillerton, J.E., and Morant, S.V. (1987). Effect on milk production of controlling muscid flies, and reducing fly-avoidance behaviour, by the use of Fenvalerate ear tags during the dry period. *Journal of Dairy Research* 54 (2): 165–171.

12 Bianchi, M.E. (2007). DAMPs, PAMPs and alarmins: all we need to know about danger. *Journal of Leukocyte Biology* 81 (1): 1–5.

13 Miller, H. (1996). Prospects for the immunological control of ruminant gastrointestinal nematodes: natural immunity, can it be harnessed? *International Journal for Parasitology* 26 (8–9): 801–811.

14 Cressler, C.E., Nelson, W.A., Day, T and McCauley, E. (2014). Disentangling the interaction among host resources, the immune system and pathogens. *Ecology Letters* 17 (3): 284–293.

15 Hart, B.L. and Hart, L.A. (1994). Fly switching by Asian elephants: tool use to control parasites. *Animal Behaviour* 48 (1): 35–45.

16 Hart, B.L., Hart, L.A., McCoy, M. and Sarath, C.R. (2001). Cognitive behaviour in Asian elephants: use and modification of branches for fly switching. *Animal Behaviour* 62 (5): 839–847.

17 Norval, R., Sutherst, R.W., Kurki, J. et al. (1988). The effect of the brown ear-tick *Rhipicephalus appendiculatus* on the growth of Sanga and European breed cattle. *Veterinary Parasitology* 30 (2): 149–164.

18 Norval, R., Sutherst, R.W., Jorgensen, O.G., et al. (1989). The effect of the bont tick (*Amblyomma hebraeum*) on the weight gain of Africander steers. *Veterinary Parasitology* 33 (3–4): 329–341.

19 Hart, B.L., Hart, L.A., Mooring, M.S. and Olubayo, R. (1992). Biological basis of grooming behaviour in antelope: the body-size, vigilance and habitat principles. *Animal Behaviour* 44 (4): 615–631.

20 Mckenzie, A.A. (1990). The ruminant dental grooming apparatus. *Zoological Journal of the Linnean Society* 99 (2): 117–128.

21 Eckstein, R.A. and Hart, B.L. (2000). The organization and control of grooming in cats. *Applied Animal Behaviour Science* 68 (2): 131–140.

22 Hart, B.L. (2005). The evolution of herbal medicine: behavioural perspectives. *Animal Behaviour* 70 (5): 975–989.

23 Huffman, M.A. and Seifu, M. (1989). Observations on the illness and consumption of a possibly medicinal plantVernonia amygdalina (Del.), by a wild chimpanzee in the Mahale Mountains National Park, Tanzania. *Primates* 30 (1): 51–63.

24 Huffman, M.A. and Caton, J.M. (2001). Self-induced increase of gut motility and the control of parasitic infections in wild chimpanzees. *International Journal of Primatology* 22 (3): 329–346.

25 Fruth, B., Ikombe, N.B., Matshimba, G.K., et al. (2014). New evidence for self-medication in bonobos: manniophyton fulvum leaf-and stemstrip-swallowing from LuiKotale, Salonga National Park, DR Congo. *American Journal of Primatology* 76 (2): 146–158.

26 Martin, S.A. and Król, E. (2017). Nutrigenomics and immune function in fish: new insights from omics technologies. *Developmental & Comparative Immunology* 75: 86–98.

27 Dezfuli, B., Bosi, G., DePasquale, J.A., et al. (2016). Fish innate immunity against intestinal helminths. *Fish & Shellfish Immunology* 50: 274–287.

28 Harris, N.L. (2017). Recent advances in type-2-cell-mediated immunity: insights from helminth infection. *Immunity* 47 (6): 1024–1036.

29 Reite, O.B. and Evensen, Ø. (2006). Inflammatory cells of teleostean fish: a review focusing on mast cells/eosinophilic granule cells and rodlet cells. *Fish & Shellfish Immunology* 20 (2): 192–208.

30 Sayyaf Dezfuli, B., Giari, L, Squerzanti, S., et al. (2011). Histological damage and inflammatory response elicited by *Monobothrium wageneri* (Cestoda) in the intestine of *Tinca tinca* (Cyprinidae). *Parasites & Vectors* 4 (1): 1–11.

31 Costa, M.M., Maehr, T., Diaz-Rosales, P., et al. (2011). Bioactivity studies of rainbow trout (*Oncorhynchus mykiss*) interleukin-6: effects on macrophage growth and antimicrobial peptide gene expression. *Molecular Immunology* 48 (15–16): 1903–1916.

32 Meeusen, E.N.T. and Balic, A. (2000). Do eosinophils have a role in the killing of helminth parasites? *Parasitology Today* 16 (3): 95–101.

33 Reinhardt, S., Scott, I., and Simpson, H.V. (2011). Neutrophil and eosinophil chemotactic factors in the excretory/secretory products of sheep abomasal nematode parasites. *Parasitology Research* 109 (3): 627–635.

34 Anthony, R.M., Rutitzky, L., Urban, J.F., et al. (2007). Protective immune mechanisms in helminth infection. *Nature Reviews Immunology* 7 (12): 975–987.

35 Capron, M. and Capron, A. (1992). Effector functions of eosinophils in schistosomiasis. *Memorias Do Instituto Oswaldo Cruz* 87: 167–170.

36 Meeusen, E.N., Balic, A., and Bowles, V. (2005). Cells, cytokines and other molecules associated with rejection of gastrointestinal nematode parasites. *Veterinary Immunology and Immunopathology* 108 (1–2): 121–125.

37 Piedrafita, D., Parsons, J.C., Sandeman, R.M., et al. (2001). Antibody-dependent cell-mediated cytotoxicity to newly excysted juvenile *Fasciola hepatica* in vitro is mediated by reactive nitrogen intermediates. *Parasite Immunology* 23 (9): 473–482.

38 Weller, P.F. (1994). Eosinophils: structure and functions. *Current Opinion in Immunology* 6 (1): 85–90.

39 Yasuda, K. and Kuroda, E. (2019). Role of eosinophils in protective immunity against secondary nematode infections. *Immunological Medicine* 42 (4): 148–155.

40 Knutie, S.A., Wilkinson, C.L., Qiu, C.W., et al. (2017). Host resistance and tolerance of parasitic gut worms depend on resource availability. *Oecologia* 183 (4): 1031–1040.

41 Vale, P.F., Wilson, A.J., Best, A., et al. (2011). Epidemiological, evolutionary, and coevolutionary implications of context-dependent parasitism. *The American Naturalist* 177 (4): 510–521.

42 Kyriazakis, I., Tolkamp, B., and Hutchings, M. (1998). Towards a functional explanation for the occurrence of anorexia during parasitic infections. *Animal Behaviour* 56 (2): 265–274.

4

Immune Response against Protozoan Parasites

Ahmed Olatunde[1,], Olalekan Ogunro[2], Habibu Tijjani[3], Shakirdeen Mayowa Obidola[4], Mustapha Abdullahi Akpaki[5], Archana Yadav[6], Manisha Nigam[7], and Abhay Prakash Mishra[8]*

[1] *Department of Medical Biochemistry, Abubakar Tafawa Balewa University, Bauchi, Bauchi, Nigeria*
[2] *Department of Biological Sciences, Kola Daisi University, Ibadan, Oyo, Nigeria*
[3] *Department of Biochemistry, Bauchi State University, Gadau, Bauchi, Nigeria*
[4] *Department of Crop Production Technology, Federal College of Forestry, Jos, Plateau, Nigeria*
[5] *Department of Biochemistry, Federal University of Technology, Owerri, Imo, Nigeria*
[6] *Department of Microbiology, Institute of Biosciences and Biotechnology, C.S.J.M. University, Kanpur, Uttar Pradesh, India*
[7] *Department of Biochemistry, Hemvati Nandan Bahuguna Garhwal University, Srinagar Garhwal, Uttarakhand, India*
[8] *Department of Pharmacology, School of Clinical Medicine, Faculty of Health Sciences, University of the Free State, Bloemfontein, South Africa*
* *Corresponding author*

4.1 Introduction

Following the entry into the host organisms, parasites first come in contact with specialized cells (antigen-presenting cells). These cells either induce phagocytosis by eating-up the whole parasitic protozoa or attack on the proteins produced by the invading protozoa. The antigen presenting cells possess intense inducible expression of major histocompatibility compounds (MHC). These compounds are involved in the presentation of engulfed proteins/antigens produced by the parasitic protozoa to the T-cells, which in turn, play important roles in immune response [1–3]. Therefore, the immune defense mechanisms, both antibody- and cell-mediated, are often stimulated by parasitic infections, and the responses that are most effective are dependent on the specific parasite and the stage of infection. Inside the T-cells, the $CD4^+$ cells, T-helper (Th) cells, display major modulatory action in the immune response. Based on the type of the invading parasitic protozoa, the $CD4^+$ cells turn into distinct subunits known as the Th1 and Th2 cells.

Parasitic Infections: Immune Responses and Therapeutics, First Edition.
Edited by Abhay Prakash Mishra and Manisha Nigam.

This process was proposed by Mosmann and co-workers and with modification stated to be a valuable paradigm [1, 4]. The Th1 $CD4^+$ cells stimulate the cell-regulated cascade of effector mechanisms. Also, an immune system stimulus involving Th1-type is associated with synthesis of interferon-gamma (IFN-γ), interleukin-12 (IL-12), and other meditators. Th1-type immune response is also linked to potent stimulation of monocytes and macrophages, however $CD8^+$ cells are involved in fighting various parasites in cells. On the other hand, the immune response involving Th2 type is linked to the production of IL-13, IL-5, IL-4 and other mediators as well as significant production of antibody. The pathway is efficacious against invading parasites found outside of cells [1].

However, protozoans such as *Plasmodium* and trypanosome have developed a variety of ways to evade host immune systems, and their structures and metabolic processes are unique to each species. To get around specific immune reactions, some parasitic protozoans have devised new ways to provoke diverse kinds of immune responses in host organisms. In contrast to fungi, bacteria, and viruses, protozoans cause distinct immune reactions [5, 6], but certain protozoans that cannot be eaten by macrophages and can survive and even multiply inside of them. For instance, *Trypanosoma brucei*, an external parasitic protozoon that can produce an immunological reaction [7]. To protect against *Leishmania* sp. infections, the host immune system relies on $CD4^+$ T lymphocytes, which recruit macrophages as effector cells and regulate them with Th1-type cytokines. Each species of *Plasmodium* has its own unique set of defense mechanisms. These range from cell-based to humoral, depending on the antigen and the protozoan's location [8].

Cytokines, important components of the immune response in host organisms, help to maintain a proper balance between the adaptive and innate immune system responses, which are critical in managing both infectious and chronic illnesses. Interferons (IFN), a broad group of cytokines, were initially discovered as a response to a viral infection. Because of their ability to prevent viral replication in mammalian cells, these antiviral factors are known as IFNs [9]. There has been a plethora of investigations since their discovery, leading to the discovery of other chemically related substances. Inflammatory factor nuclear antigens come in three varieties: type I, type II, and a newly discovered type III inflammatory factor nuclear antigens. In terms of structural characteristics and receptor molecule affinities, they fall into distinct categories [10]. Antigen presentation and natural killer (NK) cell activity can be enhanced by IFN-1. Their management is also assumed to be facilitated by antigen-specific B cells, as well as immunological memory against a variety of diseases [11]. However, IFN-1 can exacerbate disease in some situations, and these fundamental traits are important in limiting pathogen proliferation in the great majority of the time. There are number of parasitic diseases caused by protozoans, this includes trypanosomiasis, leishmaniasis, malaria, and toxoplasma, which disproportionately impact people in under-developed nations [12]. The present chapter highlights the various mechanisms used by parasitic protozoans to stimulate immune responses in host organism.

4.2 Specific Immune Response to Protozoa Infections

4.2.1 *Plasmodium* Species-Induced Immune Responses

Diseases caused by *Plasmodium* parasites are usually deadly if not treated. *Plasmodium* parasite induced immune responses is stimulated by the release of sporozoites from bites of a mosquito on the skin of a host organism. In the outer layer of the skin (dermis), the sporozoites are acted upon by innate immune system involving macrophages, neutrophils, mast cells, dendritic cells, γδ T cells, NKT and NK cells. In other parasitic protozoans, neutrophils are among the initial immune responses to mosquito sporozoites. The neutrophils can be seen in the dermis after twenty minutes of infection [13]. But, the depletion of neutrophil seen in infection induced by sporozoite does not influence parasite development and distribution in the hepatic cells, indicating redundant compensatory actions for other types of cell [14]. In the hepatic cells, sporozoites of *Plasmodium* infect the liver and transform into merozoites. The liver macrophages (Kupffer cells) are involved in vital functions, in engulfing the parasites, and reduction in Kupffer cells elevates the sporozoite invasion of the liver cells [15]. In addition, other cells of innate system including eosinophils, neutrophils, macrophages and monocytes, all attack the hepatocyte in response to sporozoite. This attack is potently linked to BALB/c mice resistance to infection induced by *P. yeolii* and *P. berghei* [16, 17]. Once the disease progresses to the red blood cell stage, major innate system mechanisms for removal depend on the potential of macrophages and monocytes to ingest and remove the infected red blood cells [18]. CD36 is a vital receptor for the clearance of infected red blood cells. Moreover, individuals that lack CD36 are more susceptible to malaria infection [19, 20]. In addition to their actions of overwhelming and inhibiting cells infected by *Plasmodium*, the innate immune cells release cytokines including IL-12, and IFN-γ, in response to molecular patterns linked to *Plasmodium*, which include hemozoin, DNA, RNA, and glycosylphosphatidylinositol [21]. Both IL-12 and IFN-γ enhance resistance against infection induced by *Plasmodium*. Although, *Plasmodium* RNA recognition via the MDA5-MAVS cascade stimulates responses of the type 1 IFN that result to protection in certain experimental conditions, Toll-like receptor (TLR)-MyD88-controlled *Plasmodium* pathogen-associated molecular pattern recognition enhances pathogenesis of *Plasmodium* parasites. Also, MyD88 experimental mice have significantly increased resistance to *P. berghei* and *P. chabaudi*, and antagonists of TLR confer protection from malaria of the cerebrum in mice

[21]. Both $CD4^+$ and $CD8^+$ cells are employed in vital functions in conferring protection against *Plasmodium* disease [22]. The $CD8^+$ T cells are vital for mediating and attacking the plasmodium parasite at the hepatocyte level [23–26], while $CD4^+$ T cells play an essential role at both blood and hepatic stages against the parasites of *Plasmodium* [27, 28]. However, a work indicated that $CD4^+$ T cells are vital in conferring protection while the $CD8^+$ T cells are not too significant [29]. Therefore, the organized coordination between T cells is vital for inducing protection against the malaria parasite [18].

4.2.2 Trypanosome-Induced Immune Responses

Innate immune cells such as monocytes, macrophages, neutrophils, dendritic cells and NK are involved in the vital role of mediating trypanosome infections such as from *Trypanosoma cruzi* in host cells. *T. cruzi* recognition by the innate immune cells is vital in the production of protective compounds such as IFN-γ, nitric oxide, tumor necrosis factor (TNF)-α, and IL-12. Experimental mice lacking IFN-γ, inducible nitric oxide synthase, and IL-12 are strongly susceptible to infection caused by *T. cruzi* [30–32]. Reduction in BALB/c mice neutrophils and macrophages is detrimental in infection caused by *T. cruzi*, showing that these mediators are vital and the same important actions for NK cells were indicated by depletion researches [33, 34]. Cytotoxic processes are not directly involved in NK cells; however, they are the main source of IFN-γ. The IFN-γ signal transduction is vital for macrophages to potently attack *T. cruzi* in cells. In another study, experimental mice deficient in MyD88 were found to be strongly susceptible to infection induced by *T. cruzi* due to defective synthesis of pro-inflammatory protein such as IL-12 and IFN-γ. Efforts to locate upstream TLRs that identify ligands of *T. cruzi* showed TLR9, TLR7, TLR4 and TLR2 as potential innate receptors [35]. TLR4 identifies glycoinositolphospholipid and TLR2 identifies *T. cruzi* glycophosphatidylinositol. Also, TLR9 and TLR7 identify DNA and RNA fragments derived from *T. cruzi* respectively. Although, TLRs deficiency leads to elevated *T. cruzi* susceptibility, the infection in such deficient experimental mice is less affected compared to $Myd88^{-/-}$ mice, indicating incomplete redundancy within the TLRs [35].

Adaptive T cells are involved in the vital role of conferring protection against infection induced by *T. cruzi*. Mice lacking either $CD8^+$ or $CD4^+$ T cells have greater susceptibility to infection caused by *T. cruzi* [36, 37]. Subsequent to activation, IFN-γ-derived T-cells ($CD4^+$ and $CD8^+$ T cells) that cause stimulation of phagocytes such as macrophages, enhance killing of the parasite [38–40]. Furthermore, $CD8^+$ T cells specifically, can phagocytize cells that are infected. However, the adaptive response of the immune system is not adequate establish a sterilizing immunity, and the parasitic organism can form potent detrimental infections in most people [18].

4.3 Immune Evasion Mechanisms of Protozoan Parasites

Malaria, leishmania, trypanosomiasis, and amoebiasis are protozoan infections that cause some of the most lethal and frequent human diseases. The ability of protozoans to infiltrate and develop inside cells makes them change antigens found on their surface, and remove their protein coating. Also, they are able to remain undetected by the host immune system for long periods of time [41]. Antigenic mimicry, which is common in parasitic infections, can cause immunosuppression directly or indirectly. Activation of a specific subset of Th cells is one of the most difficult evasion tactics. To protect the body, the host has a vast range of defense systems, from simple barriers to highly complicated devices that can recognize and eliminate pathogens. While the humoral immune response targets parasites outside the body, the cell-mediated immune response targets parasites inside the body. Parasites can survive in the host despite the host's immune and inflammatory reaction to the parasite and its large amount of antigen. This is because parasites have evolved numerous evasion mechanisms during many years of evolution [2]. The evasion methods used by parasites differ depending on their life cycle stage, the point of entry into host organisms, and the environment they find themselves.

4.3.1 Malaria

Malaria infection is caused by *Plasmodium vivax*, *P. falciparum*, *P. ovale*, and *P. malariae*. Intracellular and extracellular stages are both part of the infection's life cycle. Due to low sporozoite density and rapid migration to the liver, the host's immune response is limited in the early stages of the disease. There are two ways for a parasite to survive: first, by infecting hepatocytes and then by becoming a merozoite. After establishing itself in the host, the *Plasmodium* parasite evades the immune response of the host organism by altering the antigens on its surface throughout its life cycle [42]. Thus, each stage is associated with the expression of species- and stage-specific proteins, many of which are introduced into the red blood cells it infects or membrane of the parasite. Stage-specific proteins are challenging to immune response because of their high polymorphism or antigenic variability. Circumsporozoite protein (CSP), an antigen on malaria sporozoite surface that is susceptible to sequence polymorphism, is just one example among several *Plasmodium* proteins. Antigens with frequent repetitions can inhibit the synthesis of high-affinity antibodies and maturation of antibody isotype as superantigens of B-cells and essentially activating a polyclonal humoral response that is dependent on the thymus [43]. Antibodies are rendered useless by plasmodia, which shed their surface CSP coat while the extracellular malaria phases are still present in the bloodstream for only a short time, for example, in mammalian cells [44]. CSP has been shown to inhibit RNA translation through its RNA-binding motifs. For instance, the *P. falciparum* merozoite surface protein-1 causes blocking antibodies, which bind anywhere on merozoite surface protein-1 and prevent real inhibitory antibodies from binding [45]. Aside from CD36 and intracellular adhesion molecule-1, a protein of *P. falciparum* protein known as PfEMP-1, which is produced on infected erythrocytes surfaces, enhances vascular endothelial adhesion via CD36 and intracellular adhesion molecule-1. Based on this, infected cells are not transported to the spleen for phagocytosis because of their significantly altered appearance. Because of antibody syntheses for anti-PfEMP-1, the PfEMP-1 protein is subjected to clonal variation many times [46].

Molecular mimicry, which affects cellular immune response and cytokine release, is thought to be responsible for some of the clinical characteristics of malaria. Macrophages produce more TNF and IL-1, which induces the synthesis of acute-phase proteins and malarial fevers [47]. Two *P. falciparum* proteins with sequences like 1-thymosin have also been discovered, according to a study (a hormone from thymus that regulates T cells development). These peptides, like the human hormone, altered T-cell maturation in a comparable way, suggesting that they could interfere with the development of an effective cellular immune response. People with human malaria may develop autoantibodies because of molecular mimicry and B-cell activation [48]. Autoantibodies have been discovered in DNA from erythrocytes, carbohydrate epitopes, heat-shock protein 70, and polypeptide. The relationship between autoantibodies and malaria pathogenesis has been disputed. According to recent studies, autoantibodies may play a role in protecting against malaria. Despite the wide variety of surface proteins found on *Plasmodium*, some internal antigens appear to be substantially conserved among the many *Plasmodium* species. According to research, schizont antigens appear to initiate a cytokine cascade that is associated with most of the symptoms and pathologies linked to the infection [49, 50].

Infection with acute malaria reduces the immune response temporarily. Peripheral blood mononuclear cells' lymphoproliferative response and cytokine production are both diminished when stimulated by malaria antigens [51]. The number of T lymphocytesis also lowered. Activated macrophages' prostaglandins could be to blame for this effect, which indomethacin can partially mitigate. Releasing acute-phase proteins, which cling to lymphoid cells' surfaces, slows the development of lymphocytes. Some malaria antigens, however, have been proven to cause immunosuppression in the body on their own. In experimental mice, antigen-specific responses to *P. berghei* and *P. falciparum* schizont extracts have been found to be decreased [52]. Although parasite hemozoin accumulation has been postulated to inhibit macrophage auxiliary function, the mechanisms responsible for suppression are still speculative. Certain strains of *P. falciparum* exhibit CSP variations with T-cell epitopes identical to the original CSP, according to a recent investigation of the immune response against it.

Memory T-cell effector functions such as synthesis of lymphokine, proliferation and cytotoxicity are altered by changing T-cell epitope binding to MHC. Spirochetes need MHC-class I alleles to attach to hepatocytes during the liver stage of infection.

4.3.2 African Trypanosomiasis

Humans are infected with two strains of trypanosomes (*T. brucei rhodesiense* and *T. brucei gambiense*) that cause African trypanosomiasis, while cattle are infected with four different strains of the parasite (*T. brucei brucei*, *T. vivax*, *T. evansi* and *T. congolense*) [53]. Antibody-mediated elimination is a possibility for eradicating these trypanosomes because they are present in the circulation and do not have intracellular stages. Liver macrophages remove the vast majority of trypanosomes from the circulation. But, the remaining parasitic organisms are able to survive and spread disease because of antigenic polymorphism in the so-called variant surface glycoprotein (VSG). Many VSG genes are present on a single trypanosome, and each encodes a unique main sequence, particularly at the N terminus [54, 55]. An unexpected fact is that one VSG gene is only expressed by each parasite in turn, with the previous gene being replaced at its telomeric active transcriptional location. Thymus-independent humoral response is circumvented by trypanosomes, which can escape it through a reoccurring antigenic shift inside the VSG. This leads to a cycle of parasitemia that is similar to that of being infected by closely related but distinct diseases. As a result, it's been challenging to create a vaccine [56].

In order to understand the mechanism used by a parasitic organism to interfere with host immune system and affects cytokines and other mediators, researchers have looked into how the parasitic organism interacts with host immune system and allows the progression of the disease. B and T lymphocyte populations are changed in several ways during parasitemia. Reduction of thymus-independent humoral response against persistent *Trypanosoma* antigens, but thymus-independent humoral response against VSG surface epitopes is likely to prevent repeated parasitemias. Additionally, trypanosome products alter the pattern of cytokine release in macrophages and $CD8^+$ T cells. VSG's glycosylphosphatidylinositol group causes macrophage over-activation and the release of TNF-α by suppressor macrophages [57], which causes $CD8^+$ T cells to produce considerable IFN levels [58], which is analogous to the parasite protein T-cell-triggering factor. The proliferative T-cell response is impaired and parasite eradication is impaired when IFN levels are high enough to suppress synthesis of IL-2 and the expression IL-2 receptor. A new research shows that trypanosomes and *Leishmania* share numerous genes expressing glycoprotein 63-like proteins [59]. Even if their function is uncertain, they may be involved in a major part in the immune evasion of trypanosomes. Glycoprotein 63-like protein on the parasite's surface corresponds with its vulnerability to complement-mediated lysis in different stages of the parasite. Unlike procyclic trypanosomes, which are sensitive to the glycoprotein 63-like protein, blood trypanosomes are resistant to it.

4.3.3 American Trypanosomiasis

According to recent studies, children and neonates who are infected with the parasitic organism (*T. cruzi*) that induces American trypanosomiasis (Chagas disease) usually die. Adults, on the other hand, are more likely to acquire a chronic sickness after an initial infection, leading to a syndrome known as megacardia with mega-esophagus, mega-colon, and degeneration of the peripheral and central nerve systems [60]. Antibody defense mechanisms are in action during infection, which contributes to the long-term effects on health as well as any pathological changes. An immune response suppression is often present, which becomes more apparent as the parasitemia moves through tissues and into other organs. Surprisingly, the phagocytic activity of immunosuppressed mice was significantly increased, suggesting that the inhibition of the mononuclear phagocytic system is not the source of the immunosuppression. There was a similar condition of elevated mononuclear phagocytic action and substantial immunosuppression in *P. vinckei* experimental infections [61]. *T. cruzi* also avoids harming macrophages by infiltrating non-phagocytic cells and altering the transcriptional pattern of cytokines generated. It has been shown that the glycosylphosphatidylinositol-anchored mucin AgC10 of *T. cruzi* connects to macrophages and enhances the release of IL-1, but not TNF-α or IL-12. Both proteins are crucial in the protective response to Chagas disease. Infected macrophages also produce IL-10 and transforming growth factor, which limit IL-12's induction and effects [62]. T-cell immunosuppression has been confirmed following an experimental infection with *T. cruzi*. Lymphocyte receptor expression such as IL-2R and cytokine production in the effector phase of the immune response are negatively affected by parasite chemicals. The spleen-associated non-adherent Thy-1^+ Ly-2^+ cells may be involved. A protozoal infection, like others, causes an immune system overproduction of immunoglobulin M antibodies when *T. cruzi* makes polyclonal B-cell activation. *T. cruzi* parasite eradication in mice is hampered by the attachment of these antibodies to the surface of trypomastigote and disruption of immunoglobulin G inhibitory antibodies binding. As a result, polyclonal B-cell activation can activate B lymphocytes that recognize autoantigens [63], leading to autoimmune illness in people infected. *T. cruzi* generates chemicals that are like those found in host cells, confusing the immune system, and causing autoimmune illness. It's unclear whether parasite or host antigens cause host cell death in an autoimmune reaction. To survive in vertebrate blood, *T. cruzi* utilizes anti-complement surface molecules such T-decay-accelerating factor (DAF), glycoprotein 58/68, and glycoprotein 160, as well as the endocytic pathway to turnover surface molecules and phospholipases to breakdown anchoring glycoproteins [64].

4.3.4 Amoebiasis

The parasite *Entamoeba histolytica* causes intestinal amoebiasis. The parasite's ability to lyse its hosts is vital. The amoebic granules' aggressive components, cystine proteases and hydrolytic enzymes, lead to host cell and tissue destruction. Antigens for immunoglobulin G and A are degraded by proteases [65]. This is critical because *E. histolytica* must overcome both secretory immunoglobulin A and serum immunoglobulin G antibodies to colonize and spread outside the intestine. When entering the host's tissues, amoeba trophozoites acquire complement-mediating substances and inactivate C5a and C3a regulators, avoiding complement's lytic action and the inflammatory response. *E. histolytica* uses a sophisticated method that guides antibodies deposited on the organism's surface toward the uroid area, where they are naturally removed as molecular clumps (capping) [66]. Amoebic liver abscesses in convalescence are thought to be associated to tissue invasion by *E. histolytica*, which has been repressed by cell-mediated immunity in the extra-intestinal amoebiasis cases including amoebic liver abscesses. *E. histolytica* affects T cells (particularly Th1 response) and macrophages in a variety of ways. Uninfected cells' IL-2 production seems to be suppressed by a soluble suppressor found in the serum of infected host organism. Phorbol myristic acid acetate or IL-2 in conjunction with ionomycin can reverse it, suggesting a problem with signal transmission [67].

The cytotoxic and cytokine-secreting abilities of macrophages are decreased during the acute phase of amoebic liver abscess. *E. histolytica* trophozoites uses monocyte locomotion inhibitory factor to suppress human macrophage respiratory burst. Direct exposure to *E. histolytica* and its products appears to inhibit macrophages for the parasite. MHC molecules related to the I region (Ia molecules) were not produced in mouse macrophages *in vitro* when amoebic antigens were present. When amoebic induction is active, *E. histolytica* or macrophages release prostaglandin E2, which was the source of the impact because cyclooxygenase inhibitors reversed it [67, 68]. When prostaglandin E2 activates the macrophage phosphokinase A pathway, it raises cAMP in macrophages, which decreases the production of Ia substances on the surfaces of macrophage and the formation of Th1 factors including IFN and IL-2 by T cells. There is no TNF-α production by macrophages in the peritoneum or the blood of people with acute lymphocytic leukemia, but there is by hepatic cells and macrophages in animals with infection. In vitro proliferation of mice-derived spleen cells injected with a 220 kDa surface protein failed, despite the presence of Th2 mediators such as IL-10 and IL-4. After the parasite has invaded the liver granuloma, *E. histolytica* may alter macrophage or T cell activity and cytokine production to aid its own survival [69].

4.3.5 Leishmaniasis

It is possible to classify human leishmaniasis into three distinct categories: skin, mucosal and visceral. In order to infect cells, *Leishmania* relies on the host cell's ability to phagocytose it, as it lacks a specific method of penetration. Host cell attachment and uptake of promastigotes is facilitated by the surface-exposed protease glycoprotein 63 and a lipophosphoglycan, while macrophage proteolytic activity is suppressed [70]. The immune evasion activities of these chemcials appear to be complementary, which is surprising. Since lipophosphoglycan prevents parasites from merging with the cells' lysosomes, it helps promastigotes survive *Leishmania donovani* infection earlier on. It's possible that glycoprotein 63 blocks the degradation of the phagolysosome if the phagolysosome does form. They become amastigotes inside the macrophage because this stage of the parasite is more active in an acidic environment compared to non-acidic one. A lack of enzymes like catalase and super oxide dismutase means that promastigotes have to rely on their ability to prevent the cellular burst in order to survive. Glycoprotein 63 and lipophosphoglycan scavenge oxygen intermediates directly by preventing protein kinase C from undergoing an oxidative burst and decreasing its membrane translocation. Oxygen intermediates can be directly scavenged from lipophosphoglycan's structure, which contains repeating oxidizable phosphorylated disaccharide units [71, 72].

Apoptosis, MHC presentation, and cytokine responsiveness are all suppressed by the parasite, allowing it to persist within the macrophage. According to prior studies, stimulation of GMCSF and TNF-α is associated with a decrease in apoptosis. MHC class-1 and class-2 molecules, and the loading of peptides onto those few MHC molecules produced in macrophages driven by IFN-γ are all inhibited by *Leishmania* components like glycosylinositolphospholipids. *L. major* and *L. donovani* glycoprotein 63 proteins cleave CD4 structures on the surface of T cells, breaking the connection between T helper cells and antigen-presenting cells [73]. The antigen (which is heat-stable) and B7-1 are two examples of the macrophage co-stimulatory molecules that amastigotes suppress production of in addition to internalizing and degrading class II MHC molecules. *Leishmania*, like *E. histolytica*, secretes glycoprotein and transforming growth factor, which suppresses macrophage function. Lipophosphoglycan is used to inhibit neutrophil and monocyte chemotaxis as well as to modulate infected macrophage response by downregulation of TNF-α receptor expression. Perhaps the parasite's methods of avoiding complement are among the most complicated ever evolved by a living thing [67]. To tolerate complement, metacyclic promastigotes shed the membrane attack complex (C5b–C9) on their surface, which may be due to lipophosphoglycan structure elongation on the surface of those promastigotes. When C3b is transformed to C3bi on its surface, gp63 protects the parasite. C3, C5, and C9 are all components of the complement system that can be phosphorylated by *Leishmania* kinases, which would prevent either pathway from being activated. An immune response based on a Th1 reaction has been shown to be effective in preventing *L. major* infection, while susceptibility appears to be associated with a Th2 response. Instead of increasing IL-12 production, *L. major* metacyclic promastigotes have been reported to actively inhibit the transcription of the gene. Because interferon-gamma (IFN-γ) production is primarily stimulated by IL-12 production, and because IFN-activated macrophages can kill *Leishmania*, reducing the synthesis of IL-12 is expected to produce a significant survival benefit to *Leishmania* [74]. With these findings and their application to human patients, the future seems brighter.

4.4 Parasitic Protozoans and Their Mechanism to Induce Disruption of the Complement in Host Cells

Over the years of evolution, parasitic organisms including ectoparasites, parasitic protozoans, and helminths, have formed well-organized mechanistic actions as means of avoiding attack by the host immune system. The initial process formed was complement activation abrogation to attack the host immune clearance, mainly at the initial phases of attack. Complement system plays a vital role in the protection against infection induced by parasites, which depends on its lysis on attacking parasites via membrane attack complex formation, and serves as a connector to responses of the adaptive immune response. Parasitic organisms can execute many mechanisms that interrupt the actions of the complement in host cells as the first process to interrupt the immune attack in the host organism [75].

4.4.1 Host Regulatory Proteins Recruitment to Abrogate Complement Activation

Complement activation soluble modulator, known as C1 inhibitor (C1-INH), negatively controls the lectin and classical cascades by abrogating MBL-associated serine protease (MASP)-1, C1s and C1r, the stimulating proteases of the complement pathway [75]. *Plasmodium falciparum* (an intracellular protozoa) can aggregate and use C1-INH to abrogate activation of the complement system [76]. The invasive type of malaria (merozoites) translocate C1-INH to the surface when in-contact with human serum. pfMSP3.1 (one of the merozoite surface protein 2 family) act as a main interactive co-molecule to interact with C1-INH, to abrogate C1s, MASP2, and MASP1 [77]. Factor H (FH) from host complement system is the major inhibitor of the alternative cascade against the formation of C3-convertase. It stimulates the inhibition through promotion of the formation of iC3b from C3b. *Echinococcus granulosus* (a type of cestode that induces human cystic echinococcosis) presents a protein that aligns host FH on the wall of hydatid cyst to abrogate the complement C3b deposition [78–80]. It was later discovered that InsP6 was the binding molecule (which is a main part of the acellular laminated layer of the wall of the hydatid cyst) that interacts with host FH to abrogate the alternative cascade [81]. Also, cells of the epithelium found in the midgut of mosquito express 2 receptor proteins that interacted with host FH to abrogate deposition of C3b and impair the alternative complement cascade activation [82].

Another regulatory factor of the human complement system, DAF induces the increased decay of the C5 and C3 convertases by linking to C3b and C4b found on the membrane of cell [83, 84]. When incubated with human red blood cells, although not with erythrocytes lacking DAF, the blood fluke (*Schistosoma mansoni*), which induces schistosomiasis of the intestine, showed resistance to lysis of the complement system [85]. Another work indicated that the blood fluke obtained DAF from the red blood cell from host organism through glycosylphosphatidylinositol anchor expression on the worm surface. The potential of the *S. mansoni* worm to obtain DAF was decreased following treatment with trypsin [86]. Glycosylphosphatidylinositol-specific phospholipase D treatment promoted the interaction of DAF with the surface of schistosomula [87].

As a result of its action of inhibiting the cytolytic membrane attack complex, CD59 deceases the polymerization of C9 found on the surface of cell by intereacting with C9 and C8α. The CD59 glycosylation is associated with its complement-inhibitory action [88, 89]. The malarial parasite, *P. falciparum*, can obtain CD59 to impede invasion on the red blood cell that is infected. In addition, *P. falciparum* showed mannosyltransferase, which is involved in the synthesis of glycosylphosphatidylinositol, and then causes elevation in CD59 levels (which is a GPI-anchored protein) on the surface of cells, showing that the anchor with glycosylphosphatidylinositol is vital for CD59 expression on *P. falciparum*-infected erythrocyte surfaces [90]. *Trichomonas vaginalis*, which is an anaerobic flagellated parasitic organism, obtained CD59 from various cells in the host organism, including erythrocytes, during disease stage to inhibit the lysis of the parasite by the complement system of the host organism [91].

4.4.2 Expression of Homologous Proteins with Host Modulators of Complement System Activation

To prevent degradation controlled by the complement system, certain parasites show different types of surface complement regulatory proteins. As more genomes of parasites have been studied, helminths have been reported to show several mammalian-like receptor proteins for cytokines, growth hormones from host cells to control the development, reproduction, growth and signal cascade pathway of the parasite [92]. Furthermore, it was reported that parasites control the anti-parasite immune system of the host organism by showing host receptors of certain components of the immune system. The similarity of antigenicity between parasite- and host-expressed proteins may overwhelm the ability of the immune system of the host organism to recognize attacking parasites, and therefore prevent the elimination of the parasite [93, 94]. The host complement-orthologs are expressed by parasites to inhibit or regulate the activities of the complement system found in the host organism. A blood fluke, *Schistosoma japonicum* (which induces Asian schistosomiasis), expresses schistosome C2 receptor inhibitor trispanning (CRIT) protein that has similarities with CRIT of the host complement [95]. The expression is an indication of their presence in host organism stimulation. Furthermore, the schistosome-linked CRIT is found on the *Schistosoma* parasite tegument and it allows it to interact with C2 through a region found on the extracellular region. Subsequently, it abrogates the interaction of C2 and C4b, to disrupt the C3 convertase formation. The CRIT (a type of molecular mimicry) was showed to interact with a region that is similar to one domain found in human C4b. When C2 is interfered with, the lectin and classical complement cascade are impeded [96]. The schistosome CRIT C2 interacting region is found on 11-amino acids of the first extracellular region, which is vital in the decrease of inflammation mediated by the immune complex and classical complement cascade inhibition [97]. *T. cruzi*, the parasitic organism inducing American trypanosomiasis, also shows CRIT expression on the membrane of trypomastigotes to abrogate activation of the complement system linked to C2 [93].

Moreover, matured schistosomes show C3 receptor protein expression on the surface of their tegument. During stimulation of complement system, C3 interacts with the surface of the worm via the C3 receptor protein and activates the substitution of the tegument found on the outside region. This is then shed during the attack of the complement system [98]. The C3 receptor expression and C3-C3 receptor shedding allow parasites to consume serum C3, and then transform into non-stimulators of the alternative cascade [99, 100].

Also, the trypomastigote of *T. cruzi* expresses DAF on its virulent form to abrogate stimulation of the host complement system by C3 blocking, which is similar to DAF found in the host mammals [101–103]. Other works showed that *T. cruzi* expressed a glycoprotein gene (160 kDa) regulatory glycoprotein of the complement system on the surface of the trypomastigote. The glycoprotein 160 mRNA was found to share marked sequence of DAF gene-linked DNA homologous from the human host [104, 105]. Formation of the classical and alternative C3 convertase is inhibited by GP160 because it is part of the C4/C3 interacting group of complement mediators. The process above inhibits implication and activation of the host complement pathway on the surface of the parasitic organism [106, 107]. Another work showed that a schistosome complement inhibitor protein showed expression on the membrane of adult and larvae *S. mansoni*, and it is antigenically and functionally associated to human CD59. This protein interacts with C9 and C8 found in humans and impedes C5b-9 assembly [108]. Furthermore, other homologs of CD59 were discovered in the total genetic make-up of schistosome showing CCXXXCN as the consensus sequence at the C terminal region and around schistosome tegument found in the membrane fraction [109, 110]. A homolog of CD59 (FhCD59-1,2,3), was discovered on the surface membrane of tegument of *Fasciola hepatica* (a trematode). FhCD59-2 displayed a phylogenetic association with SmCD59-2 existing on *S. mansoni* membrane tegument [111]. But, isomers of recombinant schistosome CD59 in mammals indicated no abrogation of complement action in an *in vitro* study. This was proposed to be different from the activity of tegument-derived native proteins [112].

4.5 Cell-Mediated Killing Induced by Protozoans

Cell-mediated immunity serves as a key line of defense against protozoan infections. The effector cells immediately participating in the process include neutrophils, macrophages, and finally activated macrophages. There are huge differences in activity within this simple framework. To deal with a specific protozoan infection, effector cells from various animal species may be more or less effective, based on the species of animal and even the strain of the same species. Depending on the species, strain, and morphological shape of the protozoa in question, the sensitivity to cell-mediated death varies. They are the most vulnerable to attack from the vertebrate host if they are derived from insect vectors that have not yet acquired defense systems. *Trypanosoma promastigotes* and *Trypanosoma epimastigotes* are easily destroyed by antigen-presenting cells, whereas *Leishmania* trypomastigotes and amastigotes are far more resistant [18]. Microbic protozoa, such as *Leishmania amastigotes*, *Toxoplasma endozoites* (tachyzoids), and *T. cruzi* amastigotes with reticulotropic morphology, have developed extraordinary resistance to macrophage microbicidal activity, and this resistance has been passed on to the host cell. *T. cruzi* amastigotes that reside in muscle cells, in contrast to those from other strains, have not established this resistance to cell-mediated killing by macrophages. *T. brucei* and *Plasmodium* merozoites in the bloodstream are readily accessible protozoa that do not have the extensive resistance developed by reticulotropic protozoa. However, they do have some protection because phagocytic cells only attack them when a specific antibody is present, which is not always the case. This provides some protection. Neutrophils have a crucial role in granulocyte-mediated killing. Because of their small size, eosinophils don't seem to have much of an impact when they're mixed together with neutrophils. Among the protozoal species studied, only those of the *Trypanosoma stercorariana* genus have been demonstrated to be significantly influenced by eosinophilia. Eosinophils can kill parasite *T. cruzi* antibody-coated trypomastigotes, despite the fact that this has only been demonstrated in vitro. Even more strangely, this species appears to be the only one that can be infected by eosinophils, an extremely toxic immune system constituent that was found to be highly effective against helminth infections. Pathogens such as *Trypanosoma*, trypomastigotes of salivary, free merozoites of *Plasmodium*, promastigotes, and amastigotes of *Leishmania* are among those that neutrophils are ineffective in destroying [11].

4.6 Role of Inflammasome in Protozoan Infections

An innate immune system is critical in the quick identification and elimination of attacking organisms via various processes, including phagocytosis and the activation of inflammation [113]. Also, it is important in priming and stimulating the adaptive immunity, which leads to potent immunity. Cells of the host organism such as the myeloid and epithelial lineages, and others show several germline-encoded pattern recognition receptor proteins that identify polysaccharide, conversed protein, nucleic acid, or motifs, lipid and stimulate various inflammatory cascades. The pattern recognition receptors include TLRs, C-type lectin receptors, cytosol-based NOD-like receptors (NLRs), HIN200 proteins, and retinoate inducible gene-1-like helicase [114, 115]. To date, more than 20 NLRs in humans have been discovered. These receptors are associated with resistance to R protein of plant diseases and to apoptosis protease-stimulating factor 1 [116, 117]. They are similar to proteins with a tripartite structure containing of an N-terminal protein-protein interacting module, which can be baculovirus IAP repeat, a caspase recruitment domain (CARD), pyrin domain, and C-terminal agonist-sensing region, which in NLRs is a sequence of Leu-rich repeats, similar to TLRs [113]. The initial NLRs discovered were used determine bacterial peptidoglycan groups and stimulate inflammation by triggering MAP kinase and NF-κB cascades [118–121]. Additional work on the NLR class indicated that certain members can form inflammasome, which is a multi-protein complexes that aggregate and stimulate the activation pro-caspase-1 via adaptor proteins including apoptosis-associated speck-like protein consisting of a CARD [122].

Several inflammasomes such as members of the NLR class (NAIP5, NLRP3, NLRC4, and NLRP1) and recently, other inflammasomes such as inducible gene-1 and AIM2, have been described. Adaptor protein is needed for the recruitment of caspase-1 to the inducible gene-1 inflammasomes, NLRP3, and AIM2 whereas its action in NLRC4, NIAP5-associated complexes and NLRP1 is still speculative. After the stimulation of inflammasome, the pool of apoptosis-associated speck-like protein containing a CARD (ASC) in the cell turns into a big molecule known as pyroptosome, which is the region of activation of caspase-1. Inflammasomes are stimulated by various signals and may show redundant actions during the infective period. Impairment and or mutation or changes in mRNA encoding molecules of inflammasome were associated with human genetic diseases, such as a continuum of auto-inflammatory syndromes ranging from Muckle-Wells syndrome and familial cold urticarial to sepsis, Crohn's disease, neonatal onset multisystem inflammatory disorder, as well as susceptibility to infections induced by some pathogenic organisms [113, 123]. Caspase-1 processes several cellular substrates, which are requirement for inflammatory response induction. Specifically, pro-IL-18 and pro-IL-1β are converted into bioactive cytokine type's caspase-1 [124–126]. In addition, caspase-1 is necessary for the release of several pro-inflammatory mediators and they may not be required as substrate for caspase-1 [127]. Furthermore, excessive caspase-1 activation results to a type of cell death known as pyroptosis, which is characterized by necrosis and apoptosis [128, 129]. Among the several caspase-1 substrates, it has been indicated that the enzyme acts on enzymes of glycolysis, which may be the main mechanism to carry out cell death [113, 130].

4.6.1 Inflammasome and Protozoans

Malaria is caused by the *Plasmodium* genus, and it is endemic in sub-Saharan Africa and tropics [113]. The parasitic protozoa are transmitted by the bite of a mosquito into the host organism and then multiplies within red blood cells, following a period of incubation [131]. The breakdown of hemoglobin by the malaria parasite leads to heme removal from hemoglobin, which is thereafter turn into the crystalline molecule called hemozoin [132]. Recently, it was documented that inflammasome NLRP3 might be involved in a vital function of hemozoin identification. This crystalline product was shown to stimulate caspase-1 and active the release of IL-1β in a NLRP3 and ASC, however not NLRC4 in an in vitro model [133, 134]. This finding was not reproducible in the work of Griffith et al., [135]. In a more interesting way, NLRP3 deficiency and not caspase-1 or ASC elevated the survival of mice following injection with erythrocytes infected with *P. chabaudi adami*, a virulent strain found in mice [134]. Also, it was found that the mice induced with cerebral malaria using *P. berghei* showed a similar outcome that *Nlrp3*$^{-/-}$ mice, however not *Il1b*, *Casp1*$^{-/-}$, or *Asc*$^{-/-}$ mice, displayed more time to mortality compared to wild-type experimental animals [136]. Therefore, it was proposed that NLRP3 may possess harmful action in malaria of the cerebrum, and not be dependent on IL-1β and caspase-1 [113]. Consistent to this, it was discovered that mice with *casp*$^{1-/-}$ were similar to wild-type experimental mice in their response to the *P. berghei* and *P. chabaudi* malaria animal models) [137]. Contrarily, experimental mice lacking caspase-12 (inhibiting protein of NF-κB and caspase-1) displayed improved immune and inflammatory responses to the malaria parasitic organisms and were susceptible to malaria of the cerebrum. This phenotype was not dependent on caspase-1, but occurred from excessive activation of NF-κB and IFNγ synthesis [137]. Malaria parasites are among the most well-studied parasites in terms of activation of inflammasome and a detailed study of other helminths and protozoa may result in a clearer picture of the immune system response to these parasites and other unknown actions of the NLRs and inflammasome [113].

4.7 Conclusion

Initially, the parasites and their antigens come in contact with antigen presenting cells to aid their presentation to the T-cells, resulting in immune response stimulation. This process stimulated the formation of Th1 and Th2 cells from the CD4 cells of the T-cells. Cell-regulated cascade, monocytes, macrophages, IFN-γ, IL-12 and other cytokines are stimulated by Th1 while antibodies, specifically IL-4, IL-5 and IL-13 are activated by Th2. Parasitic protozoans such as *Plasmodium*, *Trypanosoma* and others display different mechanistic actions for evading the host organisms and this depend on the stages of the lifecycles, their location in the host organism, and others. Furthermore, these organisms use various mechanisms to avoid host immunity (such as host complement activation and others) and inflammatory reactions, which are stimulated through the innate immune system of the host organisms and various unique pattern recognition receptor proteins including TLRs, and NLRs. Therefore, *Trypanosoma*, *Plasmodium* and other protozoans induce diseases in humans and animals through the invasion of various host immune systems and responses.

References

1 Pfaff, A.W. and Candolfi, E. (2003). Immune responses to protozoan parasites and its relevance to diagnosis in immunocompromised patients. *European Journal of Protistology* 39: 428–434.
2 Schmid-Hempel, P. (2009). Immune defence, parasite evasion strategies and their relevance for 'macroscopic phenomena' such as virulence. *Philosophical Transactions of the Royal Society of London Series B, Biological Sciences* 364 (1513): 85–98. doi: 10.1098/rstb.2008.0157.
3 Mantegazza, A.R., Magalhaes, J.G., Amigorena, S., and Marks, M.S. (2013). Presentation of phagocytosed antigens by MHC class I and II. *Traffic (Copenhagen, Denmark)* 14 (2): 135–152. doi: 10.1111/tra.12026.
4 Mosmann, T.R., Cherwinski, H., Bond, M.W. et al. (1986). Two types of murine helper T cell clone. I. definition according to profiles of lymphokine activities and secreted proteins. *The Journal of Immunology* 136: 2348–2357.
5 Arsić-Arsenijević, V., Dzamić, A., Mitrović, S. et al. (2003). Karakteristike imunskog odgovora u toku infekcije protozoama [characteristics of the immune response in protozoan infections]. *Medicinski Pregled* 56 (11–12): 557–563. doi: 10.2298/mpns0312557a.

6 Mahanta, A., Ganguli, P., Barah, P. et al. (2018). Integrative approaches to understand the mastery in manipulation of host cytokine networks by protozoan parasites with emphasis on *Plasmodium* and *Leishmania* species. *Frontiers in Immunology* 9: doi: 10.3389/fimmu.2018.00296.

7 Machado, H., Bizarra-Rebelo, T., Costa-Sequeira, M. et al. (2021). *Trypanosoma brucei* triggers a broad immune response in the adipose tissue. *PLoS Pathogens* 17 (9): e1009933. doi: 10.1371/journal.ppat.1009933.

8 Caljon, G., De Muylder, G., Durnez, L. et al. (2016). Alice in microbes' land: adaptations and counter-adaptations of vector-borne parasitic protozoa and their hosts. *FEMS Microbiology Reviews* 40 (5): 664–685. doi: 10.1093/femsre/fuw018.

9 Lin, F.C. and Young, H.A. (2014). Interferons: success in anti-viral immunotherapy. *Cytokine and Growth Factor Reviews* 25 (4): 369–376. doi: 10.1016/j.cytogfr.2014.07.015.

10 Ivashkiv, L.B. and Donlin, L.T. (2014). Regulation of type I interferon responses. *Nature Reviews. Immunology* 14 (1): 36–49. doi: 10.1038/nri3581.

11 Chaplin, D.D. (2010). Overview of the immune response. *The Journal of Allergy and Clinical Immunology* 125 (2 Suppl 2): S3–S23. doi: 10.1016/j.jaci.2009.12.980.

12 Ivashkiv, L.B. (2018). IFNγ: signalling, epigenetics and roles in immunity, metabolism, disease and cancer immunotherapy. *Nature Reviews. Immunology* 18 (9): 545–558. doi: 10.1038/s41577-018-0029-z.

13 Amino, R., Giovannini, D., Thiberge, S. et al. (2008). Host cell traversal is important for progression of the malaria parasite through the dermis to the liver. *Cell Host and Microbe* 3 (2): 88–96.

14 Mac-Daniel, L., Buckwalter, M.R., Berthet, M. et al. (2014). Local immune response to injection of *Plasmodium* sporozoites into the skin. *Journal of Immunology* 193 (3): 1246–1257.

15 Vreden, S.G. (1994). The role of Kupffer cells in the clearance of malaria sporozoites from the circulation. *Parasitology Today* 10 (8): 304–308.

16 Scheller, L.F., Wirtz, R.A., and Azad, A.F. (1994). Susceptibility of different strains of mice to hepatic infection with *Plasmodium berghei*. *Infection and Immunity* 62 (11): 4844–4847.

17 Vanderberg, J.P., Khan, Z.M., and Stewart, M.J. (1993). Induction of hepatic inflammatory response by *Plasmodium berghei* sporozoites protects BALB/c mice against challenge with *Plasmodium yoelii* sporozoites. *The Journal of Parasitology* 79 (5): 763–767.

18 Gurung, P. and Kanneganti, T.D. (2016). Immune responses against protozoan parasites: a focus on the emerging role of NOD-like receptors. *Cellular and Molecular Life Sciences* 73 (16): 3035–3051. doi: 10.1007/s00018-016-2212-3.

19 Patel, S.N., Serghides, L., Smith, T.G. et al. (2004). CD36 mediates the phagocytosis of *Plasmodium falciparum*-infected erythrocytes by rodent macrophages. *The Journal of Infectious Diseases* 189 (2): 204–213.

20 Aitman, T.J., Cooper, L.D., Norsworthy, P.J. et al. (2000). Malaria susceptibility and CD36 mutation. *Nature* 405 (6790): 1015–1016.

21 Gazzinelli, R.T., Kalantari, P., Fitzgerald, K.A., and Golenbock, D.T. (2014). Innate sensing of malaria parasites. *Nature Reviews. Immunology* 14 (11): 744–757.
22 Tsuji, M. (2010). A retrospective evaluation of the role of T cells in the development of malaria vaccine. *Experimental Parasitology* 126 (3): 421–425.
23 Rodrigues, M.M., Cordey, A.S., Arreaza, G. et al. (1991). CD8+ cytolytic T cell clones derived against the *Plasmodium yoelii* circumsporozoite protein protect against malaria. *International Immunology* 3 (6): 579–585.
24 Romero, P., Maryanski, J.L., Corradin, G. et al. (1989). Cloned cytotoxic T cells recognize an epitope in the circumsporozoite protein and protect against malaria. *Nature* 341 (6240): 323–326.
25 Sano, G., Hafalla, J.C., Morrot, A. et al. (2001). Swift development of protective effector functions in naive CD8(+) T cells against malaria liver stages. *The Journal of Experimental Medicine* 194 (2): 173–180.
26 Zavala, F., Rodrigues, M., Rodriguez, D. et al. (2001). A striking property of recombinant poxviruses: efficient inducers of in vivo expansion of primed CD8(+) T cells. *Virology* 280 (2): 155–159.
27 Butler, N.S., Moebius, J., Pewe, L.L. et al. (2012). Therapeutic blockade of PD-L1 and LAG-3 rapidly clears established blood-stage *Plasmodium* infection. *Nature Immunology* 13 (2): 188–195.
28 Carvalho, L.H., Sano, G., Hafalla, J.C. et al. (2002). IL-4-secreting CD4+ T cells are crucial to the development of CD8+ T-cell responses against malaria liver stages. *Nature Medicine* 8 (2): 166–170.
29 Oliveira, G.A., Kumar, K.A., Calvo-Calle, J.M. et al. (2008). Class II-restricted protective immunity induced by malaria sporozoites. *Infection and Immunity* 76 (3): 1200–1206.
30 Graefe, S.E., Jacobs, T., Gaworski, I. et al. (2003). Interleukin-12 but not interleukin-18 is required for immunity to *Trypanosoma cruzi* in mice. *Microbes and Infection* 5 (10): 833–839.
31 Holscher, C., Kohler, G., Muller, U. et al. (1998). Defective nitric oxide effector functions lead to extreme susceptibility of *Trypanosoma cruzi*-infected mice deficient in gamma interferon receptor or inducible nitric oxide synthase. *Infection and Immunity* 66 (3): 1208–1215.
32 Torrico, F., Heremans, H., Rivera, M.T. et al. (1991). Endogenous IFN-gamma is required for resistance to acute *Trypanosoma cruzi* infection in mice. *The Journal of Immunology* 146 (10): 3626–3632.
33 Chen, L., Watanabe, T., Watanabe, H., and Sendo, F. (2001). Neutrophil depletion exacerbates experimental Chagas disease in BALB/c, but protects C57BL/6 mice through modulating the Th1/Th2 dichotomy in different directions. *The European Journal of Immunology* 31 (1): 265–275.
34 Lieke, T., Graefe, S.E., Klauenberg, U. et al. (2004). NK cells contribute to the control of *Trypanosoma cruzi* infection by killing free parasites by perforin-independent mechanisms. *Infection and Immunity* 72 (12): 6817–6825.

35 Rodrigues, M.M., Oliveira, A.C., and Bellio, M. (2012). The immune response to *Trypanosoma cruzi*: role of toll-like receptors and perspectives for vaccine development. *Journal of Parasitology Research* 2012: 507874.

36 Tarleton, R.L., Grusby, M.J., Postan, M., and Glimcher, L.H. (1996). *Trypanosoma cruzi* infection in MHC deficient mice: further evidence for the role of both class I- and class II-restricted T cells in immune resistance and disease. *International Immunology* 8 (1): 13–22.

37 Tarleton, R.L., Koller, B.H., Latour, A., and Postan, M. (1992). Susceptibility of beta 2-microglobulin-deficient mice to *Trypanosoma cruzi* infection. *Nature* 356 (6367): 338–340.

38 Antunez, M.I. and Cardoni, R.L. (2001). Early IFN-gamma production is related to the presence of interleukin (IL)-18 and the absence of IL-13 in experimental *Trypanosoma cruzi* infections. *Immunology Letters* 79 (3): 189–196.

39 Silva, J.S., Morrissey, P.J., Grabstein, K.H. et al. (1992). Interleukin 10 and interferon gamma regulation of experimental *Trypanosoma cruzi* infection. *Journal of Experimental Medicine* 175 (1): 169–174.

40 Tarleton, R.L. (2015). CD8+ T cells in *Trypanosoma cruzi* infection. *Seminars in Immunopathology* 37 (3): 233–238.

41 Zambrano-Villa, S., Rosales-Borjas, D., Carrero, J.C., and Ortiz-Ortiz, L. (2002). How protozoan parasites evade the immune response. *Trends in Parasitology* 18 (6): 272–278. doi: 10.1016/s1471-4922(02)02289-4.

42 Rénia, L. and Goh, Y.S. (2016). Malaria parasites: the great escape. *Frontiers in Immunology* 7: 463. doi: 10.3389/fimmu.2016.00463.

43 Beeson, J.G., Drew, D.R., Boyle, M.J. et al. (2016). Merozoite surface proteins in red blood cell invasion, immunity and vaccines against malaria. *FEMS Microbiology Reviews* 40 (3): 343–372. doi: 10.1093/femsre/fuw001.

44 Li, X., Chen, H., Bahamontes-Rosa, N. et al. (2009). *Plasmodium falciparum* signal peptide peptidase is a promising drug target against blood stage malaria. *Biochemical and Biophysical Research Communications* 380 (3): 454–459. doi: 10.1016/j.bbrc.2009.01.083.

45 Rubio, M., Bassat, Q., Estivill, X., and Mayor, A. (2016). Tying malaria and microRNAs: from the biology to future diagnostic perspectives. *Malaria Journal* 15: 167. doi: 10.1186/s12936-016-1222-9.

46 Sartelet, H., Garraud, O., Rogier, C. et al. (2000). Hyperexpression of ICAM-1 and CD36 in placentas infected with *Plasmodium falciparum*: a possible role of these molecules in sequestration of infected red blood cells in placentas. *Histopathology* 36 (1): 62–68. doi: 10.1046/j.1365-2559.2000.00742.x.

47 Castro-Gomes, T., Mourão, L.C., Melo, G.C. et al. (2014). Potential immune mechanisms associated with anemia in *Plasmodium vivax* malaria: a puzzling question. *Infection and Immunity* 82 (10): 3990–4000. doi: 10.1128/IAI.01972-14.

48 Thapa, P. and Farber, D.L. (2019). The role of the thymus in the immune response. *Thoracic Surgery Clinics* 29 (2): 123–131. doi: 10.1016/j.thorsurg.2018.12.001.

49 Mourão, L.C., Cardoso-Oliveira, G.P., and Braga, É.M. (2020). Autoantibodies and malaria: where we stand? insights into pathogenesis and protection. *Frontiers in Cellular and Infection Microbiology* 2020 10: 262. doi: 10.3389/fcimb.2020.00262.

50 Rivera-Correa, J. and Rodriguez, A. (2020). Autoimmune anemia in malaria. *Trends in Parasitology* 36 (2): 91–97. doi: 10.1016/j.pt.2019.12.002.

51 Angulo, I. and Fresno, M. (2002). Cytokines in the pathogenesis of and protection against malaria. *Clinical and Diagnostic Laboratory Immunology* 9 (6): 1145–1152. doi: 10.1128/cdli.9.6.1145-1152.2002.

52 Calle, C.L., Mordmüller, B., and Singh, A. (2021). Immunosuppression in malaria: do *Plasmodium falciparum* parasites hijack the host? *Pathogens (Basel, Switzerland)* 10 (10): 1277. doi: 10.3390/pathogens10101277.

53 Ponte-Sucre, A. (2016). An overview of *Trypanosoma brucei* infections: an intense host-parasite interaction. *Frontiers in Microbiology* 7: 2126. doi: 10.3389/fmicb.2016.02126.

54 Pinger, J., Chowdhury, S., and Papavasiliou, F.N. (2017). Variant surface glycoprotein density defines an immune evasion threshold for African trypanosomes undergoing antigenic variation. *Nature Communications* 8 (1): 828. doi: 10.1038/s41467-017-00959-w.

55 Moreno, C., Temporão, A., Torres, T., and Sousa Silva, M. (2019). *Trypanosoma brucei* interaction with host: mechanism of VSG release as target for drug discovery for African trypanosomiasis. *International Journal of Molecular Sciences* 20 (6): 1484. doi: 10.3390/ijms20061484.

56 McLaughlin, E.J., Rubio-Pena, K., Dujeancourt-Henry, A., and Glover, L. (2022). Transcription dependent loss of an ectopically expressed variant surface glycoprotein during antigenic variation in trypanosoma brucei. *mBio* 13 (2): e0384721. doi: 10.1128/mbio.03847-21.

57 Onyilagha, C. and Uzonna, J.E. (2019). Host immune responses and immune evasion strategies in African trypanosomiasis. *Frontiers in Immunology* 10: 2738. doi: 10.3389/fimmu.2019.02738.

58 Welsh, R.M., Bahl, K., Marshall, H.D., and Urban, S.L. (2012). Type 1 interferons and antiviral CD8 T-cell responses. *PLoS Pathogens* 8 (1): e1002352. doi: 10.1371/journal.ppat.1002352.

59 Geiger, A., Bossard, G., Sereno, D. et al. (2016). Escaping deleterious immune response in their hosts: lessons from trypanosomatids. *Frontiers in Immunology* 7: 212. doi: 10.3389/fimmu.2016.00212.

60 Montgomery, S.P., Parise, M.E., Dotson, E.M., and Bialek, S.R. (2016). What do we know about Chagas disease in the United States? *The American Journal of Tropical Medicine and Hygiene* 95 (6): 1225–1227. doi: 10.4269/ajtmh.16-0213.

61 Forthal, D.N. (2014). Functions of antibodies. *Microbiology Spectrum* 2 (4): 1–17.

62 Lopez, M., Tanowitz, H.B., and Garg, N.J. (2018). Pathogenesis of chronic Chagas disease: macrophages, mitochondria, and oxidative stress. *Current Clinical Microbiology Reports* 5 (1): 45–54.

63 Ouaissi, A., Da Silva, A.C., Guevara, A.G. et al. (2001). *Trypanosoma cruzi*-induced host immune system dysfunction: a rationale for parasite immunosuppressive factor(s) encoding gene targeting. *Journal of Biomedicine & Biotechnology* 1 (1): 11–17. doi: 10.1155/S1110724301000055.

64 Kierszenbaum, F. (1999). Chagas' disease and the autoimmunity hypothesis. *Clinical Microbiology Reviews* 12 (2): 210–223. doi: 10.1128/CMR.12.2.210.

65 Serrano-Luna, J., Piña-Vázquez, C., Reyes-López, M. et al. (2013). Proteases from *Entamoeba* spp. and pathogenic free-living amoebae as virulence factors. *Journal of Tropical Medicine* 2013: 890603. doi: 10.1155/2013/890603.

66 Begum, S., Quach, J., and Chadee, K. (2015). Immune evasion mechanisms of *Entamoeba histolytica*: progression to disease. *Frontiers in Microbiology* 6: 1394. doi: 10.3389/fmicb.2015.01394.

67 Nakada-Tsukui, K. and Nozaki, T. (2016). Immune response of amebiasis and immune evasion by *Entamoeba histolytica*. *Frontiers in Immunology* 7: 175. doi: 10.3389/fimmu.2016.00175.

68 Verkerke, H.P., Petri, W.A., Jr., and Marie, C.S. (2012). The dynamic interdependence of amebiasis, innate immunity, and undernutrition. *Seminars in Immunopathology* 34 (6): 771–785. doi: 10.1007/s00281-012-0349-1.

69 Iyer, S.S. and Cheng, G. (2012). Role of interleukin 10 transcriptional regulation in inflammation and autoimmune disease. *Critical Reviews in Immunology* 32 (1): 23–63. doi: 10.1615/critrevimmunol.v32.i1.30.

70 Walker, D.M., Oghumu, S., Gupta, G. et al. (2014). Mechanisms of cellular invasion by intracellular parasites. *Cellular and Molecular Life Sciences: CMLS* 71 (7): 1245–1263. doi: 10.1007/s00018-013-1491-1.

71 Gupta, G., Oghumu, S., and Satoskar, A.R. (2013). Mechanisms of immune evasion in leishmaniasis. *Advances in Applied Microbiology* 82: 155–184. doi: 10.1016/B978-0-12-407679-2.00005-3.

72 Gurung, P. and Kanneganti, T.D. (2015). Innate immunity against *Leishmania* infections. *Cellular Microbiology* 17 (9): 1286–1294. doi: 10.1111/cmi.12484.

73 Bhattacharya, P., Thiruppathi, M., Elshabrawy, H.A. et al. (2015). GM-CSF: an immune modulatory cytokine that can suppress autoimmunity. *Cytokine* 75 (2): 261–271. doi: 10.1016/j.cyto.2015.05.030.

74 Tegla, C.A., Cudrici, C., Patel, S. et al. (2011). Membrane attack by complement: the assembly and biology of terminal complement complexes. *Immunologic Research* 51 (1): 45–60. doi: 10.1007/s12026-011-8239-5.

75 Shao, S., Sun, X., Chen, Y. et al. (2019). Complement evasion: an effective strategy that parasites utilize to survive in the host. *Frontiers in Microbiology* 10: 532. doi: 10.3389/fmicb.2019.00532.

76 Mejia, P., Diez-Silva, M., Kamena, F. et al. (2016). Human C1-inhibitor suppresses malaria parasite invasion and cytoadhesion via binding to parasite glycosylphosphatidylinositol and host cell receptors. *The Journal of Infectious Diseases* 213: 80–89. doi: 10.1093/infdis/jiv439.

77 Kennedy, A.T. and Wijeyewickrema, L.C. (2017). Recruitment of human C1 esterase inhibitor controls complement activation on blood stage *Plasmodium falciparum* merozoites. *Journal of Immunology* 198: 4728–4737. doi: 10.4049/jimmunol.1700067.

78 Zipfel, P.F. and Skerka, C. (2009). Complement regulators and inhibitory proteins. *Nature Reviews. Immunology* 9: 729–740. doi: 10.1038/nri2620.

79 Diaz, A., Ferreira, A., and Sim, R.B. (1997). Complement evasion by *Echinococcus granulosus*: sequestration of host factor H in the hydatid cyst wall. *Journal of Immunology* 158: 3779–3786.

80 Breijo, M., Anesetti, G., Martinez, L. et al. (2008). *Echinococcus granulosus*: the establishment of the metacestode is associated with control of complement-mediated early inflammation. *Experimental Parasitology* 118: 188–196. doi: 10.1016/j.exppara.2007.07.014.

81 Irigoin, F., Laich, A., Ferreira, A.M. et al. (2008). Resistance of the *Echinococcus granulosus* cyst wall to complement activation: analysis of the role of InsP6 deposits. *Parasite Immunology* 30: 354–364. doi: 10.1111/j.1365-3024.2008.01034.x.

82 Khattab, A., Barroso, M., Miettinen, T., and Meri, S. (2015). Anopheles midgut epithelium evades human complement activity by capturing factor H from the blood meal. *PLoS Neglected Tropical Diseases* 9: e0003513. doi: 10.1371/journal.pntd.0003513.

83 Fujita, T., Inoue, T., Ogawa, K. et al. (1987). The mechanism of action of decay-accelerating factor (DAF) DAF inhibits the assembly of C3 convertases by dissociating C2a and Bb. *Journal of Experimental Medicine* 166: 1221–1228. doi: 10.1084/jem.166.5.1221.

84 Heine, H., El-Samalouti, V.T., Notzel, C. et al. (2003). CD55/decay accelerating factor is part of the lipopolysaccharide induced receptor complex. *The European Journal of Immunology* 33: 1399–1408. doi: 10.1002/eji.200323381.

85 Horta, M.F. and Ramalho-Pinto, F.J. (1991). Role of human decay-accelerating factor in the evasion of *Schistosoma mansoni* from the complement-mediated killing in vitro. *Journal of Experimental Medicine* 174: 1399–1406. doi: 10.1084/jem.174.6.1399.

86 Ramalho-Pinto, F.J., Carvalho, E.M., and Horta, M.F. (1992). Mechanisms of evasion of *Schistosoma mansoni* schistosomula to the lethal activity of complement. *Memorias do Instituto Oswaldo Cruz* 87 (Suppl. 4): 111–116. doi: 10.1590/S0074-02761992000800016.

87 Carvalho, E.M., Metz, C.N., Davitz, M.A., and Ramalho-Pinto, F.J. (1994). Binding of GPI-PLD-treated DAF to the surface of *Schistosoma mansoni* schistosomula. *Brazilian Journal of Medical and Biological Research* 27: 457–462.

88 Venneker, G.T. and Asghar, S.S. (1992). CD59: a molecule involved in antigen presentation as well as downregulation of membrane attack complex. *Experimental and Clinical Immunogenetics* 9: 33–47.

89 Ninomiya, H., Stewart, B.H., Rollins, S.A. et al. (1992). Contribution of the N-linked carbohydrate of erythrocyte antigen CD59 to its complement-inhibitory activity. *Journal of Biological Chemistry* 267: 8404–8410.

90 Kim, Y.U. and Hong, Y. (2007). Functional analysis of the first mannosyltransferase (PIG-M) involved in glycosylphosphatidylinositol synthesis in *Plasmodium falciparum*. *Molecular Immunology* 24: 294–300.

91 Ibanez-Escribano, A., Nogal-Ruiz, J.J., Perez-Serrano, J. et al. (2015). Sequestration of host-CD59 as potential immune evasion strategy of *Trichomonas vaginalis*. *Acta Tropica* 149: 1–7. doi: 10.1016/j.actatropica.2015.05.003.

92 Hu, W., Yan, Q., Shen, D.K. et al. (2003). Evolutionary and biomedical implications of a *Schistosoma japonicum* complementary DNA resource. *Nature Genetics* 35: 139–147. doi: 10.1038/ng1236.

93 Ramirez-Toloza, G. and Ferreira, A. (2017). *Trypanosoma cruzi* evades the complement system as an efficient strategy to survive in the mammalian host: the specific roles of host/parasite molecules and *Trypanosoma cruzi* calreticulin. *Frontiers in Microbiology* 8: 1667. doi: 10.3389/fmicb.2017.01667.

94 Abu-Shakra, M., Buskila, D., and Shoenfeld, Y. (1999). Molecular mimicry between host and pathogen: examples from parasites and implication. *Immunology Letters* 67: 147–152. doi: 10.1016/S0165-2478(99)00004-8.

95 Inal, J.M. (2005). Complement C2 receptor inhibitor trispanning: from man to schistosome. *Springer Seminars in Immunopathology* 27: 320–331. doi: 10.1007/s00281-005-0009-9.

96 Cestari Idos, S., Krarup, A., Sim, R.B. et al. (2009). Role of early lectin pathway activation in the complement-mediated killing of *Trypanosoma cruzi*. *Molecular Immunology* 47: 426–437. doi: 10.1016/j.molimm.2009.08.030.

97 Inal, J.M., Schneider, B., Armanini, M., and Schifferli, J.A. (2003). A peptide derived from the parasite receptor, complement C2 receptor inhibitor trispanning, suppresses immune complex-mediated inflammation in mice. *The Journal of Immunology* 170: 4310–4317. doi: 10.4049/jimmunol.170.8.4310.

98 Silva, E.E., Clarke, M.W., and Podesta, R.B. (1993). Characterization of a C3 receptor on the envelope of *Schistosoma mansoni*. *The Journal of Immunology* 151: 7057–7066.

99 Marikovsky, M., Levi-Schaffer, F., Arnon, R., and Fishelson, Z. (1986). *Schistosoma mansoni*: killing of transformed schistosomula by the alternative pathway of human complement. *Experimental Parasitology* 61: 86–94. doi: 10.1016/0014-4894(86)90138-4.

100 Rasmussen, K.R. and Kemp, W.M. (1987). *Schistosoma mansoni*: interactions of adult parasites with the complement system. *Parasite Immunology* 9: 235–248. doi: 10.1111/j.1365-3024.1987.tb00503.x.

101 Joiner, K.A., daSilva, W.D., Rimoldi, M.T. et al. (1988). Biochemical characterization of a factor produced by trypomastigotes of *Trypanosoma cruzi* that accelerates the decay of complement C3 convertases. *Journal of Biological Chemistry* 263: 11327–11335.

102 Kipnis, T.L., Joiner, K.A., da Silva, W.D. et al. (1988). Identification of membrane components of *Trypanosoma cruzi* modulators of complement system. *Memorias Do Instituto Oswaldo Cruz* 83 (Suppl. 1): 571–575. doi: 10.1590/S0074-02761988000500069.

103 Tambourgi, D.V., Kipnis, T.L., da Silva, W.D. et al. (1993). A partial cDNA clone of trypomastigote decay-accelerating factor (T-DAF), a developmentally regulated complement inhibitor of *Trypanosoma cruzi*, has genetic and functional similarities to the human complement inhibitor DAF. *Infection and Immunity* 61: 3656–3663.

104 Norris, K.A., Harth, G., and So, M. (1989). Purification of a *Trypanosoma cruzi* membrane glycoprotein which elicits lytic antibodies. *Infection and Immunity* 57: 2372–2377.

105 Norris, K.A., Bradt, B., Cooper, N.R., and So, M. (1991). Characterization of a *Trypanosoma cruzi* C3 binding protein with functional and genetic similarities to the human complement regulatory protein, decay-accelerating factor. *The Journal of Immunology* 147: 2240–2247.

106 Norris, K.A. and Schrimpf, J.E. (1994). Biochemical analysis of the membrane and soluble forms of the complement regulatory protein of *Trypanosoma cruzi*. *Infection and Immunity* 62: 236–243.

107 Norris, K.A., Schrimpf, J.E., and Szabo, M.J. (1997). Identification of the gene family encoding the 160-kilodalton *Trypanosoma cruzi* complement regulatory protein. *Infection and Immunity* 65: 349–357.

108 Parizade, M., Arnon, R., Lachmann, P.J., and Fishelson, Z. (1994). Functional and antigenic similarities between a 94-kD protein of *Schistosoma mansoni* (SCIP-1) and human CD59. *Journal of Experimental Medicine* 179: 1625–1636. doi: 10.1084/jem.179.5.1625.

109 Wilson, R.A. and Coulson, P.S. (2009). Immune effector mechanisms against schistosomiasis: looking for a chink in the parasite's armour. *Trends in Parasitology* 25: 423–431. doi: 10.1016/j.pt.2009.05.011.

110 Castro-Borges, W., Dowle, A., Curwen, R.S. et al. (2011). Enzymatic shaving of the tegument surface of live schistosomes for proteomic analysis: a rational approach to select vaccine candidates. *PLOS Neglected Tropical Diseases* 5: e993. doi: 10.1371/journal.pntd.0000993.

111 Shi, Y., Toet, H., Rathinasamy, V. et al. (2014). First insight into CD59-like molecules of adult *Fasciola hepatica*. *Experimental Parasitology* 144: 57–64. doi: 10.1016/j.exppara.2014.06.012.

112 Farias, L.P., Krautz-Peterson, G., Tararam, C.A. et al. (2013). On the three-finger protein domain fold and CD59-like proteins in *Schistosoma mansoni*. *PLOS Neglected Tropical Diseases* 7: e2482. doi: 10.1371/journal.pntd.0002482.

113 Skeldon, A. and Saleh, M. (2011). The inflammasomes: molecular effectors of host resistance against bacterial, viral, parasitic, and fungal infections. *Frontiers in Microbiology* 2: 10.3389/fmicb.2011.00015.

114 O'Neill, L.A. and Bowie, A.G. (2010). Sensing and signaling in antiviral innate immunity. *Current Biology* 20: R328–R333.

115 Takeuchi, O. and Akira, S. (2010). Pattern recognition receptors and inflammation. *Cell* 140: 805–820.

116 Belkhadir, Y., Subramaniam, R., and Dangl, J.L. (2004). Plant disease resistance protein signaling: NBS-LRR proteins and their partners. *Current Opinion in Plant Biology* 7: 391–399.

117 Proell, M., Riedl, S.J., Fritz, J.H. et al. (2008). The Nod-like receptor (NLR) family: a tale of similarities and differences. *PLoS ONE* 3: e2119. doi: 10.1371/journal.pone.0002119.

118 Chamaillard, M., Hashimoto, M., Horie, Y. et al. (2003). An essential role for NOD1 in host recognition of bacterial peptidoglycan containing diaminopimelic acid. *Nature Immunology* 4: 702–707.

119 Girardin, S.E., Boneca, I.G., Carneiro, L.A.M. et al. (2003). Nod1 detects a unique muropeptide from gram-negative bacterial peptidoglycan. *Science* 300: 1584–1587.

120 Inohara, N., Ogura, Y., Fontalba, A. et al. (2003). Host recognition of bacterial muramyl dipeptide mediated through NOD2. *Journal of Biological Chemistry* 278: 5509–5512.

121 Shaw, M.H., Reimer, T., Kim, Y.G., and Nunez, G. (2008). NOD-like receptors (NLRs): bona fide intracellular microbial sensors. *Current Opinion in Immunology* 20: 377–382.

122 Martinon, F., Burns, K., and Tschopp, J. (2002). The inflammasome: a molecular platform triggering activation of inflammatory caspases and processing of proIL-beta. *Molecular Cell* 10: 417–426.

123 Rodrigue-Gervais, I.G. and Saleh, M. (2010). Genetics of inflammasome associated disorders: a lesson in the guiding principles of inflammasome function. *The European Journal of Immunology* 40: 643–648.

124 Thornberry, N.A., Bull, H.G., Calaycay, J.R. et al. (1992). A novel heterodimeric cysteine protease is required for interleukin-1-beta processing in monocytes. *Nature* 356: 768–774.

125 Ghayur, T., Banerjee, S., Hugunin, M. et al. (1997). Caspase-1 processes IFN-gamma inducing factor and regulates LPS induced IFN-gamma production. *Nature* 386: 619–623.

126 Gu, Y., Kuida, K., Tsutsui, H. et al. (1997). Activation of interferon-gamma inducing factor mediated by interleukin-1 beta converting enzyme. *Science* 275: 206–209.

127 Keller, M., Ruegg, A., Werner, S., and Beer, H.D. (2008). Active caspase-1 is a regulator of unconventional protein secretion. *Cell* 132: 818–831.

128 Fink, S.L. and Cookson, B.T. (2005). Apoptosis, pyroptosis, and necrosis: mechanistic description of dead and dying eukaryotic cells. *Infection and Immunity* 73: 1907–1916.

129 Labbé, K. and Saleh, M. (2008). Cell death in the host response to infection. *Cell Death & Differentiation* 15 (9): 1339–1349. doi: 10.1038/cdd.2008.91.

130 Shao, W., Yeretssian, G., Doiron, K. et al. (2007). The caspase-1 digestome identifies the glycolysis pathway as a target during infection and septic shock. *Journal of Biological Chemistry* 282: 36321–36329.

131 Sachs, J. and Malaney, P. (2002). The economic and social burden of malaria. *Nature* 415: 680–685.

132 Hanscheid, T., Egan, T.J., and Grobusch, M.P. (2007). Haemozoin: from melatonin pigment to drug target, diagnostic tool, and immune modulator. *The Lancet. Infectious Diseases* 7: 675–685.

133 Dostert, C., Guarda, G., Romero, J.F. et al. (2009). Malarial hemozoin is a nalp3 inflammasome activating danger signal. *PLoS ONE* 2009 4: e6510. doi: 10.1371/journal.pone.0006510.

134 Shio, M.T., Eisenbarth, S.C., Savaria, M. et al. (2009). Malarial hemozoin activates the NLRP3 inflammasome through Lyn and Syk kinases. *PLoS Pathogens* 5: e1000559. doi: 10.1371/journal.ppat.1000559.

135 Griffith, L.M., Cowan, M.J., Notarangelo, L.D. et al. (2009). Workshop participants. improving cellular therapy for primary immune deficiency diseases: recognition, diagnosis, and management. *The Journal of Allergy and Clinical Immunology* 124 (6): 1152–60.e12. doi: 10.1016/j.jaci.2009.10.022. PMID: 20004776; PMCID: PMC2831471.

136 Reimer, T., Shaw, M.H., Franchi, L. et al. (2010). Experimental cerebral malaria progresses independently of the Nlrp3 inflammasome. *The European Journal of Immunology* 40: 764–769.

137 Labbé, K., Miu, J., Yeretssian, G. et al. (2010). Caspase-12 dampens the immune response to malaria independently of the inflammasome by targeting NF-kappaB signaling. *The Journal of Immunology* 185 (9): 5495–5502. doi: 10.4049/jimmunol.1002517. Epub 2010 Sep 27. PMID: 20876354.

5

Immune Response against Helminths

Varsha Tiwari[1,*]*, Abhishek Tiwari*[1]*, Gurdeep Singh*[2]*, and Mukesh Kr Singh*[1]

[1] *Faculty of Pharmacy, IFTM University, Lodhipur-Rajput, Moradabad, Uttar Pradesh, India*
[2] *School of Pharmaceutical Sciences, Lovely Professional University, Phagwara, Punjab*
* *Corresponding author*

5.1 Introduction

Numerous diseases that are hazardous to humans are transmitted by helminths, often known as worms. Parasitic worms can be divided into two major phyla: nematodes (or roundworms) and platyhelminthes (or flatworms), which are further subdivided into trematodes (flukes) and cestodes (tapeworms). Soil-transmitted helminths infect around 25% of the world's population, or more than 1.5 billion people. Some of the most common nematode species that infect people are *Ascaris lumbricoides*, *Trichuris trichiura*, *Necator americanus*, and *Ancylostoma duodenale*. Even exposure to a small number of these species greatly increases the risk of developing a condition that may persist for up to 20 years [1].

Macro-pathogen helminths produce compounds that act as modulators of phagocyte activation during infection by altering the microenvironment in which these cells engage in the induction and training of innate and adaptive immune responses [2]. An immune response of Type-2 is the norm in humans in the face of any pathogenic helminth. Although the protective components of the T helper (Th)-2-type response vary between parasite species and between phases of infection with the same helminth species, this response is effective against a wide range of helminths. Helminths, as a group of parasites, invade a specific ecological niche that encompasses the microenvironment in which the parasites appear and the ensuing host-parasite interactions at different points in the parasite's life cycle [3]. When it

Parasitic Infections: Immune Responses and Therapeutics, First Edition.
Edited by Abhay Prakash Mishra and Manisha Nigam.

comes to helminth infections, the symptoms might range from completely absent to debilitating chronic illness depending on the degree of the infection. Subtle symptoms, like anaemia, malnutrition, and slow physical or mental development, are experienced by the majority of people living in endemic areas [4]. The purpose of this chapter is to learn about the immunological reactions that occur in response to helminth infections, and then to use those molecular targets to find new immunomodulatory treatment drugs.

5.1.1 Variables Related to the Host that Play a Role in the Regulation of the Immune System

The production of cytokines such as interleukin-4, IL-5, IL-9, IL-10, and IL-13 as well as antibody isotypes such as immunoglobulin IgG1, IgG4, and IgE, and as increased populations of eosinophils, basophils, mast cells, type 2 innate lymphoid cells, and alternatively activated macrophages are hallmarks of the T helper immune response to all helminths [5].

Although the helminths parasite-induced Th2 response is well known, its initiation, development, and interaction with a wide range of cells, including epithelial cells, innate immune cells, lymphoid cells, dendritic cells, eosinophils, basophils/mast cells, neutrophils, T and B cells, is not as well understood. These cells include: epithelial cells, innate immune cells, lymphoid cells (Figure 5.1).

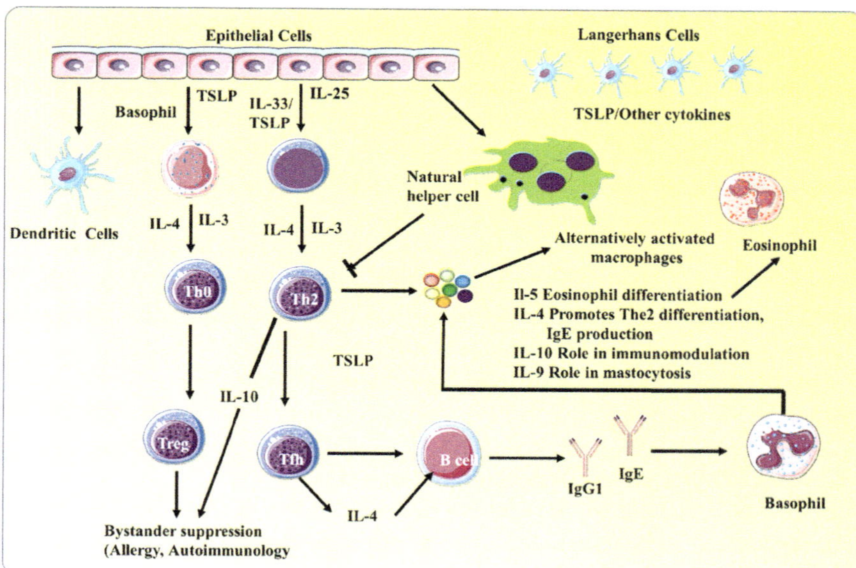

Figure 5.1 Host-related variables in the control of the immune system.

5.1.1.1 Helminths and Epithelial Cells

The bulk of helminths enter the body through the epithelial cells, which are the first barrier layer to be compromized or exposed when the barrier is broken [6]. In addition to IL-25, IL-33, thymic stromal lymphopoietin (TSLP), and IL-33, these cells are responsible for the production of the alarmins uric acid, ATB, HMGB1, and S100 proteins. They also interact with helminth parasites through nucleotide oligomerization domain-like receptors or toll-like receptors.

Intestinal epithelial cells, for example, are in continuous touch with the gut bacteria; this places them in an advantageous position for immunological monitoring within the digestive system. In addition, it has been found that epithelial cells, which can be found anywhere in the body, can send tolerogenic signals to T cells and B cells [7]. Mucus production and the bioactive compounds associated with mucus, such as mucin 5AC, trefoil factor-2, and resist in like molecule (RELM), are also factors that contribute to improved resistance to helminth infections of the digestive tract (Figure 5.1).

5.1.1.2 Helminths and Innate Lymphoid Cells

Innate lymphoid cells, also known as ILCs, are a recently discovered kind of hematopoietic effector that plays a protective role in both the homeostasis of tissue stromal cells and the innate defences against pathogenic microorganisms [8, 9].

The ILC family consists of three different members: ILC1, ILC2, and ILC3. Among these are the ILC2 cells, which are recognized for their ability to produce IL-5, IL-13, IL-9, IL-4, and IL-10. In spite of the fact that there are some similarities between ILC2s and Th2 cells, there are also variations between the two in terms of the developmental complexity, tissue tropism rules, and the specificity of the immune stimuli to which they respond [10]. Eosinophilia, the creation of goblet cell mucus, the activation of alternative activation of macrophages (AAM), muscle contraction, macrocytosis, and the repair of tissue can all be efficiently triggered by these cytokines (Figure 5.1) [11].

In addition, although their effect is not as strong as that of DCs, ILC2 have the ability to directly influence the activation of T cells by generating MHC class II molecules as well as the support molecules CD80 and CD86. In conclusion, recent studies have shown a connection between ILC2 and metabolic balance, obesity, and nutritional stress, which suggests that helminths may indirectly influence host metabolic function.

5.1.1.3 Dendritic Cells and Other Parasitic Helminths

Dendritic cells, also known as DCs, play a significant role in the detection, collection, processing, and presentation of helminth early signs (ES) to T cells, which act as mediators between innate immunity and adaptive immunity. Antigen-presenting cells, also known as DCs, are cells that are responsible for presenting antigens to T cells and initiating immune responses. DCs are also known as APCs. Despite the fact that it is known that DCs have a role in the initiation of Th1, Th17, and Treg responses, it is still uncertain how DCs influence Th2 responses. However, a number of studies have shown that DCs are essential for an appropriate Th2 response in vivo. It has been proven that the in vivo depletion of DCs places a barrier in the way of the activation of Th2 responses in response to either *Schistosoma mansoni* or *Heligmosomoides polygyrus*. Helminth products can prepare DCs for the beginning of Th2 immune responses by interacting with pattern recognition receptors (PRRs), such as Toll-like receptors (TLRs), and C-type lectin receptors (CLRs). This interaction, which is based on TLR and CLR signalling, has the potential to improve Th2 responses by lowering the amount of antigen presentation, co-stimulation, and/or the synthesis of cytokines that support Th1 responses by directly blocking these pathways. DC are hampered in their capacity to respond to a wide variety of pathogenic stimuli since it has been demonstrated that helminth antigens inhibit DC activity (Figure 5.1).

5.1.1.4 Macrophages and Other Types of Helminths

Macrophages, which are the most prevalent type of tissue-resident mononuclear phagocyte, provide a significant boost to the immune response that is mounted against helminth infections. They produce nitric oxide and other mediators, which allows them to function as effector cells in the body's defence against bacterial and protozoan infections. These macrophages are classified as alternatively activated macrophages due to their capacity to upregulate arginase, YM1, YM2, and RELM-(AAMs). It has been hypothesized that these AAMs can assist in the recovery process of wound injury brought on by the tissue migration of helminth parasites. They are known for accelerating the healing process after an injury. Because these AAMs express regulatory molecules including IL-10, TGF-, and programmed cell death, it is possible that they have a role to play in the regulation of helminth infections. 1 ligand 2. These anti-inflammatory macrophages make use of the enzymes arginase-1 and PDL2, as well as the triggering receptor expressed on myeloid cells 2 (TREM2) and RELM, in order to reduce the amount of T-cell activation, recruitment, and traditional macrophage inflammation.

When macrophages are activated in response to an infection caused by a pathogen, both pathogen-associated molecular patterns (PAMPs) and cytokines play a role in the process. For more than twenty years, based on in vitro research conducted on murine macrophages, macrophages have been classified as belonging to one of two activation phenotypes (Figure 5.1).

5.1.1.5 Both Helminths and Eosinophils are to Blame

A sizeable body of evidence lends credence to the theory that the principal function of eosinophils is to protect hosts from infection by relatively big organisms, such as parasitic helminths. This hypothesis has been backed by the aforementioned body of evidence. It succeeds in achieving the following: large numbers of eosinophils are regularly detected in close contact with both intact and injured helminths in vivo; eosinophils obviously degranulate around or onto the surfaces of helminths.

IL-5 is the factor responsible for eosinophilia in the blood and tissues, which is an indication of helminth infection. When an experimental helminth infection is carried out, eosinophils begin to migrate to the site of the infection as early as 24 hours after the initial exposure. ILC2 appears to be responsible for regulating both the accumulation of tissue following helminth infection and the basal eosinophil levels. Eosinophils are known for their ability to eliminate helminth parasites, but they are also known to play a role in tissue remodelling, metabolic balance, and anti-inflammatory cell activity by releasing cytokines that have already been generated. This is despite the fact that eosinophils are best known for their ability to eliminate helminth parasites [12]. In contrast to T and B cells, eosinophils are able to rapidly release cytokines in response to stimulation within a matter of minutes. This is possible due to the fact that the vast majority of cytokines are stored in secretory vesicles in a predetermined manner (Figure 5.1).

5.1.1.6 Mast Cells and Basophils, Helminths

Basophils play an important role in the immune system's response to infections caused by helminths [13]. Histamines, cytokines, and lipid mediators are only a few examples of the Th2 response-stimulating mediators that basophils are capable of releasing. Basophils in both humans and mice, whether they are IgE-dependent or IgE-independent, produce significant amounts of the cytokine IL-4. Basophils appear to play a vital role in the resistance to secondary infections caused by *N. brasiliense*, *H. polygyrus bakeri*, and *L. sigmodontis* (much like eosinophils do), and they also contribute considerably to the resistance to primary infections caused by *T. muris* and *T. spiralis* (via the production of IL-4 and IL-13). In addition, it has been found that basophils play a significant role as anti-parasitic cells (APCs) in helminth infection models for the purpose of triggering the development of Th2 cells. IL-4 and IL-13 are two of the most important effector cytokines, and they are responsible for mediating the defence against helminths through a variety of mechanisms [14]. In spite of the widespread belief that Th2 cells are the principal source of both cytokines, recent research has shown that basophils may also release comparable levels of IL-4 (approximately 10 femtograms per cell under a stimulation with ionomycin for 24 hours) (Figure 5.1) [15].

5.1.1.7 Helminths and Neutrophils

Despite the fact that there is evidence to suggest that neutrophils and macrophages can work together to control or remove helminth parasites, neutrophils are more commonly connected with bacterial and fungal infections [16]. However, neutrophils play a significant role in the treatment of helminth infections. When the larvae first enter the lung, local TCRgd T cells become activated and release IL-17, which attracts neutrophils to the location [17]. The acute lung damage that these neutrophils inflict is controlled and partially reversed by the type 2 immune response, which does this by lowering the amounts of IL-17 and simultaneously increasing the number of neutrophils infiltration [18]. Neutrophils also differentiate in the context of developing type-2 responses, which results in an upregulation of the expression of genes linked with type-2 responses such as IL-13, IL-33, Renta, and Chi3I3, amongst other genes. Neutrophils have an essential role in the promotion of the activation of M2 macrophages, which then contribute to tissue repair and acquired resistance during the early phases of the immune response. In conclusion, it was discovered that the absence of neutrophils led to decreased lung immunity, which ultimately resulted in greater worm burdens even during the initial infection stage. Because of this, neutrophils appear to play a role in helminth immunity that was not anticipated, and this finding warrants additional research (Figure 5.1).

5.1.1.8 B Cells and Other Head Lice

Autoimmune illnesses were the impetus for the discovery of the role of B cells in the regulation of the immune system. Helminths can interact with B cells in two different ways: by stimulating the generation of B cell cytokines, and by stimulating the creation of antibodies. Cellular interactions, most likely triggered by the stimulation of IL-10, are the primary cause of the activation of B cells, which then leads to the production of cytokines. According to the findings of a number of research, B cells enhance and sustain Th2-type immune responses during helminth infections. This is done in order to successfully combat Th-1-type immune responses. Their production of IgE is a symptom that is unique to helminth infections, as is the degranulation of mast cells and basophils, as well as the release of soluble components, which are all side effects of immunoglobulin [19].

It has been demonstrated that other antibodies, in particular IgM and IgG, are advantageous, although IgE appears to have only a limited role in the production of protective immunity. It has also been proven that maternal IgG specific for parasites, and antibodies-mediated passive immunity, protect offspring from *Trichinella spiralis* and *H. polygyrus* in a number of different helminth models. In schistosome infections, a decrease in B cells leads to a rise in immunopathology that is dependent on Th2 cells. On the other hand, B cells are essential for the generation of antibodies in conditions involving helminth infections. In the absence of B cells, a person infected with *Litomosoides sigmodontis*, *S. mansoni*, *T. muris*, or *Heligmosomoides polygyrus bakeri* is more likely to develop a secondary infection [2]. B cells, on the other hand, in addition to creating antibodies that are critical to the immune system's defence, also serve as regulators and effectors, present antigens, and co-stimulate the immune system in ways that are not dependent on antibodies [20]. Recent studies have concentrated on determining the part that regulatory B cells play in dampening the immune response to infections caused by helminths. Similar to the function of B cells, which involves the synthesis of IL-10 and IL-35, the regulatory activity of B cells in autoimmune diseases is identical to the role that B cells play (Figure 5.1).

5.1.1.9 T Cells and Other Helminths

Helminth-mediated immune control works on numerous fronts, preventing the development of adaptive immunity, the activation of effector mechanisms, and the sensitization of innate immune cells to antigens. One of the most important methods of parasite immunomodulation is the presence of regulatory T-cells (Tregs) In addition to the existence of a number of surface indicators that are important to their function, such as Tregs, which may be recognized by their expression of CD25, CTLA4, ICOS, and T-cell immunoreceptor with Ig and ITIM domains, are the cells that produce the transcription factor FOXP3, which is known as FOXP3 in humans (TIGIT). In most cases, infections caused by helminths cause a powerful Th2 reaction, which is distinguished by an increase in the synthesis of interleukin-4, interleukin-5, interleukin-9, interleukin-10, and interleukin-13 occurs in response to live parasites, parasite antigens, or mitogens [21].

In addition, the production of IL-4 and IL-13 within the Th2 cell compartment is spatially segregated. Tissues exhibit a majority of IL-13 expression, whereas lymph nodes show a predominance of IL-4 expression. In conclusion, it has been demonstrated that activating GATA-3 and lowering the expression of T-box in T cells (T-bet) are necessary steps in the transition from T-resist in-like molecular cell phenotypes to Th2 in helminth infections. Despite the fact that tissue-resident memory (TRM) T cells play a crucial role in the fight against bacterial and viral infections, relatively little is known about how they influence helminth infections. On the other hand, it has been discovered that tissue-resident Th2 cells, in response to the appropriate stimuli, engage in innate activity that is TCR-independent and IL-33-dependent. This activity acts as a defence mechanism against helminth infection. In addition, it is known that helminth infections are associated with an antigen-dependent rise in both mono- and dual-functional Th2 cells, as well as their reversal after therapy. This is the case despite the fact that multifunctionality in Th2 cells, which is defined as the capacity to create two or more cytokines, is not fully understood. In animal models of helminth infection, it has been shown that a persistent subset of parasite-induced T-bet+, GATA-3+, and Th1/Th2 hybrid T cells develops from naive progenitors and contributes to the reduction of severe inflammation. These cells come about as a result of helminth infection and develop from naive progenitors. It is believed that these cells help the parasite live longer while also protecting the host from a potentially harmful immune response. Regulatory T cells are essential for preventing autoimmunity and other types of immunological dysregulation. It is believed that these cells help the parasite survive longer. Because of this, many people who are infected with helminths do not mount an inflammatory reaction to the parasite, which is the type of reaction that would normally induce "collateral harm" in the tissues that are being impacted. It was recently demonstrated in human infections, where it was found that children who have pathology due to *S. haematobium*, exhibit stronger Th17 responses than children who do not have pathology related to *S. haematobium*. It has also been demonstrated that there is a significant association between Th17 responses and pathological responses in lymphatic filariasis. In conclusion, a subgroup of CD4 cells known as Th22 cells is responsible for the continual secretion of IL-22. Very few researchers have looked into the role of Th22 cells in helminth infections [22]. It was discovered that IL-22 was induced in the intestinal mucosa of humans after infection with *T. trichura* or *Nector americanus*. Both of these pathogens cause infections in people. In addition, it was discovered that individuals suffering with filaria had a higher number of Th22 cells compared to the healthy endemic controls. In addition, T cell clones from onchocerciasis patients that release IL-10 and TGF- have been discovered (Figure 5.1) [23].

5.2 Effector Immune Mechanisms against Helminths and Their Regulating Mechanisms

Infections caused by helminths typically show themselves through hypereosinophilia, significant production of IgE, mucosal mastocytosis, and goblet cell hyperplasia [14]. Depending on the location of the helminth, these immunological characteristics play an important part in a variety of different effector pathways.

5.2.1 The Effector Defense Mechanisms against Tissue-Dwelling Parasites and the Parasite-Developed Escape Mechanisms

There are several different types of defences that have been described in order to combat parasites that reside in tissues. Typically, these tissue-moving parasites are the larval stages of nematodes or trematodes.

Effector cells in antibody-dependent cellular cytotoxicity (ADCC), which can be mediated by IgE, IgG, or IgA antibodies, can be eosinophils, neutrophils, macrophages, or platelets. The parasitic entities that are guarded by antibodies are eliminated by cells that carry Fc fragment receptors (RFc). After being stimulated by the attachment of the antibodies to the RFc, these cells let off poisonous chemicals, which cause the worm to sustain damage (major basic protein, eosinophil cationic protein, eosinophil-derived neurotoxin, reactive nitrogen intermediates). Nematode larval stages can also be immobilized by ADCCs when they travel through the intestinal mucosa as they develop [24–27]. Nitric oxide (NO), which is a gas that is toxic to the worm, is finally released by the macrophages that had been stimulated by IFN and TNF earlier in the process. This process has been documented in the most significant way in relation to trematodes (*Schistosoma* species, *Fasciola* species) during the acute infection stage, prior to egg development in *Schistosoma mansoni* [28–30].

Tissue-dwelling parasites have devised a variety of techniques during the course of evolution in order to circumvent the effector response of their hosts. For example, *Fasciola* sp. is able to steer clear of immune responses by avoiding them in the following ways:

i) The superoxide dismutase that is produced by *Fasciola gigantica* is capable of neutralising the superoxide radicals that are damaging to young people [31, 32].
ii) *F. hepatica* is responsible for the release of chymotrypsin L-protease, an enzyme that degrades the IgE and IgG that are involved in ADCC [33].
iii) It was shown that IgM is present in flukes that are still in their juvenile stages [34]. Even though eosinophils do not have the Fc receptor, the accumulation of IgM on the fluke's tegument may inhibit eosinophils from attaching to the fluke. IgG2, which is produced during fasciolosis in susceptible sheep [35], is thought to be able to operate as an ADCC-inhibiting immunoglobulin.

5.2.2 Effector Defences against Parasites Discovered in the Duct Lumen and the Escape Mechanisms Evolved by the Parasites Themselves

Intestinal anaphylaxis with IgE-induced mast cell degranulation causes changes in the physiology of the intestine, as well as the structure and chemistry of the gut epithelium. These changes can be traced back to the immune response. Involvement of eosinophils or mast cells in the recruitment process, improvement in vascular and epithelial permeability, smooth muscle contraction, activation of fluid, electrolyte, and mucus secretion, and smooth muscle contractility are all components of these changes [36].

It is possible that this will cause the adult to be expelled, and the gastrointestinal larvae will need to be removed as rapidly as possible before they can locate their tissue niche [37]. In addition, IgA that is present on the surface of the gut mucosa helps to neutralize the metabolic enzymes that are released when strongyles are digested, which impedes the capacity of the worm to consume food [38].

When it comes to parasites that reside inside of living tissue, those that are discovered in the duct lumen have the ability to produce immunomodulatory substances, which allows them to avoid being destroyed by the immunological reactions of the host. For example, *Nector americanus* produces a metalloprotease that is responsible for the breakdown of eotaxin, which is a component of eosinophil chemotactic activity [39]. In addition, gastrointestinal nematodes are responsible for the production of the enzymes superoxide dismutase and glutathione S-transferase, both of which are capable of scavenging potentially damaging oxygen radicals [40]. A cystatin produced by *H. contortus* and *N. brasiliensis* changes the manner in which an antigen is delivered to T cells. This is accomplished by inhibiting the cysteine proteases that are utilized by antigen-presenting cells to breakdown the antigen [41, 42].

5.2.3 Controlling the Immune Response to Helminths

All of these pathways are controlled, with the exception of the ones involving traditionally activated macrophages, by cytokines that are similar to Th2 and cell types that are immunomodulatory. Interleukin-4 is responsible for influencing the isotype-switched IgE responses in B-cells. Interleukin-5 is responsible for influencing eosinophil production, and interleukin-13 is primarily responsible for influencing the development of fibrosis. Interleukin-13 has roles that are comparable to those of interleukin. T regulatory cells are responsible for producing suppressive cytokines like IL-10 and TGF. These cytokines have anti-inflammatory properties and are produced by T regulatory cells. The Th2-like responses may be the result of these cytokines. The growth of a population of Th2-activating dendritic cells, which may be induced by excretory-secretory antigens from *N. brasiliensis* [43] or soluble schistosome egg antigen, may also contribute to immunological deviation [44]. Last but not least, activated anti-inflammatory macrophages (AAMps) with powerful anti-inflammatory capabilities, accelerated Th2 cell differentiation, and the IL-4 and IL-13 cytokines can all work in concert to induce fibrosis and healing at the site of damage [45]. Because of this, an individual who is infected with helminths will experience a condition characterized by downregulation of proinflammatory reactivity, active damage repair mechanisms, and controlled development of Th2-like anti-parasite effector responses [46].

5.3 Immune Regulation and Spectrum of Disease

Immune control by parasites weakens immunity while also protecting the host from potentially harmful immunopathological reactions to the presence of parasites. This is especially true in tissue settings where the severity of the infection does not always positively correlate with pathology, such as in the cases of lymphatic filariasis, onchocerciasis, and schistosomiasis. In contrast, patients with lower levels of infection who have immunologically reactive responses and chronic pathology may not exhibit any symptoms in those with high parasite numbers and stronger regulatory responses. Long-term filaria-exposed populations provide the best examples of the disease's spectrum, with infection outcomes ranging from those who appear immune and uninfected (endemic normal) to those who range from a minority who have chronic pathology in the form of lymphatic inflammation and elephantiasis to microfilaremic (Mf), with bloodstream microfilariae, who have a large percentage of asymptomatic instances with patient infections [47, 48]. The absence of symptoms is associated with elevated levels of the immune-regulatory cytokine interleukin-10 (IL-10) [49], a suppression of key Th2 components such as IL-5 [50], as well as Th1 inflammatory cytokines such as gamma interferon (IFN), and higher concentrations of circulating T lymphocytes that express the inhibitory marker CT (cytotoxic T lymphocyte antigen 4) [51].

In contrast, cases that develop lymphatic pathology have higher levels of potent Th1 and Th17 effector components, which may be responsible for causing lymphatic inflammation against resident adult worms. However, these cases also have lower levels of regulatory T cells, which are an essential cell type (Treg). The fact that Th17 cells are also more prevalent in patients who have had their bloodstream Mf cleaned raises the intriguing question of whether Mf indirectly inhibits Th17 responses and enhances their survival as a result of shed levels of another immunosuppressive mediator, transforming growth factor (TGF), in nodules surrounding parasites. This question is raised due to the observation that nodules surrounding parasites contain shed levels of TGF [52]. It is likely that parasites and/or their released products are actively inhibiting the immune system. This is because the degree of an infection and immunological responsiveness are connected. Because a sustained granuloma formation takes place around implanted eggs in schistosomiasis and because persistent liver fibrosis is common, it might be difficult to differentiate between infections that are asymptomatic and those that have caused pathology. However, the degree of the inflammatory response to infection can vary along a continuum from mild to severe. When visitors who have not been exposed to schistosomes in the past become infected with the parasite, they experience an acute inflammatory immune reaction, with enormous granulomatous responses surrounding the parasite's implanted egg. However, persistent infection greatly reduces inflammatory cytokine responses to *Schistosoma* soluble egg antigens (SEAs). On the other hand, Th2 cytokines, such as IL-10, are enhanced [53]. Additionally, only a very small percentage of patients experience severe hepatosplenic disease.

Patients with acute schistosomiasis and severe pathology cases demonstrate considerably stronger T cell proliferative responses to SEA as compared with chronically infected individuals with minimal pathology [54]; nevertheless, the group with low pathology exhibited higher frequencies of CD4(+)CD25high [55]. Even while it is necessary for resistance, the production of Th2 cytokines, in particular IL-13, can also cause detrimental effects. This is due to the fact that it is essential for the formation of fibrosis, and chronic schistosomiasis causes damage to the liver and the spleen [56]. In the asymptomatic Mf condition, the production of IgG4 rather than IgE antibodies against filariasis antigens reaches levels that are significantly higher than typical [57, 58]. It is interesting to note that IL-4 and IL-13 promote B cells to create IgE; yet, when IL-10 and TGF- are present, it is more likely that B cells will flip to producing IgG4 [59]. As a consequence of this, it was argued that the extraordinarily powerful immune reactions could potentially be explained by studies indicating an increase in immune responses to parasite antigens following drug-induced clearance of the parasites. Immunological response to the relevant antigens is regained after treatment with medication for filaria infection [60] or individuals who have been infected with schistosomes [61]. In addition, an increased Th2 responsiveness following the administration of the medication is positively connected with infection resistance [62].

These findings reveal that helminth parasites have the ability to powerfully generate a regulatory environment that is precisely tuned to maintain a suppressed, nonresponsive condition in effector T cell levels. It is essential to note that the host's immunological competence can be restored if parasites are removed or if there is an increase in the immune response [63]. Therefore, the suppression that is generated by parasitic infection is active, and the release of mediators produced by helminths may serve as the starting point for various different pathways [64]. The hypo-proliferative condition that is seen in tissue helminth infections may be caused by a suppression of the number of antigen-presenting cells (APCs). Filarial infection has an effect on the function of monocytes as well as antigen-presenting cells (APCs): in Mf patients, there is a preponderance of circulating CD11c-CD123lo monocytic dendritic cells (DCs), which have down-regulated CCR1 and may as a result be impaired in their ability to migrate. In addition, there is a preponderance of circulating CD11c+CD123lo [65] and when compared to the norm, the production of inflammatory cytokines from microfilariae (Mf) donor monocytes is greatly suppressed by lipopolysaccharide. This is due to the fact that lipopolysaccharide acts as an immune checkpoint (LPS) [66].

After stimulation, there is a general decrease in the expression of the TLR (Toll-like receptor) family of innate pattern recognition receptors upregulated by APCs; this deficiency was also observed in B cells from Mf patients. In addition, filarial infections cause a decrease in TLR expression levels, which is part of the larger pattern of reduced TLR function that occurs during helminth infection [67]. An intriguing question that has to be answered is whether or not certain APC populations only impair their function or change into active regulatory and suppressive cell types.

5.4 Protective Immunity against Helminths

Helminths, in spite of the Th2-like immune response that is generated against them, are often able to live inside their host for an extended length of time, which might result in a chronic infection. On the other hand, premature immunity and partial immunity are also types of immunity that have been described. These types of immunity are measured by the host's resistance to reinfection and the partial clearance of established parasites. The manner in which helminth protective immunity operates is determined, in large part, by the location of the helminth infection [14]. T cells are a crucial component in the process of resisting parasites, playing a critical role in the process that mediates the expulsion of gastrointestinal (GI) nematodes. For example, mice that are deficient in T cells are unable to eliminate *T. muris*, but the T cells from mice that have immune systems that are functioning normally can be used to reestablish resistance.

The ability of mice with severe combined immunodeficiency (SCID) to receive protective immunity from the CD4 T cells of infected mice is further evidence that CD4 T cells, and not CD8 T cells, are essential for protective immunity. This evidence was provided by the fact that SCID mice were able to receive protective immunity from infected mice's CD4 T cells (lacking both B and T cells). It has also been proven that T cells are required for ejection in *N. brasiliensis* infection. Nude mice, who lack only T cells, and SCID mice, are susceptible to infection with Brugian parasites. However animals that lack CD4 T cells or CD8 T cells are not susceptible to infection with Brugian parasites. The formation of host-protective granulomas around the schistosome eggs produced by the mouse liver requires the participation of T lymphocytes. When worms are infected, the host's immunological response to the helminths may be able to control the infection; but this response may also result in tissue lesions and other symptoms, which are frequently the major causes of disease. The immunopathologic effects of *Schistosoma* infections have been the topic of a significant amount of research in recent years. Similar to what was discussed before in relation to *F. hepatica* infection, acute schistosomiasis is associated with Th1-like responses directed towards adult parasites. The Th2-like responses that are triggered as a result of the release of egg antigens are responsible for the downregulation of the production and effector activities of Th1-like mediators [68]. When Th2-like responses to the eggs were artificially repressed, a higher granuloma that was driven by Th1 and Th17 cells resulted in hepatic injury and death. This was due to the fact that Th1 and Th17 cells were overactive [69]. The function of cytokines in protective immunity has been extensively researched using mice models of both gastrointestinal helminths and tissue-invasive helminths [70].

Type 2 cytokines are responsible for activating a wide variety of downstream effector pathways. Tuft cells are a type of intestinal epithelial cell that promote the formation of goblet cells, enhance mucus secretion, and produce RELM, which is an innate effector molecule that functions directly as an anthelmintic. Mucins that can form gels are one of the most important macromolecular components of the mucus barrier, and goblet cells are able to release mucins. Two of these mucins, known as Muc2 and Muc5AC, have been proved to be necessary for the protection of the intestines against infection by nematodes. In addition, activation of IL-4R leads to increased intestinal smooth muscle hypercontractility as well as rapid epithelium turnover. These two factors, when combined with epithelial secretions, contribute to the elimination of intestinal helminths. Mast cells in the mucosa are responsible for producing proteases as part of the "weep and sweep" reflex. These proteases have the ability to breakdown epithelial tight junctions and improve fluid flow. It is also possible for AAMs in the gut to ensnare intestinal worms, which diminishes the chances of worm survival and ultimately ends in the death of the worms. Although it is common knowledge that Th2 cytokines have a role in the resistance to gastrointestinal helminth infection, considerably less is known regarding their function in the protection against tissue-invasive helminths. It is possible to establish protective immune responses in mouse models of schistosomiasis by immunising mice with irradiated cercariae and observing their subsequent behavior. This resistance is based on a Th1-mediated immune response, which is made up of endothelial cells and macrophages that are activated by TNF and IFN- and create nitric oxide. Additionally, this resistance is reliant on the fact that there is a presence of a Th1 cell. In addition to this, it is reliant on the antibodies IgG2a and IgG2b, both of which are associated with the Th1 immune response. On the other hand, studies conducted on rats as well as epidemiological study conducted on humans suggest that protective immunity is predominantly assumed to be mediated by eosinophils, IgA and IgE antibodies, as well as Th2-mediated effector pathways. It is primarily the Th2 responses in mice that are responsible for the protective immunity that mice have against filarial infections. Mice that are deficient in IL-4, IL-4R, or Stat6 are susceptible to infection with the Brugian parasite because of this. Antibodies can protect against helminth infections in some cases, although this is not the case in other instances. Passive immunity mediated by antibodies has been demonstrated in animal models of *Ascaris suum*, *A. caninum*, *Schistosoma* species, *Taenia* species, *S. ratti*, *T. muris*, *N. brasiliensis*, and *H. polygyrus bakeri*. In addition, IgM monoclonal antibodies (mAbs) that are specific for *B. malayi*, IgG or IgA mAbs that are specific for *T. spiralis*, and IgG monoclonal antibodies (mAbs) that are specific for *Fasciola hepatica* and *S. mansoni* have all been utilized to demonstrate passive immunity. IgM's significance in the host defence mechanism against *B. malayi* and *S. stercoralis* has been proven once more with the assistance of genetically modified mouse models. This protective axis seems to rely significantly on B1B cells, which are a subset of B cells that are responsible for the production of IgM. It is possible for antibodies to limit tissue damage by collecting larval helminths as they move through tissue thanks to their ability to trigger an IL-4R-independent alternative macrophage differentiation that is dependent on CD11b and FcR1. Granuloma formation, which is dependent on a Th2-like immune response, can also be observed in patients with gastrointestinal worm infections. In a recent study, the researchers found that when the Th2-like response to *Nippostrongylus brasiliensis* was experimentally downregulated, the animal's resistance to pulmonary granulomatous inflammation, fibrosis, and gastrointestinal nematode infection was significantly reduced [70]. A similar pattern can be seen with the signs of bovine ostertagiosis, which include diarrhea, loss of appetite, and weight loss. These clinical symptoms may be brought on by a contraction of the smooth muscle, an increase in the production of mucus, immediate hypersensitivity responses mediated by IgE that cause the degranulation of mast and goblet cells, the loss of specialized cells in the abomasal epitheliums, and the release of inflammatory mediators [71].

5.5 Immune Reaction-Related Pathology in Parasitic Helminth Infection

Even while certain pathological responses are the direct result of the host's response, the fact that the majority of pathological results are tied to the presence of the parasites in the tissues of the host is what differentiates one parasitic infection from another.

5.5.1 Immune Complexes

Immune complexes play a significant role as potent mediators of localized inflammatory processes in many diseases caused by parasites. This is most likely as a consequence of the constant low-dose antigen release that is characteristic of parasitic infections. Immune complexes can be found circulating in the blood of humans and animals infected with both filarial and schistosomal parasites. It has been proven that due to their deposition, they produce lymphatic inflammation as well as vasculitis in filarial infections. In addition, the outcomes of kidney biopsies performed on patients with schistosomiasis and filarial infections have revealed immune complex glomerulonephritis (ICGN), which is a common sign of immune complex-mediated illness. In patients suffering from helminth infections, two additional signs of immune complex-mediated harm have been reported. These symptoms are reactive arthritis and dermatitis.

5.5.2 Autoantibodies and Molecular Mimicry

Molecular mimicry is one of the key mechanisms that can be used by infectious organisms or chemical irritants to set off an autoimmune response. It takes place when an antigen obtained from a foreign source causes the activation of self-reactive T or B cells in a susceptible individual. This happens because foreign and self-peptides share similarities with one another. Infections with a variety of helminths, such as filarial infections, schistosomiasis, and hookworm illnesses, have been linked to autoantibodies, which are presumed to represent the polyclonal B cell proliferation that frequently follows these infections. Autoantibodies have also been linked to schistosomiasis [72]. The majority of people who have chronic schistosomiasis have autoantibodies that target nuclear material, while the majority of people who have onchocerciasis have antibodies that target human calreticulin and defensin.

5.5.3 Granulomatous Reactions

Granuloma growth is an essential part of the immune system's defence against some helminths; nevertheless, it can also have negative implications in the form of the disease if it is not controlled. Granulomatous reactions can be seen in a number of different helminth disorders, such as toxocariasis, infections caused by angiostrongyliasis, and lymphatic filariasis; however, parasitical granulomata have been studied the most in *S. mansoni* infections. In these infections, CD4 T cells are responsible for coordinating granulomatous and fibrosing reactions against tissue-trapped eggs; the fibrosis that develops as a direct result of the cellular response is the primary factor that contributes to morbidity. It has been shown that responses from Th1 and Th17 cells are associated with more severe forms of illness, whereas responses from Th2-dominant cells are associated with less severe forms of pathology. This finding holds true for both human patients and animal models used in experimental research [73]. Studies conducted with animal models of granuloma development demonstrated that TNF and IL-13 play critical roles in the development of the tumor.

5.5.4 Fibrosis

There is a strong correlation between fibrosis and repeated helminth infections, which are known to produce persistent inflammation and slow wound healing [5]. By stimulating macrophages and fibroblasts, these infections lead to the production of TGF-, platelet-derived growth factor (PDGF), and IL-1, among other chemicals. Macrophages contribute to inflammation in a number of different ways, including drawing in and boosting neutrophils and monocytes, as well as drawing in and stimulating CD4 T cells. In addition, the activation of fibroblasts results in the production of tissue inhibitors of metalloproteinases and matrix metalloproteinases (MMPs), both of which contribute to the process of fibrosis and remodelling of the extracellular matrix (TIMPs). Another impact of chronic schistosomiasis is pulmonary arterial hypertension, which has been linked to TGF-induced pulmonary vascular disease and type 2 inflammation, which is mediated by IL-4 and IL-13. In a manner that is analogous to the way in which IL-10 and IL-12 are known to govern IL-13-mediated fibrosis, IL-13–dependent fibrosis in chronic schistosomiasis patients who are deficient in IL-10, IL-12, and IL-13R advances rapidly toward cirrhosis, which is a condition that is fatal. *Wuchereria bancrofti*, one of the agents that are responsible for generating lymphatic filariasis, is associated with comparable fibrotic reactions.

Wolbachia, an endosymbiotic *Rickettsia*-like bacteria, has been connected to immunopathology in lymphatic filariasis.

5.5.5 Toll-Like Receptors

Wolbachia has been linked to immunopathology in lymphatic filariasis. It is well established that *Wolbachia* stimulates immune cells by using the proinflammatory cytokines and vascular endothelial growth factors (VEGFs) that are generated as a result of TLR2 and TLR4 interaction. as well as possibly being a contributor to lymphatic illness [74]. In addition, it has been established that the primary factor causing ocular inflammation in onchocerciasis is an interaction between *Wolbachia* and TLR4, and it has also been established that the development of pathogenic Th17 cells in *S. mansoni* infections is regulated by the TLR signalling protein IL-1 receptor-associated kinase-2. Both of these findings have implications for the treatment of onchocerciasis (IRAK-2).

5.5.6 Immediate Hypersensitivity Reactions

Immediate hypersensitivity reactions are associated with the first and/or acute stage of invasive helminth parasite infections, such as those caused by *Ascaris*, hookworm, schistosomes, or filariae. These infections can cause a variety of symptoms, including itching, hives, and rashes. Patients do show indicators that point to an allergic reaction, such as wheezing and urticaria. IgE-mediated reactions are also believed to be the cause of the signs and symptoms of clinical syndromes connected to tropical pulmonary eosinophilia, larval treatments for strongyloidiasis, and *Loa loa* infection (with associated angioedematous Calabar swellings). These syndromes include tropical pulmonary eosinophilia, larval treatments for strongyloidiasis, and *Loa loa* infection [75].

5.5.7 The Healing Process

Recent research has shown that responses of the type 2 cytokine have a close relationship with numerous aspects of the process of wound healing and tissue repair [76, 77]. Because infections caused by helminths are tightly connected to damage to tissues, according to one school of thought, the type 2 cytokine response first came into being in order to mediate resistance to helminth infections and then to initiate the wound healing mechanism in order to repair and reconstruct damaged tissue. It should come as no surprise that AAMs are involved in this process given that they are responsible for the production of MMPs, Arginase-1, insulin-like growth factor 1 (IGF1), VEGF, and TGF-. These are all factors that, when combined, promote myofibroblast activation, angiogenesis, epithelial cell turnover, and extracellular matrix deposition.

5.5.8 Lymphangiogenesis

According to the anatomical changes in the architecture of the lymphatics, which range from lymphangiectasia and granulomatous reactions to the creation of collaterals, it is possible that active lymphatic remodelling caused by early lymphatic filarial disease, which involves the growth, migration, and proliferation of endothelial cells, is a key factor. Live filarial parasites, along with their excretory and secretory secretions, have been demonstrated to stimulate lymphatic endothelial cells (LECs) to activate, proliferate, form tubes, and develop into tubelike networks.It is generally accepted that TLR-mediated pathways are the key drivers of the angiogenic and lymphangiogenic process that occurs in filarial illness [78].

5.5.9 Carcinogenesis

Infections caused by *Opisthorchis viverrini*, *Clonorchis sinensis*, and *Schistosoma hematobium* are examples of biological carcinogens that belong to Group 1. (i.e., definitive causes of cancer).

While the latter is associated with cancer of the bladder and urinary tract, the former (liver fluke) is linked to cancer of the bile duct as well as cancer of the liver (cholangiocarcinoma) (carcinoma). In addition to other cancer-causing pathways, chronic inflammation, persistent cellular proliferation, altered redox and glucose signalling, genomic instability, tumor protein destabilisation, angiogenesis enhancement, apoptosis resistance, and helminths can promote invasion and metastasis [79].

5.5.10 Epileptogenesis

In some endemic areas, neurocysticercosis is responsible for almost thirty percent of all cases of epilepsy. This condition is caused by the larval form of *Taenia solium*, which is the most common preventable risk factor for epilepsy in the entire world. The symptoms, as well as the severity of the inflammation that is caused by cyst disintegration, calcification, and/or perilesional edoema, all have a role in determining the size of the cysts, the number of cysts, and the locations of the cysts that are found in the central nervous system.

5.6 Conclusion

Helminths are able to modulate the immune system, which is now recognized as an essential survival strategy due to the passage of time and the major host–parasite relationship that is emerging. The investigation of helminth parasites is leading to a number of exciting immunological findings as well as potential therapeutic applications. Both in-depth field research and thorough laboratory testing have revealed epidemiological evidence that various diseases that are common in the industrialized world are protected against. With this new information, we can now offer methods for the management of infectious and immunopathological illnesses. These strategies can take advantage of the broad ideas that have been clarified as well as the specific modulatory mediators that have been found. Helminths are powerful agents that inhibit type 1 immune responses, which can be triggered by inflammatory diseases, secondary infections, or a combination of the two. It is currently difficult to understand the mechanisms behind this interaction and to discover possible biological targets in order to develop new methods that can prevent or postpone allergic, inflammatory, autoimmune, or metabolic illnesses in humans. This presents a challenge for the creation of new treatment options.

References

1 Motran, C.C., Silvane, L., Chiapello, L.S. et al. (2018). Helminth infections: recognition and modulation of the immune response by innate immune cells. *Frontiers in Immunology* 9: 664.

2 Harris, N. and Gause, W.C. (2011). To B or not to B: b cells and the Th2-type immune response to helminths. *Trends in Immunology* 32 (2): 80–88.

3 Nutman, T.B. (2015). Looking beyond the induction of Th2 responses to explain immunomodulation by helminths. *Parasite Immunology* 37 (6): 304–313.

4 Prasanphanich, N.S., Mickum, M.L., Heimburg-Molinaro, J., and Cummings, R.D. (2013). Glycoconjugates in host-helminth interactions. *Frontiers in Immunology* 28 (4): 240.

5 Allen, J.E. and Maizels, R.M. (2011). Diversity and dialogue in immunity to helminths. *Nature Reviews Immunology* 11 (6): 375–388.

6 Hammad, H. and Lambrecht, B.N. (2015). Barrier epithelial cells and the control of type 2 immunity. *Immunity* 43 (1): 29–40.

7 Smits, H.H., Everts, B., Hartgers, F.C., and Yazdanbakhsh, M. (2010). Chronic helminth infections protect against allergic diseases by active regulatory processes. *Current Allergy and Asthma Reports* 10 (1): 3–12.

8 Artis, D. and Spits, H. (2015). The biology of innate lymphoid cells. *Nature* 517 (7534): 293–301.

9 Spits, H., Artis, D., Colonna, M. et al. (2013). Innate lymphoid cells – a proposal for uniform nomenclature. *Nature Reviews Immunology* 13 (2): 145–149.
10 Germain, R.N. and Huang, Y. (2019). ILC2s—resident lymphocytes pre-adapted to a specific tissue or migratory effectors that adapt to where they move? *Current Opinion in Immunology* 56: 76–81.
11 Saenz, S.A., Siracusa, M.C., Perrigoue, J.G. et al. (2010). IL25 elicits a multipotent progenitor cell population that promotes TH2 cytokine responses. *Nature* 464 (7293): 1362–1366.
12 Thomas, B. and Nutman, M.D. (2015). Looking beyond the induction of Th2 responses to explain immunomodulation by helminthes. *Parasite Immunol* 37 (6): 303–313. Doi. 1111/pim.12194.
13 Karasuyama, H. and Yamanishi, Y. (2014). Basophils have emerged as a key player in immunity. *Current Opinion in Immunology* 31: 1–7.
14 Anthony, R.M., Rutitzky, L.I., Urban, J.F. et al. (2007). Protective immune mechanisms in helminth infection. *Nature Reviews Immunology* 7 (12): 975–987.
15 Mitre, E., Taylor, R.T., Kubofcik, J., and Nutman, T.B. (2004). Parasite antigen-driven basophils are a major source of IL-4 in human filarial infections. *The Journal of Immunology* 172 (4): 2439–2445.
16 Allen, J.E., Sutherland, T.E., and Rückerl, D. (2015). IL-17 and neutrophils: unexpected players in the type 2 immune response. *Current Opinion in Immunology* 34: 99–106.
17 Sutherland, T.E., Logan, N., Rückerl, D. et al. (2014). Chitinase-like proteins promote IL-17-mediated neutrophilia in a tradeoff between nematode killing and host damage. *Nature Immunology* 15 (12): 1116–1125.
18 Yap, G.S. and Gause, W.C. (2018). Helminth infections induce tissue tolerance mitigating immunopathology but enhancing microbial pathogen susceptibility. *Frontiers in Immunology* 16 (9): 2135.
19 Pennock, J.L. and Grencis, R.K. (2006). The mast cell and gut nematodes: damage and defence. *Parasites and Allergy* 90: 128–140.
20 Liu, Q., Kreider, T., Bowdridge, S. et al. (2010). B cells have distinct roles in host protection against different nematode parasites. *The Journal of Immunology* 184 (9): 5213–5223.
21 Grencis, R.K. (2015). Immunity to helminths: resistance, regulation, and susceptibility to gastrointestinal nematodes. *Annual Review of Immunology* 33: 201–225.
22 Bouchery, T., Kyle, R., Ronchese, F., and Le Gros, G. (2014). The differentiation of CD4+ T-helper cell subsets in the context of helminth parasite infection. *Frontiers in Immunology* 5: 487.
23 Doetze, A., Satoguina, J., Burchard, G. et al. (2000). Antigen-specific cellular hyporesponsiveness in a chronic human helminth infection is mediated by Th3/Tr1-type cytokines IL-10 and transforming growth factor-β but not by a Th1 to Th2 shift. *International Immunology* 12 (5): 623–630.

24 Lenzi, H.L., Pacheco, R.G., Pelajo-Machado, M. et al. (1997). Immunological system and *Schistosoma mansoni*: co-evolutionary immunobiology. What is the eosinophil role in parasite-host relationship? *Memórias Do Instituto Oswaldo Cruz* 92: 19–32.

25 Meeusen, E.N., Balic, A., and Bowles, V. (2005). Cells, cytokines and other molecules associated with rejection of gastrointestinal nematode parasites. *Veterinary Immunology and Immunopathology* 108 (1–2): 121–125.

26 Piedrafita, D., Parsons, J.C., Sandeman, R.M. et al. (2001). Antibody-dependent cell-mediated cytotoxicity to newly excysted juvenile *Fasciola hepatica* in vitro is mediated by reactive nitrogen intermediates. *Parasite Immunology* 23 (9): 473–482.

27 Weller, P.F. (1994). Eosinophils: structure and functions. *Current Opinion in Immunology* 6 (1): 85–90.

28 Cervi, L., Rossi, G., Cejas, H., and Masih, D.T. (1998). *Fasciola hepatica*-induced immune suppression of spleen mononuclear cell proliferation: role of nitric oxide. *Clinical Immunology and Immunopathology* 87 (2): 145–154.

29 Gazzinelli, R.T., Oswald, I.P., James, S.L., and Sher, A. (1992). IL-10 inhibits parasite killing and nitrogen oxide production by IFN-gamma-activated macrophages. *The Journal of Immunology* 148 (6): 1792–1796.

30 Sibille, P., Tliba, O., and Boulard, C. (2004). Early and transient cytotoxic response of peritoneal cells from *Fasciola hepatica*-infected rats. *Veterinary Research* 35 (5): 573–584.

31 Ganga, G., Varshney, J.P., and Patra, R.C. (2007). Activity of antioxidant enzymes in excretory-secretory fluid and somatic extracts of *Fasciola gigantica*. *Journal of Veterinary Parasitology* 21 (1): 51–52.

32 Piedrafita, D., Estuningsih, E., Pleasance, J. et al. (2007). Peritoneal lavage cells of Indonesian thin-tail sheep mediate antibody-dependent superoxide radical cytotoxicity in vitro against newly excysted juvenile *Fasciola gigantica* but not juvenile *Fasciola hepatica*. *Infection and Immunity* 75 (4): 1954–1963.

33 Smith, A.M., Dowd, A.J., Heffernan, M. et al. (1993). *Fasciola hepatica*: a secreted cathepsin L-like proteinase cleaves host immunoglobulin. *International Journal for Parasitology* 23 (8): 977–983.

34 Chauvin, A. and Boulard, C. (1996). Local immune response to experimental *Fasciola hepatica* infection in sheep. *Parasite* 3 (3): 209–215.

35 Hansen, D.S., Clery, D.G., Estuningsih, S.E. et al. (1999). Immune responses in Indonesian thin tail and Merino sheep during a primary infection with *Fasciola gigantica*: lack of a specific IgG2 antibody response is associated with increased resistance to infection in Indonesian sheep. *International Journal for Parasitology* 29 (7): 1027–1035.

36 Farthing, M.J. (2003). Immune response-mediated pathology in human intestinal parasitic infection. *Parasite Immunology* 25 (5): 247–257.

37 Balic, A., Bowles, V.M., and Meeusen, E.N. (2002). Mechanisms of immunity to *Haemonchus contortus* infection in sheep. *Parasite Immunology* 24 (1): 39–46.

38 Gill, H.S., Gray, G.D., Watson, D.L., and Husband, A.J. (1993). Isotype-specific antibody responses to *Haemonchus contortus* in genetically resistant sheep. *Parasite Immunology* 15 (2): 61–67.
39 Culley, F.J., Brown, A., Conroy, D.M. et al. (2000). Eotaxin is specifically cleaved by hookworm metalloproteases preventing its action in vitro and in vivo. *The Journal of Immunology* 165 (11): 6447–6453.
40 Liddell, S. and Knox, D.P. (1998). Extracellular and cytoplasmic Cu/Zn superoxide dismutases from *Haemonchus contortus. Parasitology* 116 (4): 383–394.
41 Dainichi, T., Maekawa, Y., Ishii, K. et al. (2001). Nippocystatin, a cysteine protease inhibitor from *Nippostrongylus brasiliensis*, inhibits antigen processing and modulates antigen-specific immune response. *Infection and Immunity* 69 (12): 7380–7386.
42 Newlands, G.F., Skuce, P.J., Knox, D.P., and Smith, W.D. (2001). Cloning and expression of cystatin, a potent cysteine protease inhibitor from the gut of *Haemonchus contortus. Parasitology* 122 (3): 371–378.
43 Else, K.J. (2005). Have gastrointestinal nematodes outwitted the immune system? *Parasite Immunology* 27 (10-11): 407–415.
44 Thomas, P.G., Carter, M.R., Atochina, O. et al. (2003). Maturation of dendritic cell 2 phenotype by a helminth glycan uses a Toll-like receptor 4-dependent mechanism. *The Journal of Immunology* 171 (11): 5837–5841.
45 Kreider, T., Anthony, R.M., Urban, J.F., Jr, and Gause, W.C. (2007). Alternatively activated macrophages in helminth infections. *Current Opinion in Immunology* 19 (4): 448–453.
46 Jackson, J.A., Friberg, I.M., Little, S., and Bradley, J.E. (2009). Review series on helminths, immune modulation and the hygiene hypothesis: immunity against helminths and immunological phenomena in modern human populations: coevolutionary legacies? *Immunology* 126 (1): 18–27.
47 Ottesen, E.A. (1984). Immunological aspects of lymphatic filariasis and onchocerciasis in man. *Transactions of the Royal Society of Tropical Medicine and Hygiene* 78: 9–18.
48 Pfarr, K.M., Debrah, A.Y., Specht, S., and Hoerauf, A. (2009). Filariasis and lymphoedema. *Parasite Immunology* 31 (11): 664–672.
49 Mahanty, S., Mollis, S.N., Ravichandran, M. et al. (1996). High levels of spontaneous and parasite antigen-driven interieukin-10 production are associated with antigen-specific hyporesponsiveness in human lymphatic filariasis. *Journal of Infectious Diseases* 173 (3): 769–772.
50 Sartono, E., Kruize, Y.C., Kurniawan, A. et al. (1997). Depression of antigen-specific interleukin-5 and interferon-γ responses in human lymphatic filariasis as a function of clinical status and age. *Journal of Infectious Diseases* 175 (5): 1276–1280.
51 Steel, C. and Nutman, T.B. (2003). CTLA-4 in filarial infections: implications for a role in diminished T cell reactivity. *The Journal of Immunology* 170 (4): 1930–1938.

52 Jenkins, S.J. and Allen, J.E. (2010). Similarity and diversity in macrophage activation by nematodes, trematodes, and cestodes. *Journal of Biomedicine and Biotechnology* 2010.

53 Ribeiro de Jesus, A., Magalhaes, A., Gonzalez Miranda, D. et al. (2004). Association of type 2 cytokines with hepatic fibrosis in human *Schistosoma mansoni* infection. *Infection and Immunity* 72 (6): 3391–3397.

54 Gazzinelli, G., Montesano, M.A., Corrêa-Oliveira, R. et al. (1987). Immune response in different clinical groups of schistosomiasis patients. *Memórias Do Instituto Oswaldo Cruz* 82: 95–100.

55 Carvalho, A.T., Martins Filho, O.A., Pascoal, V.P. et al. (2018). Cytokines, chemokine receptors, CD4 (+) CD25 (HIGH+) T-cells and clinical forms of human schistosomiasis. *Acta Tropica* 108: 139–149.

56 Dessein, A., Kouriba, B., Eboumbou, C. et al. (2004). Interleukin-13 in the skin and interferon-γ in the liver are key players in immune protection in human schistosomiasis. *Immunological Reviews* 201 (1): 180–190.

57 Hussain, R., Grögl, M., and Ottesen, E.A. (1987). IgG antibody subclasses in human filariasis. Differential subclass recognition of parasite antigens correlates with different clinical manifestations of infection. *The Journal of Immunology* 139 (8): 2794–2798.

58 Kurniawan, A., Yazdanbakhsh, M., Van Ree, R. et al. (1993). Differential expression of IgE and IgG4 specific antibody responses in asymptomatic and chronic human filariasis. *The Journal of Immunology* 150 (9): 3941–3950.

59 Satoguina, J.S., Adjobimey, T., Arndts, K. et al. (2008). Tr1 and naturally occurring regulatory T cells induce IgG4 in B cells through GITR/GITR-L interaction, IL-10 and TGF-β. *European Journal of Immunology* 38 (11): 3101–3113.

60 Piessens, W.F., Ratiwayanto, S., Piessens, P.W. et al. (1981). Effect of treatment with diethylcarbamazine on immune responses to filarial antigens in patients infected with *Brugia malayi*. *Acta Tropica* 38 (3): 227–234.

61 Grogan, J.L., Kremsner, P.G., Deelder, A.M., and Yazdanbakhsh, M. (1996). Elevated proliferation and interleukin-4 release from CD4+ cells after chemotherapy in human *Schistosoma haematobium* infection. *European Journal of Immunology* 26 (6): 1365–1370.

62 Joseph, S., Jones, F.M., Walter, K. et al. (2004). Increases in human T helper 2 cytokine responses to *Schistosoma mansoni* worm and worm-tegument antigens are induced by treatment with praziquantel. *Journal of Infectious Diseases* 190 (4): 835–842.

63 Taylor, M.D., van der Werf, N., and Maizels, R.M. (2012). T cells in helminth infection: the regulators and the regulated. *Trends in Immunology* 33 (4): 181–189.

64 Hewitson, J.P., Grainger, J.R., and Maizels, R.M. (2009). Helminth immunoregulation: the role of parasite secreted proteins in modulating host immunity. *Molecular and Biochemical Parasitology* 167 (1): 1–1.

65 Semnani, R.T., Mahapatra, L., Dembele, B. et al. (2010). Expanded numbers of circulating myeloid dendritic cells in patent human filarial infection reflect lower CCR1 expression. *The Journal of Immunology* 185 (10): 6364–6372.

66 Semnani, R.T., Keiser, P.B., Coulibaly, Y.I. et al. (2006). Filaria-induced monocyte dysfunction and its reversal following treatment. *Infection and Immunity* 74 (8): 4409–4417.

67 Venugopal, P.G., Nutman, T.B., and Semnani, R.T. (2009). Activation and regulation of toll-like receptors (TLRs) by helminth parasites. *Immunologic Research* 43 (1): 252–263.

68 Pearce, E.J., M. Kane, C., Sun J, J. et al. (2004). Th2 response polarization during infection with the helminth parasite. *Schistosoma mansoni. Immunological Reviews* 201 (1): 117–126.

69 Stadecker, M.J., Asahi, H., Finger, E. et al. (2004). The immunobiology of Th1 polarization in high-pathology schistosomiasis. *Immunological Reviews* 201 (1): 168–179.

70 Pesce, J.T., Ramalingam, T.R., Wilson, M.S. et al. (2009). Retnla (Relmα/Fizz1) suppresses helminth-induced Th2-type immunity. *PLoS Pathogens* 5 (4): e1000393.

71 Mulcahy, G., O'Neill, S., Donnelly, S., and Dalton, J.P. (2004). Helminths at mucosal barriers—interaction with the immune system. *Advanced Drug Delivery Reviews* 56 (6): 853–868.

72 Zandman-Goddard, G. and Shoenfeld, Y. (2009). Parasitic infection and autoimmunity. *Lupus* 18 (13): 1144–1148.

73 Wynn, T.A., Thompson, R.W., Cheever, A.W., and Mentink-Kane, M.M. (2004). Immunopathogenesis of schistosomiasis. *Immunological Reviews* 201 (1): 156–167.

74 Pani, S.P., Yuvaraj, J., Vanamail, P. et al. (1995). Episodic adenolymphangitis and lymphoedema in patients with bancroftian filariasis. *Transactions of the Royal Society of Tropical Medicine and Hygiene* 89 (1): 72–74.

75 Pritchard, D.I. (1997). The pro-allergic influences of helminth parasites. *Memórias Do Instituto Oswaldo Cruz* 92: 15–18.

76 Gause, W.C., Wynn, T.A., and Allen, J.E. (2013). Type 2 immunity and wound healing: evolutionary refinement of adaptive immunity by helminths. *Nature Reviews Immunology* 13 (8): 607–614.

77 Wynn, T.A. and Vannella, K.M. (2016). Macrophages in tissue repair, regeneration, and fibrosis. *Immunity* 44 (3): 450–462.

78 Babu, S. and Nutman, T.B. (2012). Immunopathogenesis of lymphatic filarial disease. In: *Seminars in Immunopathology* (ed. M. Stadecker), 34, No. 6. 847–861. Springer-Verlag.

79 Fried, B., Reddy, A., and Mayer, D. (2011). Helminths in human carcinogenesis. *Cancer Letters* 305 (2): 239–249.

6

Ectoparasites Host Resistance and Tolerance

Jacob Kehinde Akintunde[1,*] *and Ayodeji Mathias Adegoke*[2]

[1] *Applied Biochemistry and Molecular Toxicology Research Group, Department of Biochemistry, College of Biosciences, Federal University of Agriculture, Abeokuta, Nigeria*
[2] *Cancer Research and Molecular Biology Laboratories, Department of Biochemistry, Faculty of Basic Medical Sciences, College of Medicine, University of Ibadan, Nigeria*
[*] *Corresponding author*

6.1 Introduction

Ectoparasites are living creatures that reside on the skin of a host, from which they obtain nourishment [1]. When the larvae or maggots of ectoparasites invade the necrotic tissue of either animals or humans, a disease called myiasis may develop. The larvae or maggots of these ectoparasites that invade the necrotic tissue of both animals and humans lead to myiasis. According to Gizaw et al., [2], myiasis is a disease that often affects people and animals in tropical and subtropical regions and is characterized by itching, a feeling of movement, and stabbing pain. When the larvae are deposited in the host's eye, they penetrate the retina and sclera immediately [3], leading to ocular myiasis. The maggots of ectoparasites can also penetrate into the base of the brain to initiate meningitis and death [2]. Apparently, infestation by ectoparasites in domestic animals and animal rearing industries can result in significant financial losses for farmers due to loss of fecundity, skin infections, and death [4]. The ectoparasites may cause weight loss, depression, weakness, low productivity of milk, reduced meat (protein) production, diarrhea, and impedance. It has been reported that poor management and sanitary condition, overcrowding, and illiteracy are the indicators that support the growth rate of ectoparasites in animals [4]. Such ectoparasites include lice, sheep keds, ticks, fleas, and mange mites. The presence of ectoparasites have been linked to several health problems, such as mechanical tissue injury, irritation, inflammation, hypersensitivity, abscesses, lameness, anemia, and resultant socioeconomic

Parasitic Infections: Immune Responses and Therapeutics, First Edition.
Edited by Abhay Prakash Mishra and Manisha Nigam.

effects [2]. Ectoparasites are also responsible for the devaluation and rejection of the skin of small ruminants before slaughter, as a result of the damage that they cause to the skin of these animals [5]. Reports from tanneries indicate that in Ethiopia, 35% of sheep and 56% of goat skins are rejected because of skin abnormalities brought on by ectoparasite impacts [6]. Furthermore, ectoparasites have the capacity to transmit parasitism, a form of symbiosis in which an organism (the parasite) benefit at the detriment of another organism (the host) – typically a different species. The interaction between the host and the parasite may eventually cause harm to the host. This chapter therefore underscores the significance of prevalence, host resistance and tolerance of ectoparasites.

6.2 Epidemiology of Ectoparasites

The three main categories of parasites that are harmful to human beings include helminths, protozoans, and ectoparasites [7]. Typical human ectoparasites like lice, fleas, bedbugs, mites, and ticks dwell in the skin or skin growths for a long time [8]. Ectoparasitic infestation is prevalent in low income nations, and it has been linked to hostile environmental and poor health conditions, poor personal cleanliness, and adverse socioeconomic conditions [2]. According to reports, Ethiopia is home to about 56.71 million cattle, 29.2 million sheep, 29.3 million goats, 9.9 million horses, 1.2 million camels, and 56.9 million poultry [2]. For this reason, the number of ectoparasite infections is on the increase in this region. The presence of ectoparasites is an important variable that affects ruminant animals' productivity and output [9]. Lice, fleas, ticks, sheep keds, and mange are widespread ectoparasites in the area that are linked to detrimental effects on the health state of ruminant animals [9]. Reports have also shown that ectoparasite infections have global distribution and are documented as a major obstacle to the health of mammals, which hinders the production efficiency of ruminants. Remarkably, this reduces the economic gain of these animals in Ethiopia despite the huge livestock population.

Ectoparasites pervasiveness in the agrarian settlements of northwest Ethiopia (72.6%) is lower when compared with Brazil (76%) and also Britain (75%) but higher than in Woreta town, Ethiopia (65.7%) and Iran (67.3%). The higher pervasiveness of ectoparasites is due to deplorable living situations in agrarian areas. Studies have reported the highest prevalence of ectoparasite infestation during spring and summer in Ethiopia [9, 10]. The pervasiveness of diseases caused by ectoparasites occurs as a result of the spreading of pathogens that are vector-borne including the female *Aedes* mosquito (an ectoparasite of humans), which transmits dengue virus as it bites humans. When there is higher vector densities there is a concomitant increase in biting rates, which in turn elevates the risk of pathogen spreading. This also suggests that the number of infectious bites determines the eventual probability of infection [1].

The most common and numerous ectoparasite infections in dogs from the State of Tabasco, Mexico, were *Rhipicephalus sanguineus* and *Ctenocephalides felis*, which affected around 26.65% of the population [11]. Dog ectoparasites have also been observed in residents in underprivileged neighborhoods in South Africa's North West Province [12]. Ectoparasite infections have a history of occurring in Nigeria, which has been widely documented [12]. In an epidemiological survey by Adamu et al., [13], 81.4% of dog owners had an ectoparasitic infestation. According to the findings from other studies, southwestern Nigeria had a 98.5% prevalence rate [14], 88.6% in Ethiopia [15], and 70.7% in Brazil [10]. Several meteorological parameters (temperature, humidity, and light), which are crucial for development, transmission, as well as host intrinsic resistance, were identified in relation to the geographical variance in infestation rates [16]. According to Babamale et al., [17], the most prevalent ectoparasites found in a chicken farm in a region of Nigeria were *Gonoides gigas* (15.0%), *Menacanthus stramineus* (20.0%), *Lipeurus caponis* (25.0%), *Ceratophyllus gallinae* (25.0%), and *Amblyomma* sp. (15.0%). Globally, ectoparasites particularly mites and lice, are more prevalent in poor or developing nations than in developed nations and people with insufficient basophils, eosinophils and mast cells are highly susceptible to ectoparasites including scabies [18].

6.2.1 Ectoparasite Resistance

Ectoparasite resistance to ectoparasiticides is the adoption of a particular genetic characteristic in an ectoparasite species owing to the subjection of that species to an active substance, leading to a notable rise in the population that will be resistant to a standard dose of that chemical when used as advocated [19]. Insensitivity to one ectoparasiticide may also give rise to lack of sensitivity to some other ectoparasiticide via cross-resistance or side-resistance [20, 21]. Regulation of ectoparasite infestations is required to maintain the health of both humans and animals as well as to control the transmission of infectious diseases. The choice in the use of chemicals has inevitably resulted in resistance, except for in growth regulators and isoxazolines. The method of action and the way by which resistance occurs to organophosphates, carbamates, pyrethroids, insect growth regulators, macrocyclic lactones, phenylpyrazoles, chloronicotinyl nitroguanidines, spinosad and the isoxazolines are well documented. Also, the development of resistance has been described in individual or general insects, arthropods and mosquitoes. Resistance has not only been described in biting midges the Oestridae (gadflies, warble flies and stomach bots) and tsetse flies, but also certain species have become difficult or almost impossible to control with chemicals, e.g. the chicken red mite *Dermanyssus gallinae*. The tropical cattle tick *Rhippicephalus* (*Boophilus*) *microplus* has developed resistance to all available products while the horn fly, a very important problem in cattle in the Americas, has developed resistance to a wide variety of chemicals as well as the bed bug, *Cimex sp.*

Early on, ectoparasite control in animals was based on the utilization of mineral oils and tar distillates which include kerosene, phenols and cresol. While these have been proven to be efficacious insecticides, they have also been found to be potentially harmful to the host [22]. There was a paradigm shift to some inorganic compounds such as arsenic, boron and fluoride but the targets developed resistance to these compounds. The formulation of DDT (dichlorodiphenyltrichloroethane) as an insecticide in 1939 resulted in its global utilization for insect control. This occurred during World War II when massive doses were utilized to combat typhus and malaria vectors. In the 1970s, it was banned due to public health worries about its toxicity to both people and wildlife as well as complaints of product failure. Since then, a number of novel ectoparasiticides have been developed. However, because of the frequent and indiscriminate use of many medications, resistance has developed to many of these chemicals.

6.2.2 Common Ectoparasites and Their Resistance

The two main animal types that make up the ectoparasites that affect humans are parasitic arachnids and parasitic insects. Mites and ticks are examples of parasitic arachnids, while mosquitoes, tsetse flies, fleas, and lice are parasitic insects. Moreover, there are the blood-sucking arthropods like mosquitoes, which depend on a blood meal from human host for survival. Numerous parasites, including the malaria-causing protozoan *Plasmodium*, are spread by the mosquito [23]. Flukes, leeches, crustaceans, insects, arachinids, and arthropods are also categorized as ectoparasites.

6.2.3 Ticks

Ticks are agents or vectors of man and domesticated mammals particularly those that are warm blooded. They are persistent blood suckers and have a long-life span as well as high reproductive potential. Figure 6.1 provides a description of the life cycle of tick in a three-host life cycle.

6.3 Resistance in Ticks

Rhipicephalus microplus is the tick that displays the highest level of resistance. It is also referred to as southern or Australian cattle tick. Unlike ticks that may live on multiple hosts, cattle ticks exclusively spend their entire life cycle on one

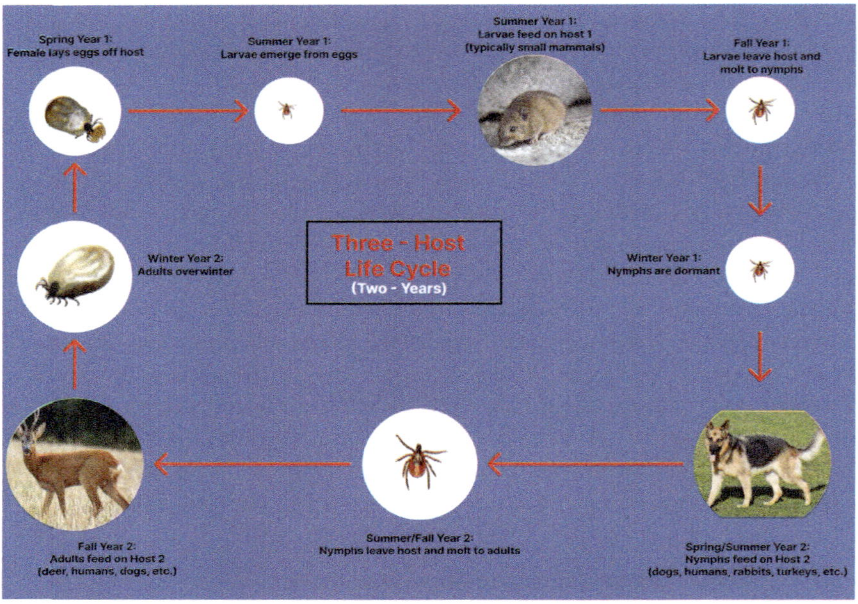

Figure 6.1 The life cycle of tick in three hosts.

species of animal. It is now widespread throughout subtropical and tropical regions of the world, but not in Europe. In the USA, it has been exterminated [20]. Brazil [24] and Mexico [25] were the two countries where *Rhipicephalus microplus* was found to be resistant to avermectins. Additionally, ten populations of brown dog ticks (*R. sanguineus*) in Mexico have shown resistance to avermectin [26]. It is noteworthy to report that there is currently no information available on cattle tick resistance in Europe, despite an in vitro study from Spain reporting a high rate of *R. sanguineus* tick resistance to deltamethrin and propoxur in dogs. *R. sanguineus* strains, however, were sensitive to amitraz [27].

6.3.1 Bed Bugs

Bed bugs are blood-feeding ectoparasites that are related to bat bugs and also minute pirate bugs, which are predators (family Anthocoridae). They are employed in integrated pest management as natural enemies. *Leptocimex boueti* in some parts of West Africa, *Cimex hemipterus* in the tropics of the Old and New Worlds, and *Cimex lectularius* in temperate and subtropical globally are three species that have been identified among humans. Figure 6.2 shows the life cycle of a bed-bug.

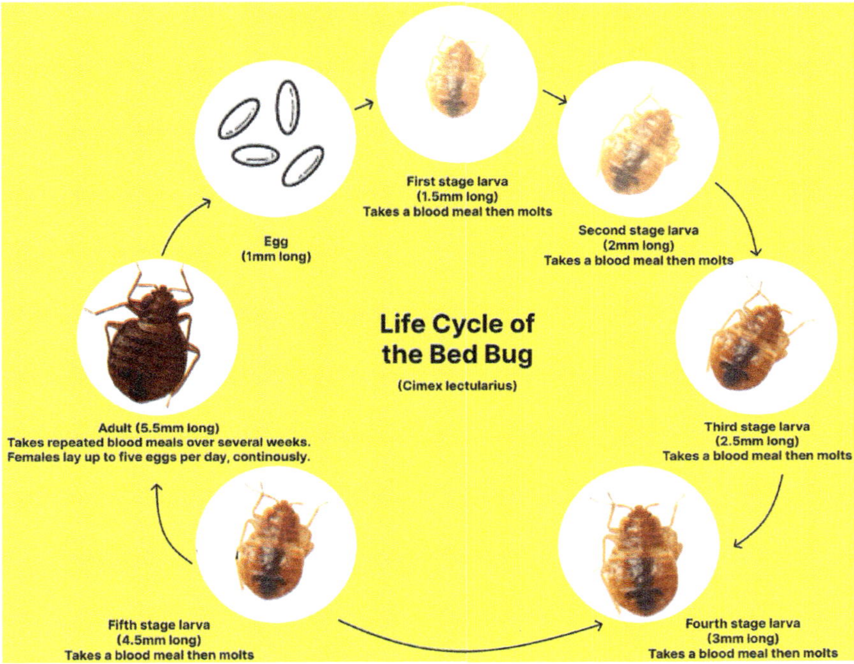

Figure 6.2 The life cycle of a bed bug.

6.4 Resistance in Bed Bugs

It has been established in recent years that bed bugs from Australia and New Zealand have developed resistance. With the introduction of the formulations of synthetic pyrethroid in Australia, decreased efficacy was seen [28]. Moulting inhibition test [29] and louse egg hatch test [30] were both also used to confirm bed bug populations' resistance to the benzoyl urea derivatives triflumuron and diflubenzuron. According to Levot and Sales [30], bed bugs had also demonstrated resistance to y-benzene hexachloride (y-BHC), aldrin, and dieldrin. In Scotland, it was also suspected that humans were infested with a population of *B. ovis* which was resistant to synthetic pyrethroids [29]. Not too long ago, a *Bovicola ocellatus* population in the UK showed signs of resistance by displaying a high level of pyrethroid tolerance [31]. Lastly, a population resistance to organophosphate insecticide by sucking *Haematopinus suis* had been reported in France, the UK and Germany [28].

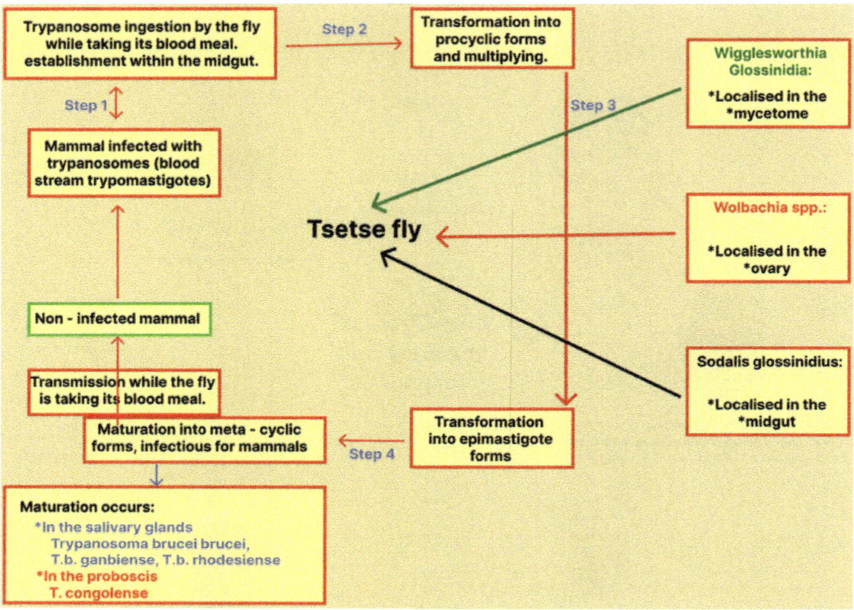

Figure 6.3 The life cycle of tsetse fly in human host.

6.4.1 Tsetse Flies

African trypanosomiasis is among the most commonly neglected tropical diseases (NTDs), wreaking havoc on both human and animal health as well as on the economy of Africa as a continent. Human African Trypanosomiasis (HAT) comes in two different types. These are the chronic form, present in West and Central Africa, which is triggered by *Trypanosoma brucei gambiense* and the acute form, present in East and Southern Africa and induced by *Trypanosoma brucei rhodesiense*. The HAT is sometimes referred to as the sleeping disease. *T. b. brucei*, which is linked to *Trypanosoma vivax* and *Trypanosoma congolense*, is the cause of the Animal African Trypanosomiasis (AAT), sometimes referred to as Nagana. When an infected tsetse fly (Diptera: Glossinidae) bites a mammalian potential host, African trypanosomes are transferred to the host. The HAT and AAT constitute major public health concerns and risks to both human and animal health. Figure 6.3 illustrates the life cycle of the tsetse fly.

6.5 Resistance in Tsetse Flies

Delta-octalactone, produced by several Bovidae, has been reported as a potential repellant of female tsetse fly attack [32]. The waterbuck repellent blend (WRB), which is a combination of geranyl acetone, guaiacol, pentanoic acid, and -octalactone, deters some tsetse flies that are acclimated to the savannah and lowers the number of riverine species that are caught in traps [33]. Aerial or ground spraying with deltamethrin can result in a drastic reduction of tsetse flies, but not elimination [32]. This may be attributed to the resistance of the flies. Additionally, *M. domestica* strains can exhibit resistance to synthetic pyrethroids and organophosphates on German farms, especially when tested with the “Fly Box” test method [32]. Furthermore, about 6–7-fold of field populations of housefly was resistant to neonicotinoids thiamethoxam and imidacloprid in European farms [34]. In the UK, eggs and larvae were reported to have developed low levels of resistance to the insect growth regulator (IGR) [35]. Insensitivity to benzoylurea and diflubenzuron were noticed in Denmark. In Turkey, field strains of *M. domestica* showed varying levels of resistance to synthetic pyrethroids from year to year, but the Antalya strain had high levels of resistance to cypermethrin [36]. A strain of *Stomoxys calcitrans* that had been treated with synthetic pyrethroid in France showed resistance in blood-engorged flies [37]. The Fly Box test method in Germany revealed that 95% of *S. calcitrans* populations were resistant to deltamethrin [38].

6.5.1 Mosquitoes

The mosquito is one of the commonest ectoparasites in developing countries and it causes a disease known as malaria. Malaria is known as a highly disastrous infectious disease in human which is caused by protozoans [39]. The malaria parasite thrives in tropical as well as subtropical regions. In the year 2015, close to 212 million cases and around 429,000 deaths were attributed to malaria. The malaria parasite spread occurs in the forests of the southern and central regions of Vietnam and throughout other Southeast Asian countries. Malaria is a public health concern in some regions of the world, especially in developing nations. The life cycle of a mosquito is shown in Figure 6.4.

6.5.2 Resistance in Mosquitoes and Sandflies

Studies have shown that mosquitoes and sandflies demonstrated resistance against commonly used chemicals [23, 40]. Deltamethrin resistance was observed in *Culex pipiens* populations from five locations in Greece (Attika, Phthiotis, Thessaloniki, Serres, and Evros), with one population in the Attika region being particularly resistant [41]. Pyrethroids showed that deltamethrin and permethrin were efficient against *Aedes albopictus* adults [42] in the eastern Mediterranean region (west of Turkey). *Phlebotomus* spp. showed resistance to deltamethrin and permethrin when insecticides were treated for a prolonged period of time [43]. It is interesting to note that a study found that three varieties of pesticides, including pyrethroids, deltamethrin, and permethrin, were potent in controlling two populations of Italian sandflies, *Phlebotomus perniciosus* and *P. papatasi* [44]. *Lepeophtheirus salmonis* (sea lice) have also become less sensitive to pyrethroids, organophosphates, and emamectin benzoate [45]. Additionally, a surveillance

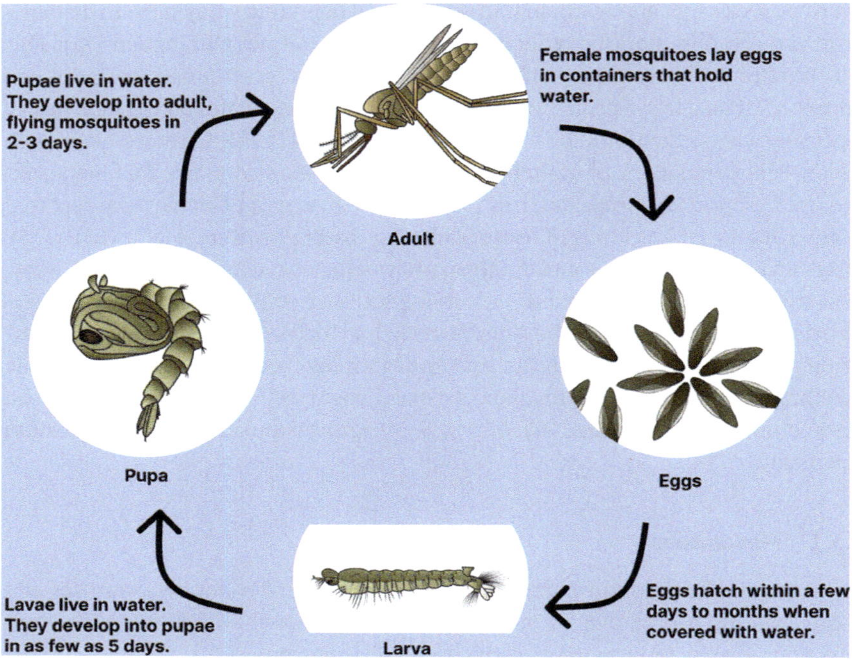

Figure 6.4 The life cycle mosquito.

study conducted in Norway found that *L. salmonis* was less sensitive to azamethiphos and deltamethrin [46].

6.6 Host Defense Mechanisms Against Ectoparasites

In view of the fact that hosts differ in their ecology, appearance, physiology, and behavior, different host species tend to employ distinct defensive tactics. Also, different host species have divergent and comparable ways of defending themselves against the same ectoparasite. According to prior research, resistance to ectoparasites is impacted by the immunological responses that various host species produce [2]. Furthermore, it is known that some ectoparasites, particularly ticks, impair the immune systems of their vertebrate hosts. This increases the susceptibility of the immunocompromized vertebrate hosts to infectious illnesses [47]. Ectoparasites are managed or stopped by the immune systems of hosts. The molecular mechanism that controls the response to immunity is the foundation of all immune evasion strategies. Immune evasion mechanisms are classified as:

A) Passive evasion parasites, which can avoid the immune system in many ways e.g., (i) ectoparasites can evade the immune system in several ways by entering immunological tissue, such as the central nervous system or by entering the eye. Additionally, certain ectoparasites lay their eggs in tissues that the host's immune system cannot easily access, such as body fat which lacks sufficient immunity. (ii) The immune system may not be able to detect some ectoparasites. This is accomplished by concealing surface elements (particularly *Plasmodium*) as soon as the host's immune system begins to opsonize them. (iii) Ectoparasites can biotransform their surface identity and thus become unrecognizable by the T-cells and antibodies of the host's immune system which has the ability to recognize specific epitopes. This causes the ectoparasites to escape detection of their antigenic surface during the infection process, whereas other ectoparasites, in particular *Plasmodium falciparum*, *Trypanosoma brucei*, etc., are capable of storing surface variants that are sequentially expressed. (iv) Lastly, some ectoparasites have the ability to temporarily lapse into quiescence – basically, bacteria – thus evading the immune system. Antibiotics that target the cell replication phase do not affect them [48].
B) Active modulation and interference. These immune evasion mechanisms show active host immune responses being interfered with by ectoparasites, which frequently disrupt the regulatory network that creates the diverse immune defense mechanisms. Additionally, ectoparasites may impair cellular processes that are vital to the host's ability to defend itself by producing chemicals that can impede or alter particular immune response steps. Through the use of specific secretion systems, like the type III secretion system, these modulatory compounds are activated by directly delivering their proteins into the cell of the host.

Generally, ectoparasites typically target a host's external surface. When an ectoparasite e.g., a tick, bites a host to have a blood meal, the infectious germs are released into the host. In order to ward off both the invading bacteria and the tick, large numbers of basophils, eosinophils, and mast cells are mobilized to the bite site. Through increasing vascular permeability, chemicals released during mast cell degranulation make it difficult for ectoparasites to find host blood vessels. Specifically, IgA and IgE antibodies stop ectoparasitic agents from attaching in the host. Toxic mediators against ectoparasites are released by mast cells, basophils, and eosinophils after IgE causes their degranulation (as shown in Figure 6.5). Unfortunately, people who don't have enough basophils and eosinophils have scabies and an itchy rash brought on by the mite *Sarcoptes scabiei* [49]. Despite the fact that research on the molecular basis of immune responses against ectoparasites is still a new concept, Th1 and Th17 responses have been shown to assist cell-mediated immunity against internal threats, whilst Th2 responses

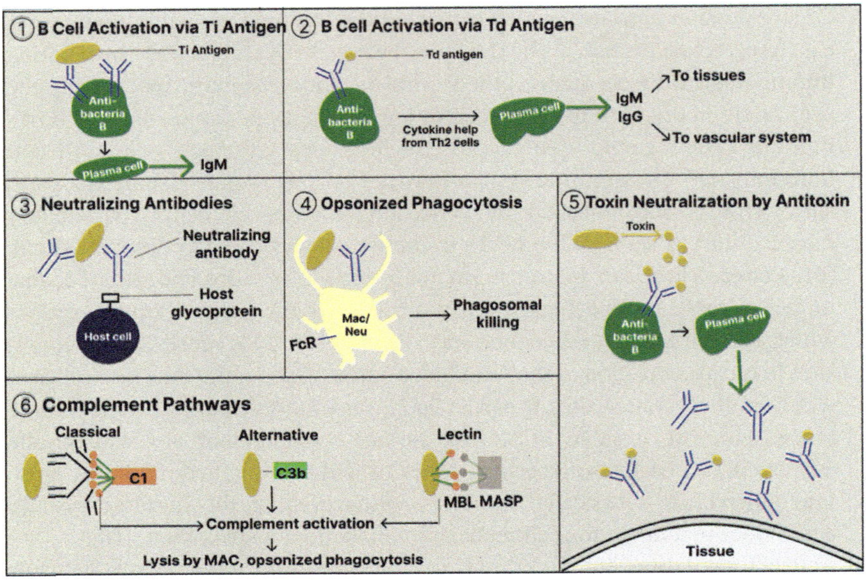

Figure 6.5 Major mechanisms of immune defense against ectoparasitic agents. (1) Ectoparasitic Ti antigens activate B cells that generate plasma cells producing antiparasitic IgM antibodies. (2) Parastic Td antigens activate additional antiparasitic B cells during Th2 responses. (3) Neutralizing antibodies recognizing parasitic components block ectoparasitic access to host cell glycoprotein receptors. (4) Antibody-bound parasites are recognized by FcRs on macrophages and neutrophils, which engulf and kill the parasite. (5) Neutralizing antitoxin antibodies bind to ectoparasite toxin molecules and prevent them from damaging cell surfaces. (6) Parasite bound by antibody plus C1, or C3b, or MBL activate complement.

reinforce humoral immunity against exterior threats. It is interesting to note that studies have shown Th2 responses to be protective against ectoparasites, as demonstrated by the increased *Demodex* skin infections in mice lacking both CD28 and STAT6 [50]. This is because STAT6 is a transcription factor necessary for Th cell production of IL-4 and CD28 is a crucial costimulator of Th cell activation [51].

Furthermore, it has been demonstrated that employing antigens obtained from salivary glands, stomach, and other organs, vaccines could be developed to immunize hosts against blood-feeding ectoparasites. In contrast to vaccines that prevent transmission by using antigens generated from pathogens, these vaccines are focused against vector arthropods and reduce their potential to spread pathogens. Also, antigen extracts have been utilized to immunize potential hosts against a variety of tick species, keds, flies, mosquitoes, fleas, and lice [52]. A recombinant antigen-based vaccine against the cow tick *Boophilus microplus* (Canestrini) has undergone field testing [52]. Mechanistically, ticks that feed on immunized hosts are impaired by an immune reaction aimed at their gut cells. While some ticks may be completely exterminated while on the host, others may swell up, consequently having their fecundity inhibited [1]. It is suggested that new vaccinations against ticks be developed using the Commonwealth Scientific Industrial Research Organization-Biotechnology Australia tick vaccine against *B. microplus* [52]. Instead of considering the time insects and ticks spend on their hosts in the development of vaccines, vaccines are developed based on the differences in the functional anatomy of gut of the parasites. The impact of vaccines on insects may be in favor of vaccines that prevent pathogens from entering vector insects. Additionally, vaccinations against mosquitoes are now being introduced to lessen their vulnerability to arboviruses [53]

6.7 Host Tolerance to Ectoparasites

In order to reduce parasite damage, hosts have evolved and created defense mechanisms. By lowering parasite fitness (resistance), hosts can lessen the harm that parasites cause. They can also do this without altering parasite fitness (tolerance). Instead of killing the ectoparasite, tolerance mechanisms like resource compensation or tissue healing enable the host to deal with the pressure from the ectoparasite [48]. The body size of the host species can also influence their susceptibility to parasitism. Hosts that are larger are better guarded and more tolerant of ectoparasites than the smaller hosts, as evidenced by the fact that larger-bodied nestling birds host species are less impacted by parasitic nest flies than smaller hosts, despite identical ectoparasite densities [54]. Smaller-bodied hosts use more energy than larger hosts because they have higher surface area to volume ratios and they also have higher metabolic rates [49]. These hosts are unable to get adequate dietary resources to generate energy for tolerance mechanisms like repairing damaged tissues or recovering lost resources like red blood cells, therefore, the cost of infection is increased.

It is anticipated that the off-host stages of temperate parasites can withstand cold weather. Due to diapause and cold acclimatization, cold tolerance is typically at its highest in the winter and at its lowest during the active summer seasons. Offspring provisioning controls cold tolerance in blood-feeding ectoparasites throughout all non-feeding, off-host stages. Large size improves cold-weather survival, although seasonal fluctuation in offspring size within a female has not been linked to offspring cold tolerance. From autumn until spring, the deer tick (*Lipoptena cervi*) reproduces on cervids. Pupae that have just emerged from the host are unaccustomed to the cold and drop off. According to reports, spring pupae fared better than winter pupae in prolonged frost exposure [55]. However, females give birth to more tolerant offspring in the winter than in the spring. Thus, when the temperature is lowest outside the host, tolerant offspring are not birthed.

6.8 Conclusion

Ectoparasites are potentiated to invade the necrotic tissue of animals and humans, producing myriad of diseases. Due to poor socioeconomic, environmental, and personal hygiene circumstances, ectoparasitic infection is more prevalent in developing nations than in developed nations. The choice in the use of many chemicals to control ectoparasites in domestic animals fosters their resistance and makes humans inevitably susceptible to their toxicity. Defense mechanisms tend to vary among host species due to distinct ecology, morphology, physiology, and behavior. Essentially, IgA and IgE antibodies hinder ectoparasitic pathogens from attaching to the host by activating mast cell, basophil, and eosinophil degranulation to produce toxic mediators against ectoparasites. Also, vaccines have been developed and directly applied against vector arthropods to mitigate the transmission of pathogens into the host. Additionally, host-tolerance mechanisms, including resource compensation or tissue repair, higher metabolic rates, body size of the hosts and climatic factors do not kill the ectoparasites but may permit the hosts to endure the ectoparasite burden. Reports have shown that findings on the molecular basis of immune responses against ectoparasites are still novel. Thus, future studies should focus on the molecular defense of the hosts against ectoparasites as well as development of tolerable natural phytotherapies from plants to control ectoparasitic infestation due to their non-toxic effects and for the fact that several synthetic chemicals are extremely toxic to humans.

References

1 McNair, C.M. (2014). Ectoparasites of medical and veterinary importance: drug resistance and the need for alternative control methods. *Journal of Pharmacy and Pharmacology* 67: 351–363.

2 Gizaw, Z., Engdaw, G.T., Nigusie, A. et al. (2021). Human ectoparasites are highly prevalent in the rural communities of northwest Ethiopia: a community-based cross-sectional study. *Environmental Health Insights* 15: 1–7.

3 Akintunde, J.K., Akintola, T.E., Hammed, M.O. et al. (2020). Naringin protects against Bisphenol-A induced oculopathy as implication of cataract in hypertensive rat model. *Biomedicine and Pharmacotherapy* 126: 110043.

4 Zewdu, S., Tsegaye, T., and Agerie, A. (2015). Ectoparasites prevalence in small ruminants in and around Sekela, Amhara regional state, Northwest Ethiopia. *Journal of Veterinary Medicine* 2015: 1–6.

5 Bahrami, A.M., Doosti, A., and Ahmady Asbchin, S. (2012). Cat and dogs ectoparasite infestations in Iran and Iraq border line area. *World Application Science Journal* 18: 884–889.

6 Leul, B., Berihun, A., and Etsay, K. (2020). Epidemiological distribution of major ectoparasites species of small ruminant in the case of chemical control campaign in welkait district, Tigray region. *Journal of Tropical Medicine* 2020: 1–9.

7 Davou, K.P., Abraham, D.G., James, T. et al. (2017). Epidemiology of ectoparasites infestation in Jos North, Plateau state, Nigeria. *Saudi Journal of Medical and Pharmaceutical Sciences* 3: 206–210.

8 Jafer, A., Danel, D., Abraham, T. et al. (2017). Prevalence of tick infestation on small ruminants in and around Dire Dawa, Eastern Ethiopia. *International Journal of Research—Granthaalayah* 5: 326–336.

9 Sisay, A., Yilkal, A., and Yacob, H. (2013). Ectoparasites of sheep and goats in North-West Amhara regional state, Ethiopian. *Ethiopian Veterinary Journal* 17: 55–56.

10 Dantas-Torres, F., Lorusso, V., Testini, G. et al. (2010). Detection of *Leishmania infantum* in *Rhipicephalus sanguineus* ticks from Brazil and Italy. *Parasitology Research* 106: 857–860.

11 Alvarado-Esquivel, C., Romero-Salas, D., Aguilar-Domínguez, M. et al. (2015). Epidemiological assessment of intestinal parasitic infections in dogs at animal shelter in Veracruz, Mexico. *Asian Pacific Journal of Tropical Biomedicine* 5: 34–39.

12 Arong, G., Shitta, K., James-Rugu, N., and Effanga, E. (2011). Seasonal variation in the abundance and distribution of ixodid ticks on mongrel, alsatian and mixed breeds of dogs (*Canis familiaris*) in Jos, in Plateau state, North-central Nigeria. *World Journal of Science and Technology* 1: 24–29.

13 Adamu, N., Adamu, J., and Salisu, L. (2012). Prevalence of ecto-, endo- and haemoparasites in slaughtered dogs in Maiduguri, Nigeria. *Review Medicine and Veterinary* 163: 178–182.

14 Agbolade, O., Soetan, E., Awesu, A. et al. (2008). Ectoparasites of domestic dogs in some Ijebu communities, Southwest Nigeria. *World Application Science Journal* 3: 916–920.

15 Tesfaye, A. and Chanie, M. (2011). Ectoparasites are major skin diseases of dogs in Gondar, Amhara National Regional state, Ethiopia. *International Journal of Animal Veterinary Advance* 3: 392–396.

16 Klimpel, S., Heukelbach, J., Pothmann, D., and Rückert, S. (2010). Gastrointestinal and ectoparasites from urban stray dogs in Fortaleza (Brazil): high infection risk for humans? *Parasitology Research* 107: 713–719.

17 Babamale, O.A., Asaba, L.C., and Ugbomoiko, U.S. (2018). Prevalence of ectoparasite infestations in owned dogs in Kwara State. *Nigeria Parasite Epidemiology and Control* 3: e00079.

18 Allen, J.E. and Sutherland, T.E. (2014). Host protective roles of type 2 immunity: parasitekilling and tissue repair, flip sides of the same coin. *Seminars in Immunology* 26: 329–340.

19 Coles, T.B. and Dryden, W.M. (2014). Insecticide/acaricide resistance in fleas and ticks infesting dogs and 823 cats. *Parasites and Vectors* 7: 8.

20 Abbas, R.Z., Zaman, M.A., Colwell, D.D. et al. (2014). Acaricide resistance in cattle ticks and approaches to its management: the state of play. *Veterinary Parasitology* 203: 16–20.

21 WHO. (2016). *Test Procedures for Insecticide Resistance Monitoring in Malaria Vector Mosquitoes*, 2e ISBN 978-9-241-51157-5 (NLM classification: WA 240).

22 Akintunde, J.K., Obisesan, O.O., and Akinsete, S.J. (2019). Diet from *Mantisa religiosa* egg case abolishes pulmonary dysfunctions triggered by sub-acute exposure to aerosolized petroleum hydrocarbons in rat model. *Clinical Nutrition Experimental* 26: 44–58.

23 Dhiman, R.C. and Yadav, R.S. (2016). Insecticide resistance in phlebotomine sandflies in Southeast Asia with emphasis on the Indian subcontinent. *Infectious Diseases of Poverty* 5: 106.

24 Klafke, G.M., Sabatini, G.A., de Alubquerque, T.A. et al. (2006). Larval immersion tests with ivermectin in populations of the cattle tick *Rhipicephalus* (*Boophilus*) *microplus* (Acari: Ixodidae) from State of Sao Paulo, Brazil. *Veterinary Parasitology* 142: 386–390.

25 Perez-Cogollo, L.C., Rodriguez-Vivas, R.I., Ramirez-Cruz, G.T., and Miller, R.J. (2010). First report of the cattle tick *Rhipicephalus microplus* resistance to ivermectin in Mexico. *Veterinary Parasitology* 168: 165–169.

26 Rodriguez-Vivas, R.I., Ojeda-Chi, M.M., Trinidad-Martinez, I., and Pérez de León, A.A. (2017). First documentation of ivermectin resistance in *Rhipicephalus sanguineus* sensu lato (Acari: Ixodidae). *Veterinary Parasitology* 233: 9–13.

27 Estrada-Pena, A. (2005). Etude de la resistance de la tique brune du chien, *Rhipicephalus sanguineus* aux Acaricide. *Revue de Medecine Veterinaire* 2: 67–69.

28 Levot, G.W. (2012). Unstable pyrethroid resistance in sheep body lice *Bovicola ovis* (Schrank), (Phthiraptera: Trichodectidae) and its implications for lice control in sheep. *Veterinary Parasitology* 185: 274–278.

29 James, P.J., Cramp A.P. and Hook S.E. (2008). Resistance to insect growth regulator insecticides in populations of sheep lice as assessed by a molting disruption assy. *Medical and Veterinary Entomology* 22: 326–330.

30 Levot, G.W. and Sales, N. (2008). Resistance to benzoylphenyl urea insecticides in Australian populations of the sheep biting louse, *Bovicola ovis* (Schrank) (Phthiraptera:Trichdectidae). *Medical and Veterinary Entomology* 22: 331–334.

31 Ellse, L., Burden, F.A., and Wall, R. (2012). Pyrethroid tolerance in the chewing louse *Bovicola (Werneckiella) ocellatus*. *Veterinary Parasitology* 188: 134–139.

32 Jandowski, A., Clausen, P.-H. Schein E. and Bauer B. (2010). Occurrence and distribution of insecticide resistance in house flies (*Musca domestica*) on dairy cattle farms in Brandenburg, Germany. *Derpraktische Tierarzt* 91: 590–598.

33 Heath, A.C.G. and Levot, G.W. (2015). Parasiticide resistance in flies, lice and ticks in New Zealand and Australia: mechanisms, prevalence and prevention. *New Zealand Veterinary Journal* 63: 199–210.

34 Kristensen, M. and Jespersen, J.B. (2008). Susceptibility to thiamethoxam of *Musca domestica* from Danish livestock farms. *Pest Management Science* 64: 126–132.

35 Bell, H.A., Robinson, K.A., and Weaver, R.J. (2010). First report of cyromazine resistance in a population of UK house fly (*Musca domestica*) associated with intensive livestock production. *Pest Management Science* 66: 693–695.

36 Akiner, M.M. and Çağlar, S.S. (2012). Monitoring of five different insecticide resistance status in Turkish house fly *Musca domestica* L. (Diptera: Muscidae) populations and the relationship between resistance and insecticide usage profile. *Turkiye Parazitol Derg* 36: 87–91.

37 Salem, A., Bouhsira, E., Liénard, E. et al. (2012). Susceptibility of two European strains of *Stomoxys calcitrans (L.)* to cypermethrin, deltamethrin, fenvalerate, λ-cyhalothrin, permethrin and phoxim. *International Journal of Applied Research in Veterinary Medicine* 10: 249–257.

38 Reissert, S.I., Bauer, B., and Steuber, S. 1009 and Clausen P-H. (2017). Insecticide resistance in stable flies (*Stomoxys calcitrans*) on dairy farms in Brandenburg, Germany. *26th International Conference of the W.A.A.V.P.*, Kuala Lumpur: 4-8.09.2017, Abstract book: 141.

39 Knutie, S.A., Owen, J.P., McNew, S.M. et al. (2016). Galápagos mockingbirds tolerate introduced parasites that affect Darwin's finches. *Ecology* 97: 940–950.

40 Salim-Abadi, Y., Oshaghi, M.A., Enayati, A.A. et al. (2016). High insecticides resistance in *Culex pipiens* (Diptera: Culicidae) from Tehran, capitol of Iran. *Journal of Arthropod-Borne Diseases* 10: 483–492.

41 Fotakis, E.A., Chaskopoulou, A., Grigoraki, L. et al. (2017). Analysis of population structure and insecticide resistance in mosquitoes of the genus *Culex*, *Anopheles* and *Aedes* from different environments in Greece with a history of mosquito borne disease transmission. *Acta Tropica* 174: 29–37.

42 Vontas, J., Kioulos, E., Pavlidi, N. et al. (2012). Insecticide resistance in the major dengue vectors *Aedes albopictus* and *Aedes aegypti*. *Pesticide Biochemistry and Physiology* 104: 126–131.

43 Karakus, M., Gocmen, B., and Özbel, Y. (2017). Insecticide susceptibility status of wild-caught sand fly populations collected from two leishmaniasis endemic areas in western Turkey. *Journal of Arthropod-Borne Diseases* 11: 86–94.

44 Maroli, M., Cianchi, T., Bianchi, R., and Khoury, C. (2002). Testing insecticide susceptibility of *Phlebotomus perniciosus* and *P. papatasi* (Diptera: Psychodidae) in Italy. *Annali dell'Istituto superiore di sanita* 38: 419–423.

45 Ljungfeldt, L.E.R., Espedal, P.G., and Nilsen, F., Skern-Mauritzen M. and Glover K.A. (2014). A common-garden experiment to quantify evolutionary processes in copepods: the case of emamectin benzoate resistance in the parasitic sea louse *Lepeophtheirus salmonis*. *BMC Evolutionary Biology* 14: 108.

46 Grontvedt, R.N., Jansen, P.A., Horsberg, T.E. et al. Annual report. The surveillance programme for resistance to chemotherapeutants in salmon lice *(Lepeophtheirus salmonis)* in Norway 2014.

47 Knowles, S.C.L., Fenton, A., Petchey, O.L. et al. (2013). Stability of within-host-parasite communities in a wild mammal system. *Proceedings of the Royal Society B: Biological Sciences* 280: 20130598.

48 Knutie, S.A. (2020). Food supplementation affects gut microbiota and immunological resistance to parasites in a wild bird species. *Journal of Applied Ecology* 57: 536–547. .

49 Grab, K.M., Hiller, B.J., Hurlbert, J.H. et al. (2019). Host tolerance and resistance to parasitic nest flies differs between two wild bird species. *Ecology and Evolution* 9: 12144–12155.

50 Chulanetra, M. (2021). Chaicumpa. revisiting the mechanisms of immune evasion employed by human parasites. *Frontiers in Cellular and Infection Microbiology* 11: 702125.

51 Zhao, L., Shao, S., Chen, Y. et al. (2017). *Trichinella spiralis* calreticulin binds human complement C1q as an immune evasion strategy. *Frontier Immunology* 8: 636.

52 Schetters, T., Bishop, R., Crampton, M. et al. (2016). Cattle tick vaccine researchers join forces in CATVAC (meeting report). *Parasites and Vectors* 9: 105.

53 Manning, J.E., Morens, D.M., Kamhawi, S. et al. (2018). Mosquito saliva:the hope for a universal arbovirus vaccine? *The Journal of Infectious Diseases* 218: 7–15.

54 McNew, S.M. and Clayton, D.H. (2018). Alien invasion: biology of *Philornis* flies highlighting *Philornis downsi*, an introduced parasite of Galapagos birds. *Annual Review of Entomology* 63: 369–387.

55 Pitts, K. and Wall, R. (2005). Winter survival of larvae and pupae of the blowfly, *Lucilia sericata* (Diptera: Calliphoridae). *Bulletin of Entomological Research* 95 (3): 179–186.

7

Microorganisms as Drivers of Host–Parasite Interactions

Rahul Negi[1]*, Munni Bhandari*[1]*, Rahul Kunwar Singh*[1,*]*, and Tribhuvan Mohan Mohapatra*[2]

[1] *Department of Microbiology, School of Life Sciences, Hemvati Nandan Bahuguna Garhwal University, Srinagar Garhwal, Uttarakhand, India*
[2] *Department of Microbiology, Institute of Medical Sciences, Banaras Hindu University, Varanasi, India*
* *Corresponding author*

7.1 Introduction

Parasitic diseases or parasitoses are the diseases triggered by parasites ranging from protozoa to worms. These diseases affect millions of people throughout the world, constitute the major portion of neglected tropical diseases declared by WHO, and are the critical contributors of mortality and morbidity, particularly in the developing world. These diseases are transmitted to humans through oral fecal route as well as with the bite of an intermediate host and vectors, usually an arthropod. The main aim of parasitology is to describe the diverse variety of parasitic pathogens, their effect on human hosts and mechanisms thereof, and develop prophylactic measures [1].

Previously, the studies on parasitic diseases were focused only on the pathogenic aspect of the parasites, and microbes associated with them were overlooked. Of late, it has been discovered that hosts are frequently connected with a variety of mutualistic symbionts that can contribute to their physiology and evolution independently or synergistically. The mutualistic microbiota play a critical role in the maintenance of the host's health and may protect the host against parasitic infection. This complexity, as well as its ramifications, is now attracting the attention of parasitologists towards the mutualistic microbiota of the host and is transforming the realm of parasitology, particularly parasitic immunology [2, 3].

Parasitic Infections: Immune Responses and Therapeutics, First Edition.
Edited by Abhay Prakash Mishra and Manisha Nigam.

Furthermore, some parasites carry their own microbiota such as bacteria or viruses, which can affect the outcome of any parasitic attack to the host. This mutualistic microbiota of a parasite may add virulence to the parasite by several means such as providing better nutrition to the parasite, activating the host's inflammatory cells and generating carcinogens. As the parasite microbiota are not commonly explored, their potential effects on pathogenesis and disease are yet to be extensively investigated [2]. Accordingly, mutualistic symbionts of both host and parasite have been considered to have significant role in the progress of parasitic diseases [4]. Any alteration to the composition of mutualistic microbiota may set off a series of events that alters the infection's fate. The mutualistic microbiota of the host or parasite forms a triangular interaction: "microbe–host–parasite" [4, 5]. Thus, these mutualistic symbionts may provide new insights for developing new medicines and remediation tactics. With this viewpoint, a new area of investigation has emerged to understand the nature of interactions between the host, parasites, and their associated microbiota [6]. Therefore, this chapter attempts to answer the questions of what role mutualistic symbionts of host and parasite play in the host–parasite interaction and parasitic disease dynamics.

7.2 Host's Symbiont

Various mutualistic symbionts typically find a home in the bodies of animals, including humans [7, 8]. The compositions of these symbionts vary greatly from one individual to another [9], and also from one anatomical site to another within the body of same person [10]. Multicellular organisms cannot function without these symbionts. This realization sparked an explosion of research into the effects of microorganisms on the health of animal hosts, revealing that microbe-mediated disease protection is common. Aside from other functions, mutualistic symbionts of the host can protect the host from parasitic pathogens. Earlier investigations have reported that the mutualistic microbiota of host affects the virulence of parasitic pathogens, which in turn influence the severity of host damage. The enhancement or weakening of *Entamoeba histolytica* virulence when co-cultivated with *E. coli* strains is an appropriate example [11]. Recent researches have focused on the role of mutualistic microbes in diseases caused by intestinal protozoans and extraintestinal *Plasmodium* parasites.

There may be several mechanisms for host protection by mutualistic symbionts that can be grouped into three main strategies: (i) competitive exclusion of the parasites, (ii) induction of host's immune system against parasites, and (iii) release of metabolites to harm the parasites. The host recognizes the parasites as well as the mutualistic symbionts through pattern recognition receptor (PPR), and pathogen-associated molecular patterns (PAMP) resulting in a systemic antiparasitic response.

7.2.1 Competitive Exclusion of the Parasites

The mutualistic microbiome of the host exhibits significant roles in several physiological processes, like nutrition, transmission of neuro signals, maturation and regulation of immune cells, regulation of intestinal permeability, and thereby provides protection to the host against pathogens.

There is likely to be competition for host's resources between the two symbionts. As a result, it is not surprising that there are numerous well-documented examples of host defense through resource depletion. Protozoan parasites, such as *Cryptosporidium*, *Giardia* etc., observed in fecal samples of Columbian children, have been linked to varied microbiota composition when compared to uninfected children [12]. Surprisingly, increased microbial diversity is linked to a healthier gut [13].

7.2.2 Release of Metabolites to Harm the Parasites

The production of toxins targeting the parasites has been linked to the protection provided by symbionts in several insects, including pea aphids (*Acyrthosiphon pisum*) protected against parasitoids by *Hamiltonella defensa* [14], and in *Drosophila neotestacea* protected against nematodes [15].

7.2.3 Induction of Host's Immune System against Parasites

Symbionts are required for the early development of the host's immune system [16]. The interaction between the mutualistic microbiome and the host's immune system has resulted in the development of a new unit that combines microbe-encoded defenses into a single word "holo-immunome" [17] or "microimmunosome" [18]. The microimmunosome can induce an immunological response in the host that suppresses parasites or diseases. For example, upregulation of mosquito immunity genes has been demonstrated to inhibit a variety of diseases, including malaria and filaria [19, 20].

7.3 Role of Host Mutualistic Symbionts in Parasite Colonization

The intestinal microbiota is important in determining susceptibility or resistance to parasite colonization. For example, the susceptibility of human subjects to *Giardia* infection differs considerably in the individuals of different ages. The susceptibility to *Giardia* infection was higher in the age group of 0.5 to 5 years compared to other age groups. It may be most probably due to alteration in mutualistic microbiota composition due to age factor and thereby host immunity [21]. It has already been proved that the mutualistic microbiota composition of children varies significantly from that of adults and advances quickly during first five years [22]. The mutualistic microbiota determines the susceptibility of host to *Giardia* infection through the interactions of the trophozoites with host's mutualistic microbiota.

Furthermore, it has been described that the *Lactobacilli* present in human vagina may impact susceptibility for infection by the parasite, *Trichomonas vaginalis* [23]. An in vitro study observed the nonadherence of the parasite trophozoites, when *Lactobacillus gasseri* ATCC 9857 was applied to human vaginal epithelia cells [24]. Though *Lactobacilli* are known to hamper the adhesion of parasite to vaginal epithelia, the exact mechanism for reduced susceptibility is still to be understood [24].

7.4 Mutualistic Symbionts and Host Immune System Interaction

During parasitic infection in humans, the immune system of the host is influenced not only by the pathogenic parasite, but also by mutualistic symbionts. This is supported by the findings of Torres et al., [25], who discovered a greater proportion of IgA in gut fluid as well as an increased proportion of IgG and IgM in serum of *Giardia*-infected conventional mice compared to germ-free mice.

Moreover, immune system modulation by mutualistic symbionts is also important in parasite pathogenesis. A study on *Giardia*-infected human subjects reported that the diversity of microbiota associated with small intestine may affect the pathogenesis as well as clinical appearances of giardiasis. Only patients with severe intestinal malabsorption caused by *Giardia* had higher concentrations of enterobacteria in the jejunum, whereas patients with mild malabsorption did not have enterobacteria overgrowth [26]. In another study, thirty-one different anaerobic microorganisms and *Giardia duodenalis* BT6 obtained from symptomatic giardiasis patients were given to germ-free mice in various combination. Although administration of only bacterial consortium was unable to recreate *Giardia* pathogenesis, the combination including *G. duodenalis* BT6 caused rise in levels of *Giardia*-specific IgA, IgM, and IgG antibodies. This result indicated that the small gut microbiome improves the host immune response against *Giardia* infection [25]. In addition, the host microbiota is also considered to exhibit antiparasitic roles against pathogenic protozoa. An in vitro study, for example, found that the *Lactobacillus johnsonii* La1 strain considerably repressed the propagation of *Giardia* trophozoites. It is a component of a healthy vaginal microbiota. In vivo study with *L. johnsonii* La1-treated gerbils confirmed the defensive role of *L. johnsonii* La1 [27]. Another in vitro investigation observed that only 20% of *Entamoeba histolytica* cells could survive, when cocultured with *Lactobacillus casei* and *Enterococcus faecium*. An in vivo study found a correlation of low *Lactobacillus* spp. levels with *Entamoeba histolytica* infections in humans [28].

The comparative 16S rRNA sequence homology analysis of fecal microbiota associated with Colombian children revealed that the subjects infected with *Giardia* alone had discrete microbiota composition compared to healthy subjects. The investigators recorded lesser values of diversity indices in the infected group compared to uninfected controls. Further, the microbiota of infected group exhibited a change in its composition from *Bacteroides* dominated enterotype I to *Prevotella* dominated enterotype II [12]. Similarly, a noteworthy increase in *Prevotella* was also noted in Argentinian patients suffering from *Giardia* infection in comparison to healthy volunteers [29].

A recent Global Enteric Multicenter Study (GEMS) investigation analyzed the data of *Giardia* infected human participants from four different countries to find out any correlation between severity of *Giardia* infection and the host's mutualistic microbiota composition. All ($N = 200$) subjects were categorized in two groups: one with mild to severe diarrhea (MSD) and another with no diarrhea. The metagenomic biomarker discovery analysis (LEfSe) of both the groups revealed that the parasitic infection caused augmentation of *Prevotella* and a decrease in *Escherichia*, irrespective of their diarrheal symptoms [30]. Further, an upsurge in cell number of *Lachnospiraceae*, *Ruminococcus*, and *Clostridiales* and diminution in cell number of *Natronobacillus*, *Lactobacillus*, and *Leuconostoc* could, to a lesser extent, also be linked to the parasitic infection.

Similar results were also recorded in the Malnutrition and Enteric Disease Study (MAL-ED) subjects in Peru. Nonetheless, increase in *Blautia*, *Faecalibacterium*, *Roseburia*, and members of Ruminococcaceae, as well as upsurge in Lachnospiraceae and *Lactobacillus* were also observed in LEfSe analysis of these subjects [30]. In general, these findings suggest a strong correlation of *Giardia* infection with significant upsurge in mutualistic microbiota of enterotype II dominated by *Prevotella*. The organisms of this group are previously identified as promoters of Th17 cell differentiation. The Th17 cell has a significant role in prophylaxis of *Giardiasis* [31].

In contrast, *Giardia* infection induced alteration in the mutualistic microbiota structure of the host, which in turn induce a pro-inflammatory response. It is evinced from a study of mice in which mice receiving microorganisms earlier exposed to *Giardia* trophozoites, exhibited augmented synthesis of pro-inflammatory cytokines like IL-6, IL-1, TNF-, and IFN-, in comparison to those mice receiving unexposed bacteria [32]. Moreover, *Giardia* infection has also been linked with initiation of Th17 type immune responses, as an increased concentration of IL-17 was recorded in Brazilian children infected with *Giardia* [33, 34].

Studies have noted that the increased production of cytokines such as IL-17 used to be induced by gut microbiota of "Candidatus Savagella" lineage [35]. These bacteria have been observed only in the *Giardia* resistant mice population, not in the susceptible population [34]. The study of T-cells and antigen presenting

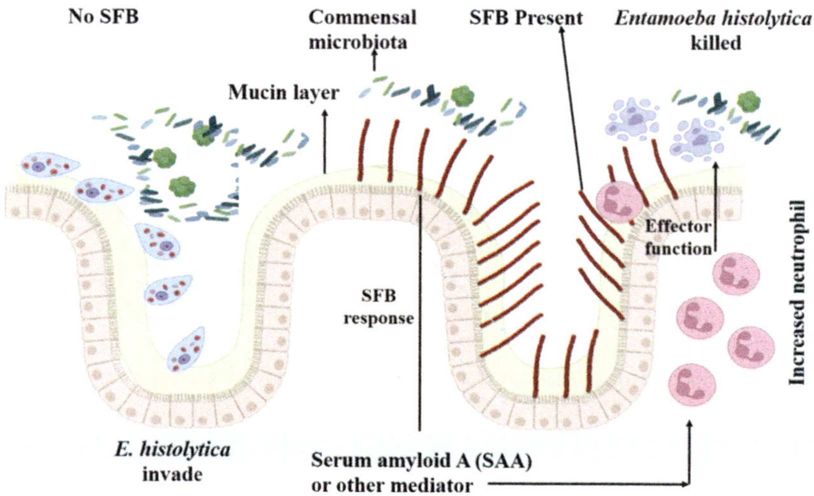

Figure 7.1 Model of bacteria ("Candidatus Savagella" lineage) mediated protection against *Entamoeba histolytica* colonization illustrated online through bioRender.com (modified from [23]).

cells (APCs) isolated from the lamina propria of germ-free mice inoculated with bacteria of "Candidatus Savagella" lineage showed the upsurged expression of IL-17 as well as fecal IgA [36]. These facts suggest that the bacteria of Candidatus Savagella lineage play critical roles in host protection against *Giardia* infection via adaptive immune responses (Figure 7.1). The intestinal tract of rodents is known to be colonized by the microbes of Candidatus Savagella lineage during initiation of weaning process, and thus are present in very low numbers in neonates leaving them susceptible to *Giardia* infection [37]. Moreover, the members of genus *Clostridium* have also been found associated to the promotion of Treg cell responses [38, 39]. A noteworthy decrease in clostridial population have been noted in *Giardia*-infected mice resulting in very weak Treg response [35].

7.5 Host's Mutualistic Symbionts Interaction with Parasites

The intestinal parasitic pathogens inhabiting the gut mucosa remain in vicinity of the mucosa-associated microbiota. Thus, the interactions between the parasite, the mucosa associated microbiota, and the defense system of the host determines the consequences of intestinal parasitic infections [40]. For example, *Entamoeba histolytica* infection in the human gut has been considerably linked with patient's fecal microbiome. The composition of human gut mutualistic symbionts was found able to diagnose the infection by *Entamoeba* species in a particular Cameroon population with approximately 80% precision. The major organisms used for diagnosis of *Entamoeba* infection belonged to family Prevotellaceae [41]. Another study in Bangladesh also observed the upsurged level of *Prevotella copri* in the intestine of patients suffering from diarrhea caused by *Entamoeba histolytica* infections (Table 7.1). The increased level of *P. copri* in the human gut have been

Table 7.1 Specific components of the microbiota and their effect during human protozoan infection.

S.	Protozoan	Microbiota	Effect	Reference
1.	*Entamoeba histolytica*	*Escherichia coli*	Augment and attenuate parasite virulence	[58]
2.	*E. histolytica*	*Lactobacillus casei, Enterococcus faecium*	Coculture inhibits parasite	[5]
3.	*E. histolytica*	Prevotellaceae	Predicted infection	[23, 41]
4.	*E. histolytica*	*Prevotella copri*	Predicted diarrhea	[23, 42]
5.	*Cryptosporidium*	*Proteobacteria, Firmicutes, Escherichia coli* CFT073, *Bacillus* spp., *Clostridium* spp.	Increased relative abundance in *Cryptosporidium*-negative subjects	[23, 45]
6.	*Giardia*	*Lactobacillus johnsonii* La1	Inhibit proliferation of *Giardia* trophozoites	[5; 27]
7.	*Giardia duodenalis*	*Bifidobacterium*	Increased relative abundance in *Giardia*-positive subjects	[47; 59]
8.	*Blastocystis*	*Clostridia, Enterobacteriaceae*	Increased *Clostridia* levels but lower Enterobacteriaceae levels in blastocystis-positive subjects	[23]
9.	*Trichomonas vaginalis*	*Lactobacilli, Mycoplasma, Parvimonas, Sneathia*	Decreased lactobacilli and increased *Mycoplasma, Parvimonas*, and *Sneathia* abundances in *T. vaginalis*-positive subjects	[23, 5]
10.	*Trichomonas vaginalis*	*Lactobacillus gasseri* ATCC 9857	Parasite adhesion inhibition	[5, 24]
11.	*Plasmodium falciparum*	*Bifidobacterium, Streptococcus*	Higher proportion of Bifidobacterium and *Streptococcus* organisms in a low-infection-risk group	[23, 46]

linked earlier with severe inflammation, colitis and risk of autoimmune diseases [42]. Thus, these investigations reveal that the composition of mutualistic symbionts in the human gut play critical roles in *Entamoeba* infection and highlight the pro-inflammatory properties of intestine that may change the consequences of parasitic infection [40, 42].

Moreover, the infection of another parasitic pathogen, *Giardia*, in the human intestine results in invasion by mutualistic symbionts along with the parasite due to depolymerization of the mucus matrix, as shown by several in vivo studies [32, 43]. Further, the mucus matrix depolymerization also results in the alteration of intestinal mutualistic microbiota including mucolytic bacteria. This alteration causes changes in characteristic properties of mucus, e.g., permeability and glycosylation patterns, and may results in mucus degradation [44]

Chappell et al., [45] compared the abundance of mutualistic bacterial taxa found in adult human intestine before or within two days of *Cryptosporidium* infection and thereafter. A greater concentration of *Bacillus* spp., *Clostridium* spp., and *E. coli*, whereas lower concentration of *Bacteriodes* spp., *Prevotella bryantii* and *Akkermansia muciniphila* were recorded in the intestine of uninfected individuals as compared to the infected ones (Table 7.1). The difference in abundance of seven bacterial groups was as high as 2.5- fold.

Recently, intestinal mutualistic microbiota composition of a Malian population was determined by their stool sample analysis and evaluated for risk of *P. falciparum* infection. A substantial correlation was observed in composition of mutualistic symbionts and the prospective risk of malarial infection. The individuals, who got ill with malaria, had lower proportion of *Bifidobacterium* spp. and *Streptococcus* spp. in their intestine as compared to those who did not get infected [46].

Another investigation of a Côte d'Ivoire population infected with intestinal parasites using temporal temperature gel electrophoresis (TTGE) classified the individual in four different categories: 1) those positive for *Giardia* infection, 2) those positive for *Entamoeba* and *Blastocystis* infection, 3) those positive for all three protozoans, and 4) those negative for protozoan infection. Further, the investigators analyzed the mutualistic microbiota composition in the intestine of individuals belonging to these four groups and observed the considerable upsurge in the number of *Bifidobacterium* in the first group [47].

In addition to the intestinal parasitic diseases, the mutualistic microbiota may also affect the outcome of extraintestinal disease. This might be due to the changes in acquired immunity and increase in the magnitudes of humoral and cell mediated responses [48]. Although the exact mechanisms behind the extraintestinal effect of intestinal microbiota has not been studied fully, some microbial metabolites, e.g., small fatty acids, or factors of host origin, induced by the mutualistic symbionts, might be involved in these outcomes [23, 49].

7.6 Parasite-Associated Mutualistic Symbionts

Mutualistic symbionts associated with hosts have been discovered in many systems, whereas mutualistic symbionts of parasites have not been studied very often [2]. *Wolbachia*, for example, is necessarily associated with filarial nematodes in a mutualistic relationship [50]. In humans, seditious responses linked with various diseases are induced by this mutualistic symbiont of nematodes. *Wolbachia*-induced swelling is accountable for impaired vision and sightlessness caused by *Opisthorchis volvulus* and lymphatic vessel inflammation caused by *Wuchreria bancrofti* infection. Similarly, *Opisthorchis viverrini*, popularly recognized as the liver fluke, and a common cause of cholangiocarcinoma, harbors many symbionts [51]. Few of these symbionts are found spread in the host, often abundantly in tumors. Thus, there is need to evaluate their involvement in tumor development [52]. Moreover, *Trichomonas vaginalis* harbors a viral symbiont, *Trichomonas*. An analysis of 34 *T. vaginalis* clinical isolates, revealed the presence of RNA virus, *Trichomonas*, in almost half of the clinical samples tested, as well as their correlation to sexually transmitted diseases, infertility, miscarriage, and pre-term delivery. The worsening of symptoms following successful antibiotic therapy is due to the *Trichomonas* virus as the dying parasites release its viral burden into the host vaginal canal [53].

Similarly, an endosymbiont, *Wolbachia*, associated with the parasite, *Onchocera volvulus*, plays critical role in the disease oncocerciasis. *Wolbachia* and its metabolites reemitted by nematodes interact with the host defense arrangements, inducing responses in different immune cell types, as observed in one more filarial species, *Brugia malayi*. The phagocytic cells like DCs and macrophages get activated using TLR2-MyD88 pathway upon exposure with *Wolbachia* as demonstrated by increased synthesis of co-stimulatory surface particles and production of pro-inflammatory cytokines [53]. A subsequent stimulation can homo- and heterotolerize macrophages, and contributes to the repression of immune response that characterizes the common filarial diseases [53]. Similarly, peripheral polymorphonuclear leukocytes (PMN) and stromal cells of the cornea may react to *Wolbachia* through the TLR2-MyD88 pathway, synthesize CXC chemokines and trigger the parasite's inflammatory responses [54]. *Wolbachia* stimulates mast cells via TLR2 to degranulate and rise in vascular permeability to enable infection [54]. This interaction was replicated in vitro in a mouse model of onchocercal keratitis by means of *Onochocerca volvulus* extracts comprising *Wolbachia* cells, demonstrating that the *Wolbachia* was required for PMN leukocytes-mediated inflammation, opaqueness, and corneal mist via different mechanisms of action.

Wolbachia released in the cornea after microfilaria death contributes to corneal edoema and opaqueness by persuading PMN and macrophage penetration and instigation in the corneal stroma, which is reliant on TLR2-MyD88 gene activation and CXC chemokine production. In the corneal stroma, keratocytes and bone marrow-derived cells may induce this action, which is then sustained by inflammatory cells [54]. After treatment with filaricidal drugs, lot of *Wolbachia* bacterial cells are unconfined from microfilariae causing cutaneous and systemic adverse reactions such as fever, tachycardia, hypotension, lymphadenopathy, and pruritus. Neutrophils are the first cells to be recruited and activated in the skin, causing dermal inflammation. The adverse events at systemic level are related to *Wolbachia* DNA and levels of *Wolbachia* cells in blood, pro-inflammatory cytokines, neutrophilis, and antibacterial peptides like calprotectin and calgranulin. *Wolbachia* is linked to PMN penetration in the cornea, skin, and onchocercomas [53, 54].

The following sequence of events is consistent with *O. volvulus/Wolbachia*-induced keratitis, which was studied using an ocular onchocerciasis mouse model: *Wolbachia*, which is released by vanishing microfilariae in the corneal stroma, triggers indigenous cells like fibroblasts, dendritic cells, and macrophages. These cells secrete proinflammatory and chemotactic cytokines, such as macrophage-inflammatory protein 2 (MIP-2) and predominately keratinocyte-derived chemokine (KC) [54]. Predominately KC causes CXCR2-dependent PMN relocation to the corneal stroma, where *Wolbachia* activates PMN via TLR2/6. These leukocytes secrete other chemokines, primarily MIP-2, which induces additional relocation and degranulation of PMN, as well as the exudation of cytotoxic molecules such as nitric oxide, myeloperoxidase, and oxygen radicals, which exhibit a cytotoxic consequence on indigenous corneal cells such as fibroblasts and endothelium, ensuing corneal edema and additional harm to corneal clarity [54].

Helminths go through dramatic phenotypic variations and severe alterations in the environment during their transit to other host species, each with its particular set of immunological responses. Thus, the mechanism of reproduction in these organisms could create serious limitations in the propagation of their mutualistic symbionts.

The particular characters of these microbial symbiont–parasite interactions not only offer the new options for developing novel treatments against parasitic disorders [55, 56], but it also makes the host–parasite interactions and ecology more complex [57]. For example, the entire microbiota of parasitic helminths at all stages of helminth's development need to be understood.

7.7 Conclusion

Microorganisms associated with the human host as well as pathogenic parasites are significantly involved in host–parasite interactions, resulting in a multifaceted web of interactions with mutual feedback. This results in a complex web of interconnected organisms formed by the multidimensional linkages between the host, parasite, and host-parasite-associated microbiota in which changes in any of these components trigger a chain of counter-responses in the others. The progress of infections caused by either mucosal or systemic parasites are also determined with particular microorganisms as part of mutualistic symbionts. On the other hand, the composition of mutualistic symbiont community and thereby the host immunity is altered by parasitic infection. This alteration in mutualistic microbiota composition is very specific and may be unique in response to the infection by a particular parasite. Further, this alteration can moderate both innate and adaptive immunity; however, the precise mechanism behind host immunity modulation is not completely understood. Moreover, the parasite associated symbionts also play critical roles in modulating the microbiota composition and immunity of the human host. Thus, it is apparent that both the host and parasite's mutualistic symbionts act as drivers for host–parasite interaction and play meaningful roles in parasitic disease consequences. Detailed knowledge about the role of mutualistic symbionts in driving host and parasite interactions will enable us to utilize the mutualistic symbiont for prophylaxis and treatment of parasitic diseases.

Acknowledgement

Authors are thankful to their respective employers for providing necessary facilities.

References

1 Parija, S.C. and Chaudhury, A. (2022). *Textbook of Parasitic Zoonoses*. Singapore: Springer.

2 Dheilly, N.M., Bolnick, D., Bordenstein, S. et al. (2017). Parasite microbiome project: systematic investigation of microbiome dynamics within and across parasite–host interactions. *MSystems* 2 (4): e00050–17.

3 Itoh, H., Jang, S., Takeshita, K. et al. (2019). Host–symbiont specificity determined by microbe–microbe competition in an insect gut. *Proceedings of the National Academy of Sciences* 116 (45): 22673–22682.

4 Hahn, M.A. and Dheilly, N.M. (2016). Experimental models to study the role of microbes in host–parasite interactions. *Frontiers in Microbiology* 7: 1300.

5 Mumcuoğlu, I. (2019). Interactions between parasites and human microbiota. European Journal of Therapeutics 25(1): 6–11.

6 Leung, J.M., Graham, A.L., and Knowles, S.C. (2018). Parasite–microbiota interactions with the vertebrate gut: synthesis through an ecological lens. *Frontiers in Microbiology* 9: 843.

7 Gerardo, N.M. and Parker, B.J. (2014). Mechanisms of symbiont-conferred protection against natural enemies: an ecological and evolutionary framework. *Current Opinion in Insect Science* 4: 8–14.

8 Ford, S.A. and King, K.C. (2016). Harnessing the power of defensive microbes: evolutionary implications in nature and disease control. *PLoS Pathogens* 12 (4): e1005465.

9 The Human Microbiome Project Consortium. (2012). Structure, function and diversity of the healthy human microbiome. *Nature* 486: 207–214.

10 Smith, M.I., Yatsunenko, T., Manary, M.J. et al. (2013). Gut microbiomes of Malawian twin pairs discordant for kwashiorkor. *Science* 339 (6119): 548–554.

11 Galván-Moroyoqui, J.M., Domínguez-Robles, M.D., Franco, E., and Meza, I. (2008). The interplay between entamoeba and enteropathogenic bacteria modulates epithelial cell damage. *PLoS Neglected Tropical Diseases* 2 (7): e266.

12 Toro-Londono, M.A., Bedoya-Urrego, K., Garcia-Montoya, G.M. et al. (2019). Intestinal parasitic infection alters bacterial gut microbiota in children. *PeerJ* 7: e6200.

13 Grace-Farfaglia, P., Frazier, H., and Iversen, M.D. (2022). Essential factors for a healthy microbiome: a scoping review. *International Journal of Environmental Research and Public Health* 19 (14): 8361.

14 Oliver, K.M., Degnan, P.H., Hunter, M.S., and Moran, N.A. (2009). Bacteriophages encode factors required for protection in a symbiotic mutualism. *Science* 325 (5943): 992–994.

15 Ballinger, M.J. and Perlman, S.J. (2017). Generality of toxins in defensive symbiosis: ribosome-inactivating proteins and defense against parasitic wasps in *Drosophila*. *PLoS Pathogens* 13 (7): e1006431.

16 Gollwitzer, E.S. and Marsland, B.J. (2015). Impact of early-life exposures on immune maturation and susceptibility to disease. *Trends in Immunology* 36 (11): 684–696.

17 Dheilly, N.M. (2014). Holobiont–holobiont interactions: redefining host–parasite interactions. *PLoS Pathogens* 10 (7): e1004093.

18 Dietert, R.R. (2017). The microbiome-immune-host defense barrier complex (microimmunosome) and developmental programming of noncommunicable diseases. *Reproductive Toxicology* 68: 49–58.

19 Pan, X., Zhou, G., Wu, J. et al. (2012). *Wolbachia* induces reactive oxygen species (ROS)-dependent activation of the Toll pathway to control dengue virus in the mosquito *Aedes aegypti*. *Proceedings of the National Academy of Sciences* 109 (1): E23–31.

20 Rancès, E., Ye, Y.H., Woolfit, M. et al. (2012). The relative importance of innate immune priming in *Wolbachia*-mediated dengue interference. *PLoS Pathogens* 8 (2): e1002548.

21 Harvey, T.V., Tang, A.M., da Paixao Sevá, A. et al. (2020). Enteric parasitic infections in children and dogs in resource-poor communities in northeastern Brazil: identifying priority prevention and control areas. *PLoS Neglected Tropical Diseases* 14 (6): e0008378.

22 Lynch, S.V. and Pedersen, O. (2016). The human intestinal microbiome in health and disease. *New England Journal of Medicine* 375 (24): 2369–2379.

23 Burgess, S.L., Gilchrist, C.A., Lynn, T.C., and Petri, W.A., Jr (2017). Parasitic protozoa and interactions with the host intestinal microbiota. *Infection and Immunity* 85 (8): e00101–17.

24 Phukan, N., Parsamand, T., Brooks, A.E. et al. (2013). The adherence of *Trichomonas vaginalis* to host ectocervical cells is influenced by lactobacilli. *Sexually Transmitted Infections* 89 (6): 455–459.

25 Torres, M.F., Uetanabaro, A.P., Costa, A.F. et al. (2000). Influence of bacteria from the duodenal microbiota of patients with symptomatic giardiasis on the pathogenicity of *Giardia duodenalis* in gnotoxenic mice. *Journal of Medical Microbiology* 49 (3): 209–215.

26 Tomkins, A.M., Wright, S.G., Drasar, B.S., and James, W.P. (1978). Bacterial colonization of jejunal mucosa in giardiasis. *Transactions of the Royal Society of Tropical Medicine and Hygiene* 72 (1): 33–36.

27 Humen, M.A., De Antoni, G.L., Benyacoub, J. et al. (2005). *Lactobacillus johnsonii* La1 antagonizes *Giardia intestinalis* in vivo. *Infection and Immunity* 73 (2): 1265–1269.

28 Verma, A.K., Verma, R., Ahuja, V., and Paul, J. (2012 December). Real-time analysis of gut flora in *Entamoeba histolytica* infected patients of northern India. *BMC Microbiology* 12 (1): 1-1.

29 Mejia, R., Damania, A., Jeun, R. et al. (2020). Impact of intestinal parasites on microbiota and cobalamin gene sequences: a pilot study. *Parasites & Vectors* 13 (1): 1–9.

30 Berry, A.S., Johnson, K., Martins, R. et al. (2020). Natural infection with *Giardia* is associated with altered community structure of the human and canine gut microbiome. *Msphere* 5 (4): e00670–20.

31 Calcinotto, A., Brevi, A., Chesi, M. et al. (2018). Microbiota-driven interleukin-17-producing cells and eosinophils synergize to accelerate multiple myeloma progression. *Nature Communications* 9 (1): 1–3.

32 Beatty, J.K., Akierman, S.V., Motta, J.P. et al. (2017). *Giardia duodenalis* induces pathogenic dysbiosis of human intestinal microbiota biofilms. *International Journal for Parasitology* 47 (6): 311–326.

33 Cascais-Figueiredo, T., Austriaco-Teixeira, P., Fantinatti, M. et al. (2019). Giardiasis alters intestinal fatty acid binding protein (I-FABP) and plasma cytokines levels in children in Brazil. *Pathogens* 9 (1): 7.

34 Singer, S.M., Fink, M.Y., and Angelova, V.V. (2019). Recent insights into innate and adaptive immune responses to *Giardia*. *Advances in Parasitology* 106: 171–208.

35 Barash, N.R., Maloney, J.G., Singer, S.M., and Dawson, S.C. (2017). *Giardia* alters commensal microbial diversity throughout the murine gut. *Infection and Immunity* 85 (6): e00948–16.

36 Atarashi, K., Tanoue, T., Ando, M. et al. (2015). Th17 cell induction by adhesion of microbes to intestinal epithelial cells. *Cell* 163 (2): 367–380.

37 Ericsson, A.C., Hagan, C.E., Davis, D.J., and Franklin, C.L. (2014). Segmented filamentous bacteria: commensal microbes with potential effects on research. *Comparative Medicine* 64 (2): 90–98.

38 Keselman, A., Li, E., Maloney, J., and Singer, S.M. (2016). The microbiota contributes to CD8+ T cell activation and nutrient malabsorption following intestinal infection with *Giardia duodenalis*. *Infection and Immunity* 84 (10): 2853–2860.

39 Fink, M.Y. and Singer, S.M. (2017). The intersection of immune responses, microbiota, and pathogenesis in giardiasis. *Trends in Parasitology* 33 (11): 901–913.

40 Burgess, S.L. and Petri, W.A. (2016). The intestinal bacterial microbiome and *E. histolytica* infection. *Current Tropical Medicine Reports* 3 (3): 71–74.

41 Morton, E.R., Lynch, J., Froment, A. et al. (2015). Variation in rural African gut microbiota is strongly correlated with colonization by *Entamoeba* and subsistence. *PLoS Genetics* 11 (11): e1005658.

42 Gilchrist, C.A., Petri, S.E., Schneider, B.N. et al. (2016). Role of the gut microbiota of children in diarrhea due to the protozoan parasite *Entamoeba histolytica*. *The Journal of Infectious Diseases* 213 (10): 1579–1585.

43 Allain, T., Chaouch, S., Thomas, M. et al. (2018). Bile-salt-hydrolases from the probiotic strain *Lactobacillus johnsonii* La1 mediate anti-giardial activity in vitro and in vivo. *Frontiers in Microbiology* 8: 2707.

44 Amat, C.B., Motta, J.P., Fekete, E. et al. (2017). Cysteine protease-dependent mucous disruptions and differential mucin gene expression in *Giardia duodenalis* infection. *The American Journal of Pathology* 187 (11): 2486–2498.

45 Chappell, C.L., Darkoh, C., Shimmin, L. et al. (2016). Fecal indole as a biomarker of susceptibility to *Cryptosporidium* infection. *Infection and Immunity* 84 (8): 2299–2306.

46 Yooseph, S., Kirkness, E.F., Tran, T.M. et al. (2015). Stool microbiota composition is associated with the prospective risk of *Plasmodium falciparum* infection. *BMC Genomics* 16 (1): 1–5.

47 Iebba, V., Santangelo, F., Totino, V. et al. (2016). Gut microbiota related to *Giardia duodenalis*, *Entamoeba* spp. and *Blastocystis hominis* infections in humans from Côte d'Ivoire. *Journal of Infection in Developing Countries* 10: 1035–1041.

48 Netea, M.G., Joosten, L.A., and Latz, E. (2016). Trained immunity: a program of innate immune memory in health and disease. *Science* 352: aaf1098.

49 Richards, J.L., Yap, Y.A., McLeod, K.H. et al. (2016). Dietary metabolites and the gut microbiota: an alternative approach to control inflammatory and autoimmune diseases. *Clinical & Translational Immunology* 5 (5): e82.

50 Dudzic, J.P., Curtis, C.I., Gowen, B.E., and Perlman, S.J. (2022). A highly divergent *Wolbachia* with a tiny genome in an insect-parasitic tylenchid nematode. *Proceedings of the Royal Society B* 289 (1983): 20221518.

51 Plieskatt, J.L., Deenonpoe, R., Mulvenna, J.P. et al. (2013). Infection with the carcinogenic liver fluke *Opisthorchis viverrini* modifies intestinal and biliary microbiome. *The FASEB Journal* 27 (11): 4572–4584.

52 Osborne, L.C. and WegenerParfrey, L. (2016). Liver flukes and the microbiota in cancer. *EBioMedicine* 8: 12–13.

53 Tamarozzi, F., Halliday, A., Gentil, K. et al. (2011). Onchocerciasis: the role of *Wolbachia* bacterial endosymbionts in parasite biology, disease pathogenesis, and treatment. *Clinical Microbiology Reviews* 24 (3): 459–468.

54 Manoj, R.R., Latrofa, M.S., Epis, S., and Otranto, D. (2021). *Wolbachia*: endosymbiont of onchocercid nematodes and their vectors. *Parasites & Vectors* 14 (1): 1–24.

55 Castiglioni, P., Hartley, M.A., Rossi, M. et al. (2017). Exacerbated leishmaniasis caused by a viral endosymbiont can be prevented by immunization with its viral capsid. *PLoS Neglected Tropical Diseases* 11 (1): e0005240.

56 Jenkins, T.P., Brindley, P.J., Gasser, R.B., and Cantacessi, C. (2019). Helminth microbiomes–a hidden treasure trove? *Trends in Parasitology* 35 (1): 13–22.

57 Dheilly, N.M., Martínez Martínez, J., Rosario, K. et al. (2019). Parasite microbiome project: grand challenges. *PLoS Pathogens* 15 (10): e1008028.

58 Fernández-López, L.A., Gil-Becerril, K., Galindo-Gómez, S. et al. (2019). *Entamoeba histolytica* interaction with enteropathogenic *Escherichia coli* increases parasite virulence and inflammation in amebiasis. *Infection and Immunity* 87 (12): e00279–19.

59 Fekete, E., Allain, T., Siddiq, A. et al. (2021). *Giardia* spp. and the gut microbiota: dangerous liaisons. *Frontiers in Microbiology* 11: 618106.

8

Neglected Parasitic Infections

History to Current Status

Sarmistha Debbarma[1,*], *Jupi Talukdar*[2], *Prabhakar Maurya*[3], *Luit Moni Barkalita*[4], *and Anupam Brahma*[5]

[1] *ARDD, Govt of Tripura, Tripura*
[2] *Mahisah Perspicientia LLP, Assam*
[3] *CEHTRA, New Delhi*
[4] *Department of Animal Biotechnology, CVSc, AAU Assam*
[5] *SRF, Krishi Vigyan Kendra, Dudhnoi, Assam*
[*] *Corresponding author*

8.1 Introduction

Parasitic diseases, simply put, are diseases and their manifestations caused by helminths arthropods and protozoa [1]. Parasitic diseases affect humans and animals alike and many parasitic diseases are zoonotic in nature i.e., they can jump species and can affect humans even though they originate in animals. The public health authorities, governments, NGOs, civil society organizations and other interested stakeholders all over the world, especially the World Health Organization, have taken cognizance of a broader classification of diseases, which includes neglected parasitic diseases as well as some other tropical diseases, that are of grave public health concern and need collaborative efforts and strategy to combat them. These are termed neglected tropical diseases or NTDs. These are a set of transmissible diseases of public health importance that are caused by various worms which are parasitic in nature, bacteria, and protozoa [2]. Neglected parasitic infections are considered as a subclass of diseases under neglected tropical diseases. At present, 20 such infectious parasitic diseases are classified under this category and place a heavy burden on over 1.7 billion people around the world.

Neglected parasitic and tropical diseases cause immeasurable suffering; they debilitate, disfigure and can often be fatal to the host. NTDs mostly affect the

Parasitic Infections: Immune Responses and Therapeutics, First Edition.
Edited by Abhay Prakash Mishra and Manisha Nigam.

vulnerable populations around the world who often live in remote communities, thus creating a cycle of disease and hence poverty causing social and economic burden on developing nations while costing billions of dollars every year in economic losses [2–4]. In developing nations, neglected parasitic and tropical diseases are a major cause of human suffering as well as mortality, thus pose a devastating challenge to the health for millions of people [2]. South Asia is burdened by large proportion of NTD [5]. The poorest section from rural areas, urban slums and geopolitically disturbed zones are the worst sufferers and yet the policy makers, drug firms and research sectors have failed to reach out to the sufferers leaving a major gap in treatment needs for many parasitic and tropical diseases. Also, there are unmet treatment needs for patient groups that are neglected due to economic constraints where, in spite of availability of treatment, the high cost prohibits its affordability [5]. The neglected tropical diseases alone result in greater than 500,000 deaths annually [6] and loss of more than 50 million life years due to adjusting to disability [7]. These diseases are most prevalent in countries with poor economic and hence underdeveloped health infrastructure, thus affecting populations in low- and middle-income countries and geographically mostly African and Asian nations, and in terms of DALYs (Disability-Adjusted Life Years), they are listed as in the top ten for causing global health burden [5, 7, 8]. Considering the burden of this group of diseases and their socio-economic impact, it is very important to have a proper control strategy. In this chapter, we intend to shed light into the evolution of the NTDs diseases by highlighting their past and present status including the impact on global public health.

8.2 Historical Perspectives of Parasitic Infections

The accurate history of parasites is impossible to chart out, but owing to years of knowledge contributions by scholars, historians, scientists, and researchers, a seemingly accurate chronology is available which can help to understand the evolution of parasites, their association with humans and animals, as well as its emergence and development in the field of parasitology (Figure 8.1). Parasites seemed to have first evolved long before humans evolved, as it is quite evident from evolutionary processes that single celled organisms gradually gave way to multi-cellular organisms. From the times when art and writing developed and events circling life, culture and society were documented in some form or other, some records can be seen to shed light on medicine and diseases and many queries around them [9]. The eggs of helminths and cysts of protozoa in coprolites and preserved bodies from the artefacts in the past have been able to shed some light about the prevalence of parasitic disease [10]. Roundworms and tapeworms are mentioned in certain early Egyptian literature [9].

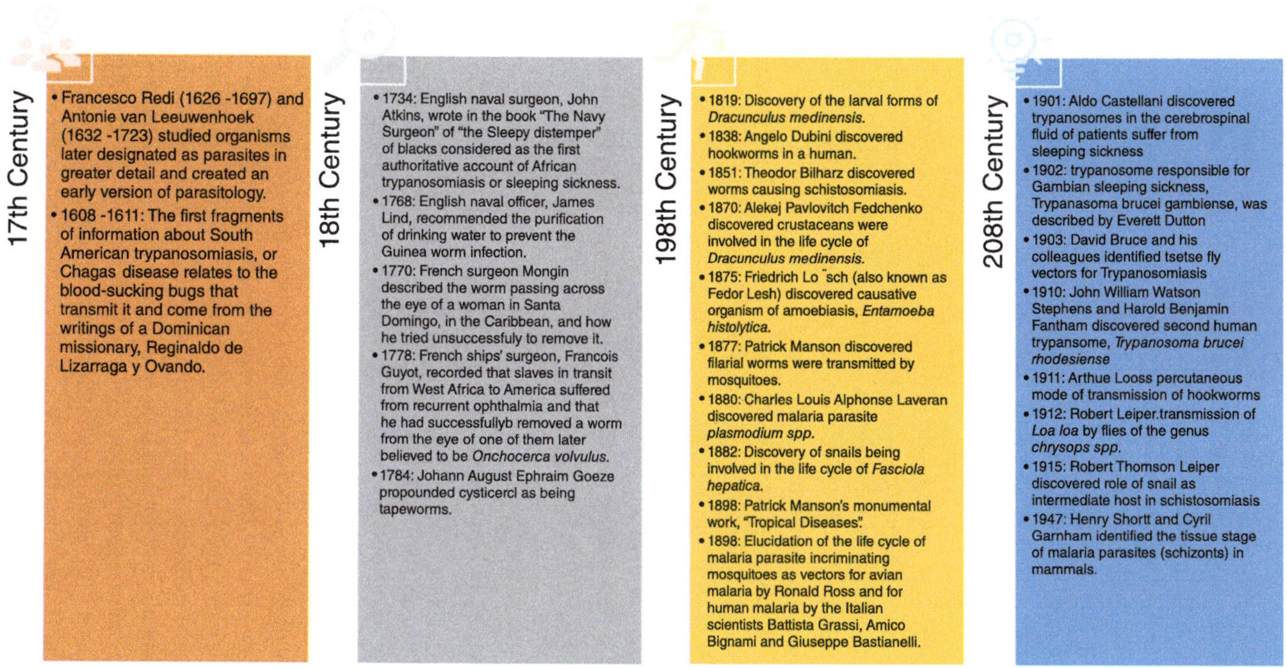

Figure 8.1 Landmarks in the development of parasitic diseases. *Source:* Sarmistha Debbarma et.al., 2022.

The earliest discovered records of human parasites refer to those of eggs of the lung trematode (or flukes) and date back to sixtieth century BCE having been found in the Atacama region from human coprolites. There is also evidence of disputed records dating back to 5000 BCE from Brazil about hookworm eggs, and eggs of *Ascaris lumbricoides* in coprolites dating 2330 BCE from Peru. While the records are disputed, it gives a fair notion that human civilization is replete with the occurrences of common gastrointestinal helminths. The well-preserved mummies from Egypt by embalming have been found to contain eggs of *Taenia* spp. dating to 2000 BCE and *Schistosoma haematobium* dating between 1250 and 1000 BCE, while the presence of the *Dracunculus* spp. in Egypt has been confirmed in a well-preserved female worm and a calcified worm in mummies [10]. In the seventeenth century, scientists such as Francesco Redi and Antonie van Leeuwenhoek began to study organisms later designated as parasites in greater detail and created an early version of parasitology [9].

With the inception and circulation of Johannes Japetus Smith Steenstrup's resourceful book *On the Alteration of Generations*, the initial theory on "Theory of spontaneous generation" gradually began to slip into oblivion and was totally dead by mid nineteenth century. The new advent in the field of research of parasites began due to breaking of the dogma of spontaneous generation and promulgation of the germ theory of disease resulting from Louis Pasteur's work in the 1860s, and for the first time in history scientists and the researchers were able to recognize that parasitism was not merely an association between two organisms but sometimes related to infections causing a relationship of a harmful nature. It initiated a new and logically sound understanding of relationships between parasites and their hosts thus creating enthusiasm for the study of parasites [11].

The liberation from earlier dogmas opened up many avenues and led to rapid and comprehensive discoveries and many observers witnessing the developments made in tropical medicine specially in the last few decades of the nineteenth century, which has been regarded as the the golden age of the subject, and it was considered as equivalent to the study of diseases or infections due to parasitic organisms. One of the earliest books, (considered a monumental work in the field of parasitology), is *Tropical Diseases* published in 1898 and written by Patrick Manson. However, the birth of parasitology, if not limited to tropical medicine, can also be credited with the useful insights provided by the systematic studies of many eminent scientists and zoologists on helminths in Britain and Germany before shedding of light on microorganisms. The knowledge was gathered over years of research and understanding and later became an integral part of university courses of zoological sciences and assimilated along with coelenterates and annelids, and such practices persisted up to the later part of the twentieth century. The later part of the nineteenth century is considered as the golden age in the field of parasitology, as it was the time during which research led to the elucidation of life-cycles of many parasites, which together with insights provided by previous centuries, had created coherent subject matter [11].

During the twentieth century, parasitology emerged as a part of zoological underpinning for tropical medicine. The subject matter of parasitology became well defined by the curriculum of tropical medicine and botany-zoology divergence and the way tropical parasites were compatible with the "type"-based zoological curriculum. During world wars the study of and research into tropical diseases took a relative backseat and research was more focused on vector control. The post-war period resulted in a very schematic and uneven development of five major fields of study: parasitology, tropical medicine, medical research generally, zoology, and veterinary science over a period of 50 years, embracing many countries. The history of parasitology in incredibly intertwined with the history of science and medicine as an interdisciplinary field and has to offer a lot in the fields of current scientific, veterinary, medical and public health research [12].

8.3 Classification Systems of Neglected Parasitic Diseases

The common names of parasites may owe their origin to the local language or to the place of origin or the place of discovery or even based on their symptoms, and hence not only it may be confusing in terms of understanding but also may find difficulty in scientific acceptance world-wide. Hence, to avoid such problems, a standard and widely accepted system of binomial terminology exists that consists of the italicized forms of genus and species respectively. Classification systems are developed to indicate the common relationship among the various parasitic species, and such divisions of nomenclature have been based mostly on life cycles and morphometry of the organism (Figure 8.2).

There are many different ways in which parasites have been classified, but no single system is correct or false. Some of the common ways in which parasites have been classified are: based on location of the parasite in the host i.e., ectoparasite (external to the body) and endoparasite (internal to the body); based on number of cells i.e. single celled (protozoa) or multi-celled (metazoa), which are divided again into arthropods and helminths; based on mode of locomotion i.e. Mastigophora: locomotion by use of flagella, Sporozoa: devoid of flagella and cilia, apply self-distorting movements to locomote; Sarcodina: amoeboidal locomotion and Ciliophora: use of cilia for locomotion; and also based on the relationship with the host i.e., obligate parasite: fully dependent on host for a specific developmental stage of their life-cycle or the entire stages of their lives, facultative parasite: they are not completely dependent on the host for their survival, accidental/incidental parasites: they are those that inadvertently parasitize on unnatural hosts, and erratic parasites: those that localize in inconducive body organs. Although new

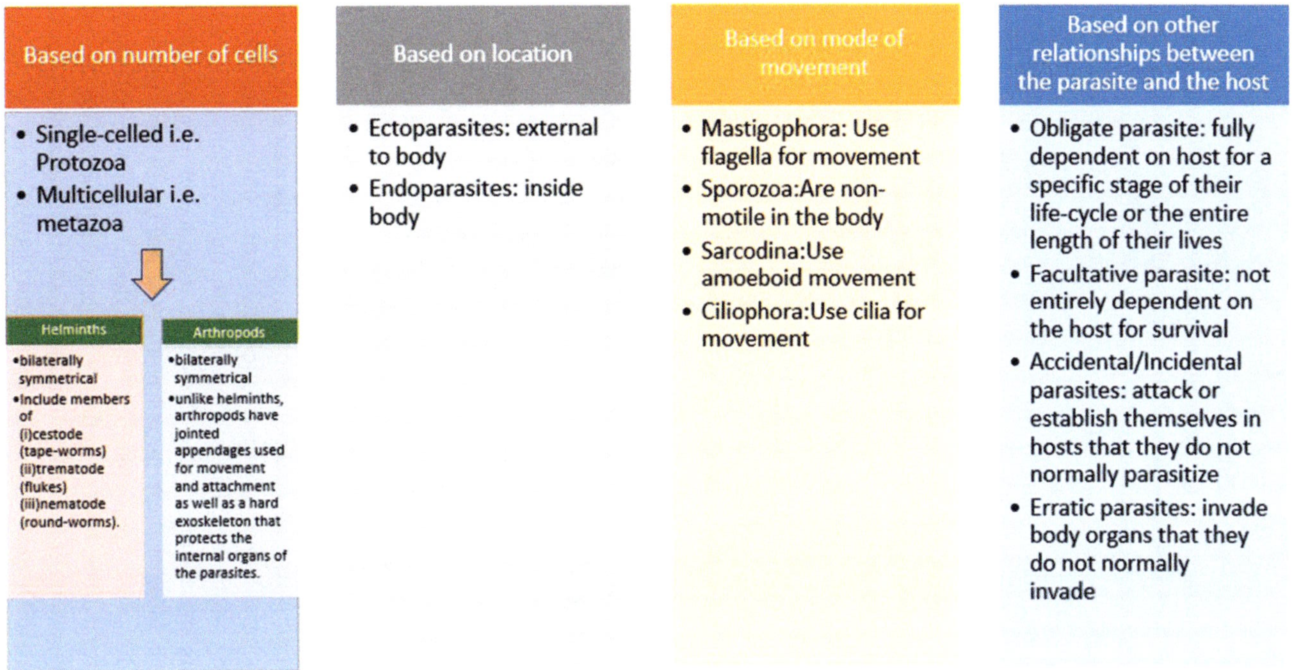

Figure 8.2 Classification systems of parasites. *Source:* Sarmistha Debbarma et al., 2022.

classification names are proposed frequently and there may be some disagreement among scientists [13]. The International Code of Zoological Nomenclature (ICZN) is the organization which governs the classification of organisms for categorization as "parasite" in the medical, veterinary and public health realms [14].

There have been different classifications of neglected tropical diseases but the most widely accepted classification to date is by World Health Organization, 2005 which included 13 diseases, and since then seven more categories have been included as of 2022 [5, 7, 15]:

(i) Bairns dale ulcer (Buruli) (ii) American trypanosomosis or Chagas disease (iii) Dengue and Chikungunya (iv) Guinea worm disease (v) Hydatidosis (vi) Trematodiases (foodborne) (vii) African trypanosomiasis (viii) Aleppo boil or Leishmaniasis (ix) Hansens disease or Leprosy (x) Lymphatic filariasis (xi) Chromoblastomycosis, Mycetoma, and other mycoses (xii) Onchocerciasis (xiii) Schistosomiasis (xiv) Scabies and other ectoparasites (xv) Hydrophobia or Rabies (xvi) Soil-transmitted helminthiases (xvii) Snakebite envenoming (xviii) Taeniasis or Cysticercosis (xix) Egyptian ophthalmia and (xx) Yaws (endemic treponematoses). Out of these neglected tropical diseases, 12 are neglected parasitic diseases, where eight are caused by helminths, three are caused by protozoa and one by ectoparasites as explained in Figure 8.3.

8.4 Impact and Interventions: Neglected Parasitic and Tropical Diseases

Neglected parasitic and neglected tropical diseases related interventions resort to addressing socio-environmental health determinants as well drug treatments. Drug treatments consisting of the application of essential medicines are some of the best actions currently available against neglected tropical diseases [4, 15]. Despite significant improvements attained in health parameters when utilizing available drugs in current times [5, 16–18], fewer compounds are being researched or developed for neglected parasitic or tropical diseases [4, 19, 20]. The major institutions in neglected parasitic or tropical diseases drug development i.e., governmental bodies, pharmaceutical industries, universities and research organizations, are involved in scaling up with the research and development requirements and addressing the challenges of neglected parasitic and tropical diseases [21]. Some of these initiatives include forging of new research based around public–private partnership (PPP) models, building global research and development facilities, lower regulatory bottlenecks to speed up the approvals, broad collaborations of scientific community, and harmonization of governance structures around neglected parasitic or tropical diseases.

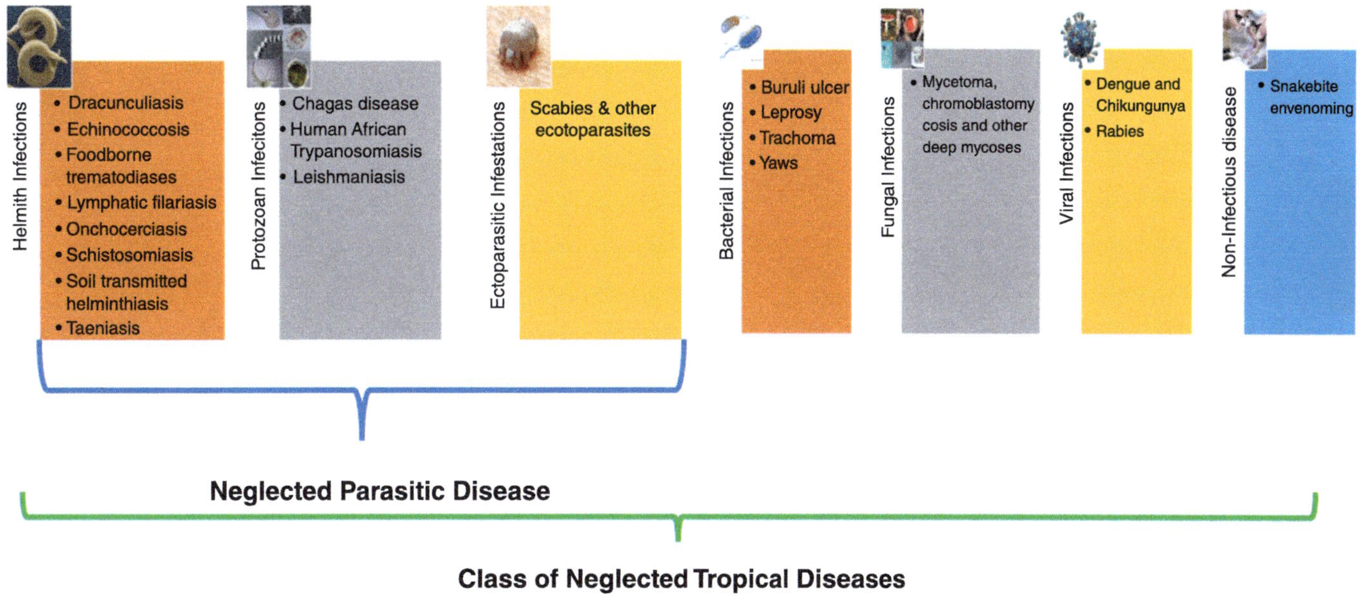

Figure 8.3 List of neglected parasitic diseases and neglected tropical diseases (as per WHO's classification of neglected tropical diseases).
Source: Sarmistha Debbarma et. al., 2022.

In some instances, the treatments are relatively economical [3], while as per an estimate in 2010, financial assistance to the tune of 2–3 billion US$ was required towards neglected parasitic and tropical diseases over the next 5–7 years up to 2015–2017 [4]. To help in the cause of, control of, and elimination of neglected parasitic and tropical diseases, some pharmaceutical majors have committed towards providing a supply of drug therapies for free, and due to this, huge drug administration programs have been successfully introduced across nations [15].

But it has been observed that measures aimed at prevention are more effective in the developed countries compared to developing and poor nations [16]. Within developed countries, neglected parasitic and tropical diseases affect the severely impoverished sections of society, although many of the similar issues put populations at risk in developing nations as well [18]. A total of twenty neglected parasitic and tropical diseases have been on priority list of World Health Organization, but other organizations may have varying definitions or priorities for these neglected parasitic and tropical diseases [19]. These diseases are found to be similar in about 149 countries, and combinedly they affect a population of more than 1.4 billion including about 500 million children [20] while costing billions every year to developing economies [22].

In the beginning of the 2000s, along with the development of Millennium Development Goals, an impetus was provided for the first time towards neglected tropical diseases as part of 6th Millennium Development Goal 6 (MDG-6). The initial classification of neglected tropical diseases focused on diseases that cause infections of a chronic nature and that rendered feebleness, and because the majority of them are parasitic diseases they possessed one common metric of being prevalent in the impoverished population sets [7]. The first road map for neglected tropical diseases by WHO was aimed to speed up the integrated approach of 2005 for the then 17 neglected tropical diseases, by targeted systems which were not only clear but also quantifiable and in agreement of all the collaborating partners. The London Declaration of 2012 was designed to gain a strong commitment in the form of partner support aimed at the elimination of ten major diseases that it was most likely to be feasible to deal with. A very strategic approach by WHO caused the inception of alliances with committed global multipartner support systems wherein the pharmaceutical sector has also been brought in and it has led strategic accomplishments in a shorter timeframe reinforcing the validity and operability of such planning [18].

8.4.1 World Neglected Tropical Diseases Day

World Neglected Tropical Diseases Day has been launched as a worldwide movement to end neglected tropical diseases. In the 2019 Reaching the Last Mile Forum, Abu Dhabi Crown Prince Court proposed the idea of World Neglected

Tropical Diseases Day to spread awareness and consolidate commitment and efforts towards neglected parasitic and tropical diseases. The main goal of institutionalizing NTD is to mobilize political activity and securing binding commitments to eliminate these diseases in line with 2021–30 NTD roadmap of WHO [4]. The neglected tropical diseases were given official recognition in May 2021 after a proposal by UAE, Brazil and Oman with the support of other member states of the WHO. In 2022, Uniting to Combat NTDs launched the 100% Committed movement on 27 January, just ahead of World NTD Day celebrated on 30th Jan 2022. Worldwide commitments, both financial as well as political, are being secured by 100% Committed movement. The Kigali Declaration is a political declaration and it provides a platform to mobilize the political will, display the community commitment, consolidate and direct resources and action needed to end unnecessary suffering from neglected parasitic and tropical diseases [3].

8.4.2 Organizations Involved in Combating Neglected Parasitic and Tropical Diseases

Many organizations, governments, pharmaceuticals, NGOs, civil society organizations, foundations, academia, research organizations, individuals and other bodies all over the world have united towards prevention, control or elimination of neglected parasitic and tropical diseases. The sustained and multi-collaborative efforts of these bodies and individuals are commendable and some of such major bodies are:

(a) World Health Organization

World Health Organization (WHO) efforts to combat neglected tropical diseases culminated in the WHO's 2011 "Accelerating work to overcome the global impact of NTDs: A roadmap for implementation" and Word Health Assembly (WHA) resolution WHA66.12 [6], which was adopted by the 66th WHA in May 2013. The roadmap for implementation timeline was 2012–2020, overlapping with the WHO's 12th and 13th General Programme of Work (GPW). The WHO's Neglected Tropical Disease Programme aligns to the needs of Member States as identified in the roadmap for implementation, WHA resolutions and Sustainable Development Goals (SDGs), and their own country strategies. The London Declaration on NTDs and the Roadmap for implementation were instrumental in increasing the profile of NTDs globally, regionally and nationally [2].

(b) Centre for Disease Control

Centre for Disease Control (CDC) works with many nations and also through broad regional collaborations to ameliorate existing interventions towards combating neglected parasitic and tropical diseases, like mass drug administration

(MDA) campaigns. It includes various programs and stakeholders across continents and counties. Centre for Disease Control actively supports most of the neglected tropical disease programs globally which are focused on locating areas of endemicity across nations where such neglected tropical or parasitic diseases are prevalent. The main programs covering NTDs are: (i) Global control efforts such as Global Programme to Eliminate Lymphatic Filariasis (GPELF) started in 2000 by WHO, (ii) Global Efforts to Eliminate Onchocerciasis, which was launched as a regional initiative in 1992 named as Onchocerciasis Elimination Program for the Americas (OEPA) and has supported onchocerciasis elimination through a partnership with the endemic countries, (iii) Global Efforts to Control Schistosomiasis, where depending on disease prevalence, school going children, or those within such age depending on the requirement, or even the whole community, is annually dosed with the drug praziquantel, and (iv) global efforts to mitigate trachoma, through SAFE: surgery, antibiotics, facial cleanliness and environmental educational efforts and which has shown very effective results. An antibiotic called Zithromax has been in use to assist in trachoma control. The CDC is ardently working for Guinea Worm Eradication Program and Global Guinea Worm Eradication Program, which is aimed at eradicating dracunculosis and is run by The Carter Center in the USA [7].

(c) United States Agency for International Development

United States Agency for International Development (USAID) works on many issues all over the world and is a donor agency for many causes worldwide. USAID has supported and aided many developing countries in their efforts to mitigate diseases and also helped in successful elimination of five of the prevalent neglected parasitic and tropical diseases i.e., river blindness, bilharziasis, elephantiasis, intestinal worms and trachoma. USAID works across the world with governments in developing countries, partners with public as well as private sectors for the mapping of diseases, treatment of at-risk or vulnerable populations, determining the achieved progress, and building an in-country response. To contribute to the cause of USAID in combating neglected parasitic or tropical diseases, pharmaceutical partners have also pledged and donated billions of dollars [7].

(d) Drugs for Neglected Diseases initiative

Drugs for Neglected Diseases initiative (DND*i*) is a patient-needs driven organization and it runs as not-for-profit, working in areas of drug research and development. Since its inception, it has worked towards the development of safe and effective treatments for neglected diseases and making such drugs and treatment affordable to the needy population groups [5].

(e) Global Antibiotic Research & Development Partnership

Global Antibiotic Research & Development Partnership (GARDP) was established in May 2016 as a joint initiative of parent organization DND*i* and the World Health Organization. DND*i* was set up with the mission to address the burning issue of antimicrobial resistance along with developing antibiotics with improved therapeutical and pharmaceutical potential, while ensuring their sustainable access. GARDP became an independently functioning organization in 2019.

8.5 Current Status of Neglected Parasitic Diseases

8.5.1 Helminth Infections

(a) Dracunculiasis

Dracunculiasis is also known as thread worm or Guinea worm disease and is soon to be eliminated with only 54 occurrences reported in 2019, 27 human cases reported in 2020 and five cases reported up to May 2021. At present it is endemic to only a few countries in Africa having been eradicated or controlled in other parts of the world [23].

It is a neglected parasitic disease affecting humans who drink unfiltered water containing copepods (small crustaceans) that are infected with larvae of *D. medinensis*. Once inside the body, the larvae multiply. Over the due course of next 300–400 days in the body, female larvae have been reported to develop into a metre long worm. Gradually, these Guinea worms start emerging outside skin forming painful blisters on feet or legs. This expulsion of worms outside the skin is also accompanied by other symptoms such as pyrexia and nausea [24].

Once a mature worm emerges outside the body, it needs several weeks to be cautiously and slowly removed. Wounds inadvertently caused during this process often may develop into secondary bacterial infections [24].

(b) Echinococcosis

Echinococcosis (hydatid disease) is a zoonotic parasitic disease and it is caused due to infection by tiny *Echinococcus* tapeworms. >1million people are affected at any one point of time in the world [23]. One can become infected with hydatid disease through the ingestion of food and water contaminated with the eggs of the parasite which are passed through the feces of dogs [25].

Many different canine species can act as definitive hosts harboring the adult worms in the gastro-intestinal tract and excreting the eggs via feces. On ingestion of eggs, potent larvae are developed inside different organs of the individual, though the main affected organs are lungs and liver. Both cystic and alveolar echinococcosis have grave prognoses and can result in death [25].

(c) Foodborne Trematodiases

Foodborne trematodiases are caused by parasitic trematode worms. 70+ countries have reported cases with approximately 200,000 cases annually resulting in around 7000 deaths [23]. Humans can get infected from foodborne trematodiases due to consumption of raw or improperly cooked fish, shellfish and vegetables containing the cercarial or metacercarial stages of the trematode. Initial early and light infections may not be conspicuous and therefore go unnoticed as they often are symptomless [26].

However, at later stages when infections become severe, it can adversely affect the several organs such as *ductus biliaris*, cholecyst (gall bladder) and lungs. Symptoms may range from general dejection to severe gut and acute pectoralgia, pyrexia, inflammation, fibrosis, colic pain and icterus [26].

(d) Elephantiasis or Lymphatic filariasis

Lymphatic filariasis is prevalent in mostly human populations of South America, West Indies (some parts), Western Pacific, Africa and Asia. It is a chronic condition that occurs in about 36 million people. The worldwide at risk population in 2019 was about 859.5 million [23]. It is caused by filarial worms and transmitted by mosquitoes and affects the lymphatic circulatory system [27].

Lymphatic filariasis can cause severe lymphedema along with episodes of fever. Persons affected from lymphedema are also susceptible to secondary infections by bacteria causing reduced movements or mobility due to thickening and hardening of skin. In men, other symptoms also can occur such as swelling of the scrotum i.e., hydrocele [27].

Elephantiasis has been seen to be prevalent mostly in extremely poor communities, thus impeding individuals from living a fully functional and productive social or working life, further pushing them into the vicious cycle of poverty. Seventeen countries have been able to eliminate it since 2012 [27]:

In 2020: Malawi, in 2019: Kiribati and Yemen, in 2018: Wallis and Futuna, Vietnam, Palau and Egypt, in 2017: Thailand, Tonga, Togo and Marshall Islands, in 2016: Vanuatu, Niue, Cook Islands, Cambodia, Sri Lanka and Maldives.

(e) Onchocerciasis

Onchocerciasis is also known as river blindness and currently has prevalence in American and sub-Saharan Africa affecting about 21 million people (potential people at risk is 218 million with 99% of the infected population is in sub-Saharan Africa) [23]. Onchocerciasis is a parasitic infection which is transmitted by blackflies or *Simulium* sp. The larvae produced by worms progress to eyes, skin and other body organs leading to severe itching which may progress to disfiguring skin conditions and also vision loss that may also include irreversible blindness. Due to such a grave prognosis, it adversely impacts the lives of affected individuals and affects their social, mental and economic capabilities. Preventive chemotherapy for river blindness has taken a significant positive shift in the aftermath of London Declaration of 2012 [28].

Since 2012 four nations have been able to eliminate river blindness namely: Guatemala in 2016, Mexico in 2015, Ecuador in 2014 and Colombia in 2013.

(f) Schistosomiasis

Schistosomiasis or bilharziasis is endemic in 51 countries spread across the continents of South America, Africa and Asia, leading to treatment being required by about 237 million people worldwide [23]. It is caused in humans by close proximity with water contaminated with furcocercous cercariae of schistosomes released by intermediate hosts i.e., snail species. These parasites have the ability to penetrate through skin (by keratolysis) and affect the individual while migrating towards specific organs inside the body. Schistosoma mainly affects the urinary or intestinal system, and prognosis can be severe to grave causing chronic sickness and, as well as mortality. School children are an at risk-population for bilharzia due to unhygienic or poor hygienic conditions along with water-based activities that may include swimming and fishing. The infection with bilharzia may result in malnutrition, and absenteeism as well as impaired intellectual development in young children [29].

Children affected with chronic and severe bilharzia infections are also susceptible to other secondary pathological conditions. For women suffering from this disease other symptoms such as female genital schistosomiasis (FGS) can occur. Preventive chemotherapy for bilharzia has taken a significant positive shift in the aftermath of London Declaration of 2012 [29].

(g) Soil-transmitted helminthiases

Soil transmitted helminthiases are caused by a group of helminths that transmit infective stage gastrointestinal helminths from soil. They are mostly endemic to South-East Asia, China, America and sub-Saharan Africa where the at risk population is about 1.05 billion children (82% live in Africa and South-East Asia) [23]. Intestinal worms are prevalent in climatic conditions where soil is warm and humid but poor sanitary conditions exist. Humans can get infected with intestinal worms on exposure to soil contaminated with eggs of the parasite [30].

Intestinal worms can cause reduction of ability to absorb and/or utilize the nutrients and vitamins causing malnutrition and may lead to other debilitating conditions such as anaemia and stunted growth. This infection is associated with poverty and also linked to a broader community development challenge, since it can severely limit and hamper the ability of infected individuals to live fructuous and productive lifestyle [30].

The WHO roadmap of Neglected Tropical Diseases comprehensively aims towards combating the various neglected tropical diseases by 2020, including intestinal helminths. Intestinal worms target: To achieve control by 2020 wherein treatment is available for 75% of pre-school and school-aged children in need [30].

(h) Taeniaisis/Cysticercosis

Taeniasis is an infection of intestines and is caused by two species of tapeworms in humans. Over 75 countries are endemic for taeniasis and about 5.5 million people are infected worldwide [23]. Humans can become infected with taeniasis on consumption of infected beef or pork, which has not been adequately cooked [31].

Taeniasis usually causes symptoms which are mild and nonspecific. But fully developed tapeworms, which may take about eight weeks to develop after ingestion of larvae, can result in severe excruciating abdominal pain, nausea, diarrhea or constipation [31].

8.5.2 Protozoan Infections

(a) American trypanosomiasis or Chagas disease

American trypanosomiasis or Chagas disease is caused by the metacyclic trypomastigote stage of the protozoan *Trypanosoma cruzi*. It is endemic in 21 Latin American countries affecting around seven million people, with 75 million people at risk of infection [23]. One can get infected with Chagas disease on coming in contact with feces of blood-sucking stercorarian group of bugs known as kissing bugs infected with the protozoa [32].

Chagas disease can also be transmitted in other ways such as eating foot contaminated by bug excretions, transfusion of blood or transplantation of organs, as well as from mother to children during birth. Initially the symptoms are mild for a few weeks with symptoms such as pyrexia, body pain, rash of body parts, diarrhea or vomiting and it is very much possible to be unaware of being infected for long time. Only about 30–40% of the infected population progresses to symptomatic forms which can result in death [32].

(b) African trypanosomiasis (Human)

African trypanosomiasis or sleeping sickness is endemic in 18 countries in sub-Saharan Africa and there were 992 cases worldwide in 2019 [23]. It is caused by being bitten by infected tsetse flies (*Glossina*). Sleeping sickness disease can manifest either in its chronic form which is caused by *Trypanosoma brucei gambiense* or acute form of infection which is caused by *Trypanosoma brucei rhodesiense*.

In the first stage of the disease, which manifests due to multiplication of parasites in the body, there are symptoms of itching, joint pain, headaches and fever, and during the second stage of the disease termed as sleeping sickness, the protozoa enter the central nervous system, leading to behaviorial changes, lack of coordination, schizophrenia, and insomnia [33].

(c) Visceral Leishmaniasis

Visceral leishmaniasis (VL) is also known as kala-azar mostly affects people in Latin America Asia and Africa. In 2020 there was under 13,000 cases reported [23]. Visceral leishmaniasis is transmitted by being bitten by infected sandflies. On progression of visceral leishmaniasis, immunocompetence is compromised and bone marrow as well as internal organs are affected with physiological disintegration of the spleen and liver, and it can cause irregular bouts of pyrexia, emaciation and anaemia. Without treatment, visceral leishmaniasis can be 100% fatal rate within 24 months [34].

Cutaneous leishmaniasis (CL) mostly affects people in the Mediterranean region, America, Middle East and Central Asia. It is the most commonly occurring form of leishmaniasis and it causes skin lesions and ulcers on the parts of the body which normally remain exposed and hence cause social stigma. In the year of 2020, the new cases of this disease were mostly limited to only 10 nations with more than 85% of total incidence in world occurring in Tunisia, Syrian Arab Republic, Peru, Pakistan, Libya, Iraq, Colombia, Brazil, Algeria and Afghanistan [34]. It is estimated that there are 600,000 to one million new cases annually [23].

8.5.3 Ectoparasitic Infestations

8.5.3.1 Scabies and Associated Ectoparasitic Infections

Scabies is an ectoparasitic infection caused by microscopic mite *Sarcoptes scabiei*. The female mite creates a winding burrow parallel to the surface of skin and once eggs laying occurs, it leads to an immune reaction causing intense itching as well as a skin rash. It can be further complicated by secondary infections by bacteria leading to the development of skin wounds, which may further be causing other medical complications. Scabies is usually transmitted through close skin contact from person-to-person. Scabies patients typically suffer from severe forms of itching, burrows (linear) and vesicles near finger webs, wrist area, limbs and belt area [35]. It is estimated that there are around 200 million cases at any one time [23].

8.6 Conclusions

In our efforts to eradicate NTDs, we are making slow but steady progress. Forty-three countries have eliminated at least one neglected parasitic or tropical disease so far, and at present about 600 million people are out of significant vulnerability. This reinforces the fact that eradication of neglected parasitic or tropical diseases is possible [16]. Mass Drug Administration (MDA) or integrated MDA has led to a huge drop in occurrences of number of NTDs, such as lymphatic filariasis and onchocerciasis, and trachoma. The evaluation of WHO's Neglected Tropical Diseases Programme [2, 6], from 2019, revealed the evidence of great success for the NTD medicine donation programme, with about 1.762 billion treatments provided to about 1 billion people across five preventive chemotherapy (PC) diseases (i.e., lymphatic filariasis, onchocerciasis, soil-transmitted helminthiasis, schistosomiasis, and trachoma). In many cases, WHO guidelines are in place to support the prevention, control and treatment of these parasitic and tropical diseases, but they are focused on a few selected diseases. Again, the funding relationships are extremely important given the competitive environment for funding to meet the SDGs (Sustainable Development Goals). The WHO NTD programme is reliant on donor funding for 85% of its budget, which can challenge the programme's autonomy.

The epidemiology of neglected parasitic and tropical diseases is more often not linked to environmental factors and the NTDs are prevalent mainly in rural resource-poor environments. So combating these will require multi-pronged and cross-sectoral approaches, as well as proper understanding of the associated social issues [7]. There is also a need to have a strong and integrated disease management, water, sanitation and hygiene (WASH), vector control, as well as the health systems in place for effective control of the NTDs. Control and eradication of some of the NTDs might need new drugs or strategies and research is progressing in such areas. To achieve a rapid rate of control and worldwide prevention and elimination, it is vital to have increased social awareness, collaborative research and political will from leaders around the globe [20].

References

1 Carrion, M.N., Hamer, D.H., and Evans, D. (2017). Parasitic diseases, an overview. In: *International Encyclopedia of Public Health*, 2e. 399–408. Elsevier Inc. doi: 10.1016/B978-0-12-803678-5.00241-1.

2 Dyke, E., McKay, B., Ly, M., and Peeling, R. (2019 July). Evaluation of the WHO neglected tropical diseases programme volume 1: report [Internet]. *Canada: WHO Evaluation Office*. Available from: https://www.who.int/docs/default-source/documents/evaluation/evaluation-ntd-report.pdf (Accessed 26 January 2022).

3 WHO *World neglected tropical diseases day: WHO calls for equitable health services for all [Internet].* Available from: https://www.who.int/news/item/30-01-2022-world-neglected-tropical-diseases-day-who-calls-for-equitable-health-services-for-all (Accessed 30 January 2022).

4 World NTD Day (2022). [Internet]. Available from: https://worldntdday.org (Accessed 30 January 2022).

5 DNDi in India (2019). Strengthening partnerships to respond to the needs of neglected patients [Internet]. *New Delhi: 2019.* Available from: https://dndi.org/wp-content/uploads/2019/01/DNDi_in_India_2019.pdf (Accessed 25 January 2020).

6 WHA66/2013/REC/1 (2013). Resolution WHA66.12 neglected tropical diseases. in: sixty-sixth world health assembly, Geneva, 20-28 May 2013 [Internet]. *Resolutions and decisions. Geneva: World Health Organization.* Available from: http://apps.who.int/gb/ebwha/pdf_files/WHA66-REC1/A66_REC1-en.pdf#page=25 (Accessed 26 January 2022).

7 Centre for Disease Control (2022). Neglected tropical diseases [Internet]. *Global Health, Division of Parasitic Diseases and Malaria.* Available from: https://www.cdc.gov/globalhealth/ntd/global_program.html (Accessed 25 January 2022).

8 Pedrique, P., Strub-Wourgaft, N., Some, C. et al. (2013). The drug and vaccine landscape for neglected diseases (2000-11): a systematic assessment. *The Lancet Global Health* 1: e371–79. doi: 10.1016/S2214-109X(13)70078-0.

9 biomedcentral.com [Internet] *Historical aspects of parasitology* Available from: https://www.biomedcentral.com/collections/HistParasitol#:~:text=Parasites%20first%20evolved%20long%20before,an%20early%20version%20of%20parasitology (Accessed 20 January 2022).

10 Cox, F.E. (2004). History of human parasitic diseases. *Infectious Disease Clinics of North America* 18 (2): 171–188.

11 Cox, F.E. (2009). George Henry Falkiner Nuttall and the origins of parasitology and Parasitology. *Parasitology* 136 (12): 1389–1394.

12 Worboys, M. (1983). Emergence and early development of parasitology. In: *Parasitology* (eds K.S. Warren and J.Z. Bowers), 1–18. New York, NY: Springer.

13 Garcia, L.S. (1999). Classification of human parasites, vectors, and similar organisms. *Clinical Infectious Diseases* 29 (4): 734–736.

14 Mathison, B.A. and Pritt, B.S. (2019). Medical parasitology taxonomy update, 2016–2017. *Journal of Clinical Microbiology* 57 (2): e01067–18.

15 Hotez, P.J., Aksoy, S., Brindley, P.J., and Kamhawi, S. (2020). What constitutes a neglected tropical disease? *PLOS Neglected Tropical Diseases* 14 (1): e0008001. doi: 10.1371/journal.pntd.0008001.

16 World Health Organization *NTD progress dashboards 2011–2020 [Internet]* Available from: https://www.who.int/teams/control-of-neglected-tropical-diseases/overview/progress-dashboard-2011-2020 (Accessed 25 January 2022).

17 Elphick-Pooley, T. and Engels, D. (2022). World NTD day 2022 and a new Kigali declaration to galvanise commitment to end neglected tropical diseases. *Infectious Diseases of Poverty* 11: 2. doi: 10.1186/s40249-021-00932-2.

18 Engels, D. and Zhou, X.N. (2020). Neglected tropical diseases: an effective global response to local poverty-related disease priorities. *Infectious Diseases of Poverty* 9 (1): 10. doi: 10.1186/s40249-020-0630-9.

19 World Health Organization *Ending the neglect to attain the sustain—able development goals: a road map for neglected tropical diseases 2021–2030 [Internet]* Available from: https://www.who.int/teams/control-of-neglected-tropical-diseases/ending-ntds-together-towards-2030 (Accessed 25 January 2022).

20 Hotez, P.J. and Lo, N.C. (2020). Neglected tropical diseases: public health control programs and mass drug administration. In: *Hunter's Tropical Medicine and Emerging Infectious Diseases* 10th edition (eds E.T. Ryan, D.R. Hill, T. Solomon et al.), 209–213. Elsevier. doi: 10.1016/b978-0-323-55512-8.00027-2.

21 Hotez, P.J., Molyneux, D.H., Fenwick, A. et al. (2006). Incorporating a rapid-impact package for neglected tropical diseases with programs for HIV/AIDS, tuberculosis, and malaria. *PLOS Medicine* 3: e102.

22 Molyneux, D.H., Hotez, P.J., and Fenwick, A. (2005). "Rapid-impact interventions": how a policy of integrated control for Africa's neglected tropical diseases could benefit the poor. *PLOS Medicine* 2 (11): e336. doi: 10.1371/journal.pmed.0020336. PMID: 16212468.

23 World Health Organization (2022). THE GLOBAL HEALTH OBSERVATORY-explore a world of health data indicators index [Internet]. Available from: https://www.who.int/data/gho/data/indicators/indicators-index (Accessed 15 February 2022).

24 Uniting To Combat Neglected Tropical Diseases (2022). Guinea worm disease [Internet]. *Uniting To Combat Neglected Tropical Diseases.* Available from: https://unitingtocombatntds.org/ntds/guinea-worm-disease (Accessed 28 January 2022).

25 Uniting To Combat Neglected Tropical Diseases (2022). Echinococcosis [Internet]. *Uniting To Combat Neglected Tropical Diseases.* Available from: https://unitingtocombatntds.org/ntds/echinococcosis (Accessed 28 January 2022).

26 Uniting To Combat Neglected Tropical Diseases (2022). Foodborne trematodiases [Internet]. *Uniting To Combat Neglected Tropical Diseases.* Available from: https://unitingtocombatntds.org/ntds/echinococcosis (Accessed 28 January 2022).

27 Uniting To Combat Neglected Tropical Diseases (2022). Elephantiasis (lymphatic filariasis) [Internet]. *Uniting To Combat Neglected Tropical Diseases.* Available from: https://unitingtocombatntds.org/ntds/lymphatic-filariasis (Accessed 28 January 2022).

28 Uniting To Combat Neglected Tropical Diseases (2022). River blindness (onchocerciasis) [Internet]. *Uniting To Combat Neglected Tropical Diseases.*

Available from: https://unitingtocombatntds.org/ntds/onchocerciasis (Accessed 28 January 2022).

29 Uniting To Combat Neglected Tropical Diseases (2022). Bilharzia (schistosomiasis) [Internet]. *Uniting To Combat Neglected Tropical Diseases.* Available from: https://unitingtocombatntds.org/ntds/schistosomiasis (Accessed 28 January 2022).

30 Uniting To Combat Neglected Tropical Diseases (2022). Intestinal worms (soil-transmitted helminths) [Internet]. *Uniting To Combat Neglected Tropical Diseases.* Available from: https://unitingtocombatntds.org/ntds/soil-transmitted-helminths(Accessed 28 January 2022).

31 Uniting To Combat Neglected Tropical Diseases (2022). Taeniasis and cysticercosis [Internet]. *Uniting To Combat Neglected Tropical Diseases.* Available from: https://unitingtocombatntds.org/ntds/taeniasis-and-cysticercosis (Accessed 28 January 2022).

32 Uniting To Combat Neglected Tropical Diseases (2022). Chagas disease [Internet]. *Uniting To Combat Neglected Tropical Diseases.* Available from: https://unitingtocombatntds.org/ntds/chagas-disease (Accessed 28 January 2022).

33 Uniting To Combat Neglected Tropical Diseases (2022). Sleeping sickness [Internet]. *Uniting To Combat Neglected Tropical Diseases.* Available from: https://unitingtocombatntds.org/ntds/human-african-trypanosomiasis (Accessed 28 January 2022).

34 Uniting To Combat Neglected Tropical Diseases (2022). Leishmaniasis cutaneous and visceral [Internet]. *Uniting To Combat Neglected Tropical Diseases.* Available from: https://unitingtocombatntds.org/ntds/leishmaniasis (Accessed 28 January 2022).

35 Uniting To Combat Neglected Tropical Diseases (2022). Scabies [Internet]. *Uniting To Combat Neglected Tropical Diseases.* Available from: https://unitingtocombatntds.org/ntds/scabies (Accessed 28 January 2022).

9

Molecular Techniques for the Study and Diagnosis of Parasite Infection

Syed Muhammad Mukarram Shah[1], Saira[2,], and Fida Hussain[3]*

[1] Professor Department of Pharmacy, University of Swabi, Swabi, Pakistan
[2] Assistant Professor, Department of Zoology, University of Swabi, Swabi, Pakistan
[3] Lecturer, Department of Pharmacy, University of Swabi, Swabi, Pakistan
* Corresponding author

9.1 Introduction

9.1.1 Parasite

An organism (literally derived from *para* means besides and *sites* means food) that is not able to survive independently and gets benefit from other living organisms and harms the host [1]. Parasitic organisms are responsible for causing the most dangerous and pandemic diseases worldwide. Parasites consist of protozoa such as sporozoans, amoebae, ciliates, flagellates, and coccidians [2]. Parasitic infections are common in animals that evolved with the human species for millions of years [3]. Parasites associated with humans are divided into several groups (Table 9.1).

The groups consist of the protozoans group (amoebae, flagellates, ciliates, sporozoans, and coccidia), the platyhelminths or flatworms group (cestodes, trematodes), thorny-headed worms or the acanthocephalan group, roundworms or the nematode and the arthropod group (spiders, ticks, mites, insects) [4].

9.1.2 Protozoa Parasites

Protozoan parasites consist of a cell-like unit structure that can perform all the functions of the living organism. Parasites consist of protoplasm which is divided into nucleoplasm and cytoplasm. The cytoplasm is divided into two layers the inner granular endoplasm and the outer hyaline ectoplasm. The ectoplasm helps

Parasitic Infections: Immune Responses and Therapeutics, First Edition.
Edited by Abhay Prakash Mishra and Manisha Nigam.

Table 9.1 Brief outline classification of the common human protozoan parasite, helminths parasites, and ectoparasite.

Protozoan Parasites	Helminth Parasites	Arthropod Parasites
Sarcodina 1) Genus: *Entamoeba* (*E. histolytica*) 2) Genus: *Endolimax* (*E. nana*) 3) Genus: *Iodameba* (*I. butchlii*) 4) Genus: *Dientmeba* (*D. fragilis*)	**Class: Trematodea** 1) Genus: *Schistosoma* (*S. mansoni*) 2) Genus: *Fasciola* (*F. hepatica*)	**Class: Arachnida** *A. sanguineu* *D. gallinae* *O. bacoti* *P. ventricossus* *S. scabiei* *T. alfreddugesi*
Class: Mastigophora (Flagellates) 1) Genus: *Giardia* (*G. lamblia*) 2) Genus: *Trichomonas* (*T. vaginalis*) 3) Genus: *Trypanosoma* (*T. brucci*) 4) Genus: *Leishmania* (*L. donovani*)	**Class: Cestoda** 1) Genus: *Diphylobotrium* (*D. latum*) 2) Genus;: *Taenia* (*T. saginata*) 3) Genus: *Echinococcus* (*E. granulosus*)	**Class: Insecta** *P. capitis* *P. humanus* *A. senegalensis* *C. americana* *C. americana* *C. hominivorax* *C. anthropophaga* *D. hominis* *C. lectularius* *C. hemipterous* *P. irritans* *T. penetrans* *X. cheopis*
Class: Sporozoa 1) Genus: *Plasmodium* (*P. falciparum*) 2) Genus: *Toxoplasma* (*T. gondi*) 3) Genus: *Cryptosporidum* (*C. parvum*) 4) Genus: *Isospora* (*I. beli*) **Class: Ciliates** *Balantidium coli*	**Class: Nemathelminthes** 1) Intestinal nematodes (*A. lumbricoides*) 2) Somatic nematodes (*W. bancrofti*)	

Source: Adapted from Assafa et al., 2006 [4].

in movement, respiration, ingestion of food, and excretion. Vacuoles present in the cytoplasm are involved in the storage of food material and its digestion as well as excretion of the resulting waste products. The nucleus is the involved in reproduction and the maintenance of life. The protozoan parasite can shift from an active to inactive phase, in which the motility is diminished and the parasite is enclosed in a thick wall. In this inactive phase the cellular structure of the parasite is called a cyst. The cyst is the resistant stage of the parasite and is infectious to humans (Figures 9.1 and 9.2 and Table 9.2) [5].

9.1.3 Helminth Parasites

The helminth parasites are multicellular organisms that are symmetrical bilaterally and composed of three germ layers. Clinically important helminths are divided into three main groups. In general, helminths are large having size of (> 1 mm long) whereas some of them are very large in size (> 1 m long) [5]. All have well-developed organs. The parasitic body is either enclosed in a cell membrane, such as in flatworms, or covered with cuticle such as roundworms. The parasites may be hermaphrodites or individual sexes. The most severe helminthic infections occur in poor subtropical and tropical areas but they may also be acquired in developed countries; less serious infections are distributed worldwide. Susceptibility to infections is influenced by diet, hygienic condition, contact with vectors, and climatic condition.

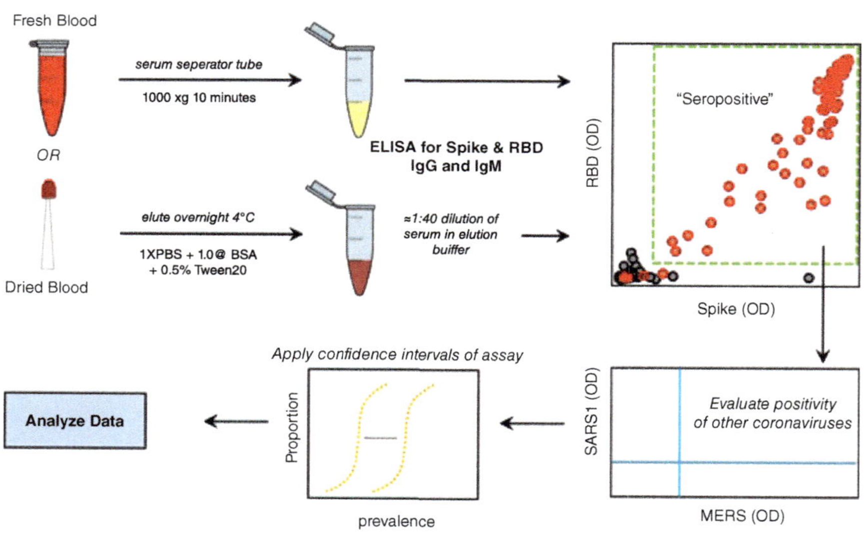

Figure 9.1 Serology testing protocol. (*Source:* adapted from Klumpp-Thomas et al., 2021 [57]).

Table 9.2 Some important parasites and their associated infections.

Parasite	Infection
Protozoans	Trypanosomiasis
T. cruzi	Chagas disease
T. b. gambiense	West African trypanosomiasis
T. b. rhodesiense	East African trypanosomiasis
Leishmania spp.	Leishmaniasis
Giardia lamblia	Giardiasis
Coccidium	Coccidiosis
Plasmodium *falciparum*	Malaria (Black-water fever)
Balantidium coli	Balantidiasis
Babesia spp.	Babesiosis
Helminths	
Wuchereria bancrofti	Filariasis
Brugia malayi	Filariasis
Dirofilaria spp.	Filariasis
Mansonella perstans	Filariasis
Mansonella streptocerca	Filariasis
Mansonella ozzardi	Filariasis
Trichura spp.	Trichuriasis
Trichinella	Trichinosis
Strongyloides stercoralis	Strongyloidiasis
Schistosoma spp.	Schistosomiasis
Onchocerca volvulus	Onchocerciasis
Loa loa	Loiasis
Dipylidium caninum	Dog tapeworm infection
Hymenolepis diminuta	Rat tapeworm infection
Hymenolepis nana	Dwarf tapeworm
Fasciola hepatica	Fascioliasis
Enterobius vermicularis	Enterobiasis
Ectoparasites/Arthropods	
Cimex lectularius	Cellulitis
Sarcoptes scabiei	Scabies

Source: Adapted from Assafa et al., 2006 [4].

Strong immunity of the host can eradicate latent infections; some infections can extend for longer periods of time, even for years. Whereas most infections are asymptomatic severe symptoms can occur. As helminths are mostly of a relatively large size they can cause direct injury to the host tissues as they move through the host body. Indirect damage to the host tissues occurs to the immune mechanisms of the host (Figures 9.1 and 9.2 and Table 9.2) [6].

9.1.4 Arthropod Parasites

Arthropods are the largest group of animals in the Kingdom Animalia. They are characterized by segmented bodies that are bilaterally symmetrical with grouped appendages. Arthropods have hard exoskeletons that protect the inner soft organs. The circulation system is open with or without the heart located dorsally which pumps hemolymph through arteries to different organs of the body. Hemolymph returns to the pump through spaces called hemocoel. A nervous system, an respiratory system, and an excretory system are present in arthropods. Arthropods can act either as direct agents of the disease or act as carriers for the transmission of the disease.

Arthropods that are pathogenic in nature belong to the classes, Arachnida, Insecta, and Crustacea. The body of the insect class consists of the head, thorax and, abdomen having two antennae. Malaria, trypanosomiasis and onchocerciasis are primarily caused by insects (Figure 9.1 and Table 9.2) [5].

Arthropods generally reside on the host and are referred to as ectoparasites. Ectoparasites can attack a large number of living beings including humans, plants, and other insects.

Most of the time infestations of parasites cause very little trouble to the host organism and do not need any treatment, but occasionally these ectoparasites can cause severe disease either directly or indirectly as vectors, or leading to secondary infection [7]. Infestations of ectoparasites in human beings are viewed as diseases of the developing world. This is an incorrect notion as contagious ectoparasites are very common in developed countries, examples being bed bugs, lice, head louse, mites, and scabies. These infestations are not fatal but symptoms are attached to social stigma (Figure 9.2) [8].

Numerous infectious parasites distributed widely cause diseases in humans. Acute severe diseases are caused by only a small number of organisms. Chronic infections cause severe morbidity in only a few cases. Several chronic infestations are very common throughout the world but serious morbidity occur in only few of them.

These characteristics of parasitic diseases are some of the most interesting, challenging and important infectious diseases faced by clinicians, scientists, and public health officials in the twenty-first century [9].

9.2 Diagnostic Techniques

In addition to many laboratory tests parasitic infections are identified and diagnosed based on geographical area, travel history, clinical history and clinical symptoms of the patient.

The diagnostic tests used recently have undergone very few changes since the invention of the microscope. Most of the tests used today are unable to differentiate between recurrent, past, acute, and latent infections and are not able to monitor the reaction of treatment and follow the prognosis of the disease [5].

Preference has been given to rapid diagnosis to prevent fatalities and to initiate therapy as soon as possible. Efficient diagnostic techniques can limit transmission and ensure active surveillance. Sadly, only a few advances have been made in techniques used in the diagnosis of parasitic infections. No significant efforts have been made in this regard most of the diagnostic techniques used today, such as microscopy, are time-consuming and labor-intensive. The new diagnostic assays and techniques should give quick results. Such issues could be resolved by reducing the number of steps to be carried out in a process and making results clearer so avoiding significant operator-dependent biases. An optimal analysis method would bear positive characteristics while still sustaining the highest level of sensitivity and specificity. Diagnostic examinations currently used are unable to distinguish between existing and former infections and diagnostic tests that can differentiate between past and current infections are required for the assessment of the disease prevalence, and optimum treatment, and to evaluate the outcome of the treatment.

Presently, reference and diagnostic laboratories use many methods comprising microscopy, molecular assays, and serological analysis. Every diagnostic protocol has its own pros and cons. Most of the research centers are trying to upgrade the old diagnostic protocols and develop new diagnostic techniques. A special focus has been given to the development of molecular diagnostic techniques [5, 10]. RT-PCR methodologies are constantly optimized for the identification of several parasites. LAMP has achieved much attention and is viewed as a future molecular tool [11, 12].

The current need is to develop multiplex real-time PCR protocols. These techniques can help in the analysis of mixed infections [13]. For many years serology-based diagnostic methods have been complementary to microscopy; nevertheless, for most infectious diseases they cannot replace microscopic diagnosis entirely. Serology tests are based on the detection of antigens or antibodies from the patient's blood samples. Most popular serology-based assays are used as faster assessment tests [14–17]. They are the center of many studies because of their potential to be used in field settings. Rapid tests have been introduced which led to the advancement in the diagnosis of mixed infections [5,18–21]. Just like the multiplex PCR processes, these serology-based examinations are of vital importance as these can diagnose mixed infections, which is a growing problem in many areas. New diagnostic tools are continuously developed but many of the protocols need standardization, optimization, and, validation [22].

This chapter highlights different diagnostic techniques that are in use currently, and those tools which are under examination, for the diagnosis of parasite infections. For expediency, the currently used diagnostic tests and those under examination along with those used in the past are listed in Table 9.3.

9.2.1 Microscopy

Microscopy is the most economical technique for the examination of biological samples and serves as a reference standard in parasitology. Generally fluorescent or light microscopes are used by experienced operators for the observation of vaginal swabs, wet samples of phlegm, tissue biopsies, abscesses, urine, duodenal aspirates, and sigmoidoscopic material [24, 25].

Samples in iodine or saline are useful for spotting *G. duodenalis* trophozoites [25, 26], *E. histolytica* [22] *N. fowleri* [27], *T. vaginalis* [28], *B. coli* [29], and *Acanthamoeba* [30, 31], eggs, larvae like *Strongyloides* spp. and adults of protozoan and helminths (oo)cysts like those of *E. histolytica*, and *G. duodenalis* [25, 32], whereas oocysts of *I. belli*, *C. cayetanensis*, or *Cryptosporidium* spp., can be envisaged in severe infections [33]. The saline mounts of surface skin nicks are valuable to diagnose *O. volvulus* species whereas wet mount analysis of the blood is used to diagnose microfilariae of many nematodes and trypanosomes [24].

Detection and imagining of parasitic life stages can be enhanced by staining biological samples with fluorescent or chemical dyes [32]. For example, acid-fast stains are utilized for the detection of *Isospora*, *Cryptosporidium*, and *Cyclospora* spp., in stool samples [26].

In a similar way, a fast protocol is provided by Giemsa and Wright stains for the diagnosis of *Plasmodium*, *Leishmania*, Microfilariae, and *Trypanosoma* in thin and thick blood films [34], *T. gondii* in imprints of the lung, brain, and other tissues, trypanosomes in lymph node imprints, *E. histolytica* and *G. duodenalis* in imprints of gastrointestinal biopsy samples [35] and *Leishmania* spp., in aspirates of mucocutaneous or cutaneous lesions in and bone marrow aspirates. Fluorescent stains are also utilized in parasitic diagnosis.

An acid-fast stain called auramine-rhodamine attaches to mycolic acid present in protozoa for example *Cryptosporidium*. Auramine-rhodamine stain is comparatively more economical when used as rapid screening test tool although it is not as specific as Ziehl-Neelsen stain.

Table 9.3 Diagnostic techniques for parasitic diseases.

Tool/ Techniques	Intestinal parasitic diseases	Blood-borne parasitic diseases
Microscopy **Serology-based Assay** ELISA FAST ELISA Dot ELISA (Dipstick) RIPA ELISA DHA or IHA DFA or IFA PRISM RDT **Molecular-based Assay** PCR RT-PCR QT-NASBA RT-QB-NASBA LAMP PCR-ELISA Luminex OC-PCR **Proteomics** Mass Spectrometry (LDMS, MALDI-TOF, SELDI-TOF)	**Protozoa** **Cryptosporidiosis** *(C. hominis, C. parvum,)* **Trematodes** **Fasciolosis** *(F. hepatica, F. gigantica)* **Schistosomiasis** *S. mansoni* **Cestodes** **Taeniasis/ Cysticercosis** *T. solium* **Hydatidosis** *E. multilocularis* *E. granulosus* **Nematodes** **Filariasis** *W. bancrofti,* *B. malayi* *B. timori* *L. loa* **Strongylodiasis** *S. stercoralis*	**African trypanosomiasis** *T. brucei* **Babesiosis** *B. microti* **Chagas disease** *T. cruzi* **Leishmaniasis** *Leishmania* spp. **Malaria** *Plasmodium* spp. **Toxoplasmosis** *T. gondii*

ELISA: enzyme-linked immunosorbent assay; FAST-ELISA: Falcon assay screening test; RIPA-ELISA: radioimmunoprecipitation assay; IFA: immunofluorescent assay, IHA or DHA: indirect or direct hemagglutination assay; IFA or DFA: indirect or direct immunofluorescence assay; LDMS: laser desorption mass spectrometry; LAMP: loop-mediated isothermal amplification; MALDI-ToF: matrix-assisted laser desorption/ionization-time of flight; OC-PCR: oligochromatography polymerase chain reaction; PCR: polymerase chain reaction; QT-NASBA: quantitative nucleic acid sequenced-based amplification; RT-QT-NASBA: real-time quantitative nucleic acid sequenced-based amplification; RDT: rapid diagnostic test; RT-PCR: Real-time PCR; SELDI-TOF: surface-enhanced laser desorption/ionization-time of flight.
Source: Adapted from Ndao et al., 2009 [23].

For identification of *Acanthamoeba* and *Leishmania*, acridine orange is used [34]. Calco flour white has also been used for the detection of *Acanthamoeba and Naegleria*, which binds selectively to chitin and cellulose present in the fungal cell wall [30, 32, 36].

Calco flour white, which is carefully attached to chitin and cellulose in the fungi cell wall, has also been cast off to diagnose *Naegleria and Acanthamoeba*. These diagnostic techniques are relatively economical and they are of increasing importance in less developed countries where more expensive techniques such as molecular and immunological kits cannot commonly be used.

9.2.1.1 Disadvantages of Conventional Methods

The best way to control the disease is the identification of the exact parasite. The main drawback of the conventional microscopic method is the identification of different stages of life of the parasites. Microscopic methods of diagnosis require extensive knowledge of parasites [24, 30, 36–44] and the operator must have enough knowledge to assess the differences in morphologies and the life stages [45]. These techniques are time-consuming, labor-intensive, and have low sensitivity and reproducibility. These diagnostic methods frequently need multiple samples taken independently in order to reduce the variations caused due to changes in life-cycle stages [46, 47]. Another drawback of the conventional methods is that these methods are not capable of differentiating parasitic species of the same genera for example *Cryptosporidium* [48], *Cyclospora* [49], and *Entamoeba* [19], For example, they incapable of differentiating between *Entamoeba histolytica* and *Entamoeba dispar* [19]. In addition, the collection of samples for tissue parasites is extensively invasive leading to the inefficacy of the methods [50, 51].

9.2.2 Serology-Based Assays

When tissue specimens or biological samples are not available serology is the most important method for parasite infection diagnosis. Serology diagnostic techniques are categorized into two main groups: antibody-detection assays and antigen-detection assays. (the key variance between antigen and antibody is that one identifies the virus in the body at its most contagious stage, while the antibody test detects either the body has developed a defense against the virus.) [52, 53]. Serology-based tools consist of EIA also known as ELISA, and ELISA-derived tests such as the dot-ELISA and the FAST-ELISA. More types of assays consist of IFA or DFA tests, the HA test, immunoblotting, and RDTs.

Serology-based tests take a similar time for completion of the test and have a similar comfort of usage as microscopy although they are essentially very specific and more sensitive than microscopy. Serology tests are especially valuable in those patients in whom differentiation among *Plasmodium* and *Babesia* [54] is required or in low parasitemia patients or asymptomatic patients such as Chagasic patients [55]. If an infected person is found the transmission of the parasite may occur through blood transfusions during the transplantation of organs. Serology tests are significant for the detection of chronic fascioliasis where the production of the egg is low or occasional [56]. Lastly, these tests are used to monitor the eradication of parasites from patients after therapy (Table 9.1).

9.2.2.1 Falcon Assay Screening Test ELISA (FAST-ELISA)

FAST-ELISA utilizes recombinant and artificial peptides for evaluation of the antibody response to an antigen [58]. This method has been used for studying malaria [59], fasciolosis [56], and taeniasis [60]. Nevertheless, this tool has similar limitations to serology-based tests. Antibodies produced beside a specific peptide from a particular parasitic protein may have cross-reaction with another specie's proteins. Further, antibodies produced counter to a specific peptide may retort in several assays. Recently, no investigation has been available on the practice of this technique to diagnose parasitic infections.

9.2.2.2 Rapid Antigen Detection System (RDTS)

Immunochromatographic antigen detection forms the basis of RDTs. RDTs have been used as an addition to microscopy in most laboratories for the detection of malaria. In RDTs soluble proteins are captured by complexation with antibodies which are rooted on a strip of nitrocellulose. This assay involves the application of collected blood samples on the nitrocellulose strip and eluted with a small number of drops of the buffer which contains the labeled antibody. From the membrane, the antigen-antibody complex is then observed directly [13]. In the 1990s, after the advent of the first RDTs, key developments have been made in the tool, to make the use of RDTs feasible in countryside endemic areas. Advancement in RDTs technique makes it speedy, constant at 40 °C, easy to execute, and less costly thus giving numerous benefits compared to conventional microscopic techniques [61]. RDT techniques are beneficial to detect *P. vivax and P. falciparum* infections, but are not suitable to diagnose *P. ovale* and *P. malariae* infections [13]. Moreover, it is unusable for diagnosis of lowest-density infections where PCR-based methods are the optimal technique. In excess of 80 RDTs are available for the diagnosis of HRP specific to *P. falciparum* species-specific isotypes of LDH [62] but, for international marketing, only 23 have come across the WHO's standards (Table 9.2) [61].

Recently, in African countries, malaria RDTs have been made available to help in the prevention of incorrect diagnoses of malaria infections and to decrease the exercise of possible cure [49]. As a matter of fact, the propensity to deal with slide-negative specimens through antimalarials is still a common phenomenon. Such kind of

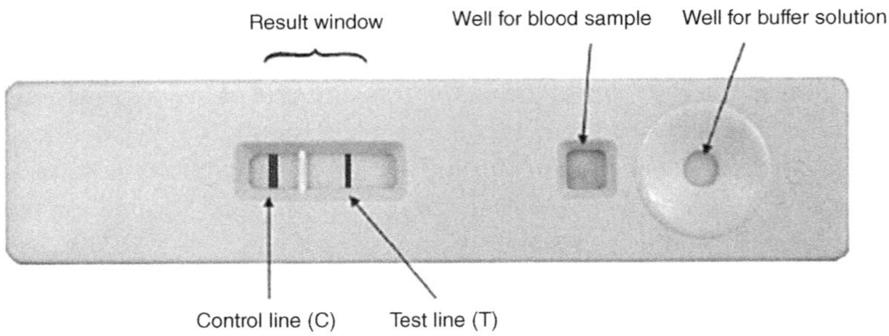

Figure 9.2 Typical RDT for malaria diagnosis. (*Source:* World Health Organization (WHO), 2011 [62]).

exercise causes unease in the health care of patients and also the expense it indicates by recommending the very costly antimalarial sulfadoxine/pyrimethamine and artemisinin-based combinations [61]. In conclusion, the uncontrolled use of antimalarials can potentially cause the advent of drug-resistant strains.

9.2.2.3 Luciferase Immunoprecipitation System (LIPS)

LIPS is an altered ELISA-based assay having serum-containing antigen-specific antibodies that could be recognized through determining the production of light. In mammalian cells selected antigen is bonded to the enzyme reporter Renilla luciferase (Ruc) and expressed as a Ruc-fusion to permit mammal precise post-translational alterations. Then extracted raw protein is incubated with the protein A/G beads and test serum. In the course of the incubation, on the A/G beads, the Ruc-antigen union is immobilized, which allows the antigen-specific antibody to be measured through washing the beads and then adding up coelenterazine substrate and measuring the production of light (Figure 9.3) [63]. Recently, the effective use of LIPS was reported for the recognition of sera specimens infected by *S. stercoralis* by means of Ruc-NIE fusion [64] and *L. loa* by means of Ruc-LlSXP-1 fusion [11]. Several benefits of the LIPS tools are their speed and accuracy in identifying infected patients. In the case of mammalian cells, sensitivity is enhanced that yields merging antigens free of infecting bacterial proteins. Furthermore, matched to the ELISA low backgrounds are produced. This significantly enables differentiation between positive and negative samples. Moreover, the *Strongyloides* LIPS centered on the NIE antigen revealed better precision than the ELISA as no feedback was perceived through serum samples of filarial-infected patients [11]. A LIPS assay could executed in two hours and thirty minutes. Burbelo et al., [66] reported 100% sensitivity and specificity by using LIPS assay on the basis of *Loa loa* SXP-1 antigen with insignificant cross-reactivity with a small number of patient sera infected from *W. bancrofti* and *O. volvulus*. By reducing the times of incubation

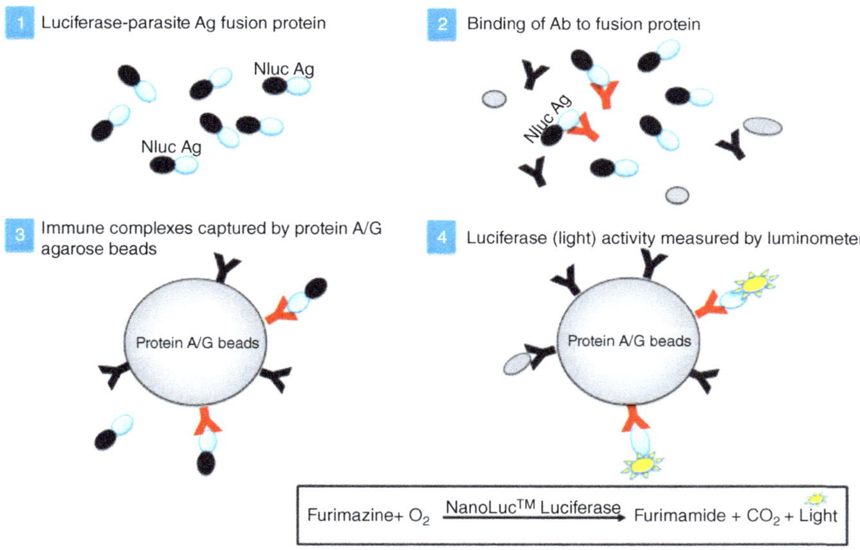

Figure 9.3 Assessment of LIPS for serodiagnosis of toxoplasmosis. (*Source:* Aye et. al., 2018. [69] With permission of Elsevier).

of a usual LIPS assay, it was possible to minimalize feedback. With the LIPS assay, several of the sera samples of *O. volvulus* tested positive, these were negative by means of LIPS assay (15-minutes) also known as QLIPS. Further, in-field investigations would be valuable in evaluating and authenticating the accuracy and possible efficacy of the QLIPS and LIPS assays. As discussed, immunodiagnostic tests have some drawbacks. For diagnosis of parasitic diseases, like amoebiasis, malaria, cryptosporidiosis, cysticercosis, schistosomiasis, African trypanosomiasis, giardiasis, and filariasis, there are no commercially certified antibody diagnosis tests. Results from investigations were too variable because of the kind of antigen construction used, such as adult worm, crude, egg, recombinant purified, and also due to the usage of non-standardized test processes. Cross-reaction causes false positives and diagnosis is too difficult, particularly in areas where several parasites are endemic [65]. Similarly, in Africa, cross-reactivity occurs among filarial and some other helminth antigens [66]. Also vital is the inability of antibody-diagnostic tests to distinguish among previous and present on-the-go infections [67]. Moreover, tests of antibody detection cannot be applied to a parasitic infection which is not able to develop a substantial antibody reaction (e.g., this has been seen in subjects having *Echinococcus* cysts) [68].

Likewise, while diagnosing African trypanosomiasis, these tests are inadequate as seroconversion happens after the beginning of clinical signs [70]. Therefore, further improvement is required in available up-to-date detecting methodologies.

Ever since the development of the PCR, parasitologists have been focusing on molecular-based methodologies in order to find better diagnosis techniques.

9.2.3 Molecular-Based Methodologies

9.2.3.1 Nucleic Acid-Based Methods

Shortcomings associated with serology-based assays and microscopy have guided parasitologists to practice amplification of gene techniques based on PCR. In addition to the nested and multiplexed traditional PCR, RT-PCR has been used for the diagnosis of various parasite diseases.

LAMP and Luminex-based assays are new-fangled techniques used for the observation of parasites in specimens. Nucleic acids-based molecular methods have more sensitivity and specificity than microscopy and serology-based approaches. These techniques can detect parasites in samples that have very low levels of parasites and samples of asymptomatic patients [71]. Furthermore, in the same reaction tube, multiplexed PCR is used to uncover multiple sequences showing benefits in the detection of numeral parasitic infections concurrently [72].

9.2.3.2 Real-Time Polymerase Chain Reaction (RT-PCR)

Compared to conventional PCR, the RT-PCR system can quantify the original concentration by using FRET, Sybergreen, Taqman probes, and Scorpion primers [6]. With the association of standard curves, concentration can be measured. It exterminates the necessity to envisage the amplicons through gel electrophoresis thus significantly decreasing the menace of adulteration and the establishment of false positives. The use of this technique permits the increased yielding of diverse sequences in one single-closed tube reaction (Figure 9.4) [73]. By means of this method, in a single reaction tube, Shokoples et al., find the four types of *Falciparum, plus P. vivax, P. ovale*, and *P. malariae* [13]. Multiplex assay decreased cost as well as the time for completion of the test, taking only three hours to complete, making it more useful than microscopy which has a slow turnaround time. Multiplexed RT-PCR is also helpful in distinguishing drug-sensitive strains of malaria [55] which is very useful for the selection of suitable drugs for the treatment of malaria.

Diez et al., [74] detected a *T. cruzi* infection using PCR after a heart transplant. This detection led to the prompt initiation of treatment before Chagas disease reactivation.

These examples determine that effective and timely detection can openly affect the care of patients and that PCR-based methods can lead to a better choice for a cure.

Even though DNA-based approaches have revealed tremendous sensitivity and specificity, in daily laboratory practice the use of these techniques is still unusual, particularly in rural endemic areas. Additionally, as seen with various PCR-based

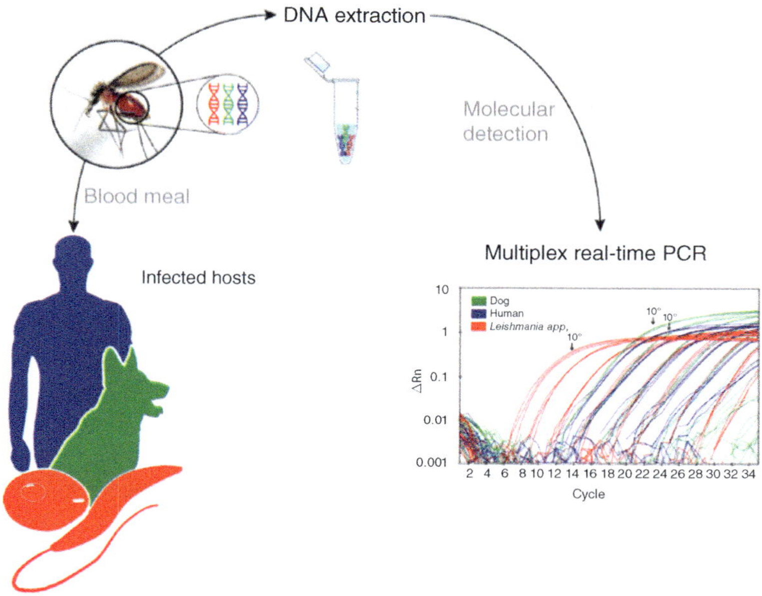

Figure 9.4 Multiplex real-time PCR. (*Source:* Adapted from Sales *et al.*, 2020 [76]).

methods, serology-based assays, are also affected because of absence of the calibration [75]. Several factors like DNA isolation, primer selection, and the use of different amplification protocols could cause this variation in findings [76]. The addition of an automated DNA isolation step would aid PCR assays in detecting parasitic infections.

9.2.3.3 Loop-Mediated Isothermal Amplification (LAMP)

LAMP is a magnification technique with very high specificity and sensitivity capable of distinguishing a single nucleotide variance [77]. This is described through using six different primers especially planned to identify eight different locations on a gene of interest, with intensification only arising if primers are attached and produce a product [78]. Earlier, LAMP has been effectively used to quickly diagnose both RNA and DNA viruses like the West Nile [11] and SARS viruses [12]. In recent times, for the detection of a number of parasitic infections (*Entamoeba*, *Trypanosoma*, *Taenia*, *Plasmodium*, and *Cryptosporidium*, *Plasmodium, and Dirofilariaimmitis*) [79–87], parasitologists have applied the LAMP technique. Most of these investigations have shown the several benefits of this technique over the common PCR method. LAMP is executed at 60–65 °C (constant temperature). Moreover, the procedure can be conducted without isolating the isolated DNA as presented such as in RIME, an unfree retroelement observed in *T. b. rhodesiense* and *T. b. gambiense* [88]. Within a short period of time (35 minutes), RIME LAMP was capable to diagnose *T. b. rhodesiense and T. b. gambiense* directly from serum, blood, and CSF samples. LAMP reactions are also stress-free to set up, and findings can readily be evaluated. The sample of interest is mixed with primers, substrates, and a DNA polymerase capable of strand displacement in a microcentrifuge tube. In the course of the reaction, huge numbers of pyrophosphate ions are formed, which causes the synthesis to precipitate [89]. The indicated opacity is proportionate to the quantity of DNA produced, the reaction can evaluate by real-time measurement of opaqueness or, essentially, via the naked eye Figures 9.5 and 9.7. Due to all these ins and outs, the implementation of LAMP in the future as a detecting technique for parasitic diseases in countryside endemic areas shows potential. Moreover, as the application of LAMP is used in the field of parasitology, it will help to develop LAMP-modified assays that meet particular diagnosis requirements. In a recent investigation on *B. babesia* [82], a mLAMP assay was established to diagnose *B. bigemina and B. bovis* from blood (DNA). Likewise, in a study applied a LAMP assay based on the 18S rRNA gene for the diagnosis of *Plasmodium* spp., i.e., *P. ovale*, *P. falciparum*, *P. vivax*, and *P. malariae* [90]. LAMP had the same sensitivity and more specificity matched to nested PCR, obtaining like findings but in a shorter time.

9.2.3.4 Luminex xMAP Technology

The Luminex tool is a bead-based flow-cytometric assay that permits the diagnosis of many targets at the same time. The beads act as probes that could be covalently bound to antibodies, antigens, or oligonucleotides. About 100 microspheres are used, and when stimulated by a laser each radiates distinctive fluorescent

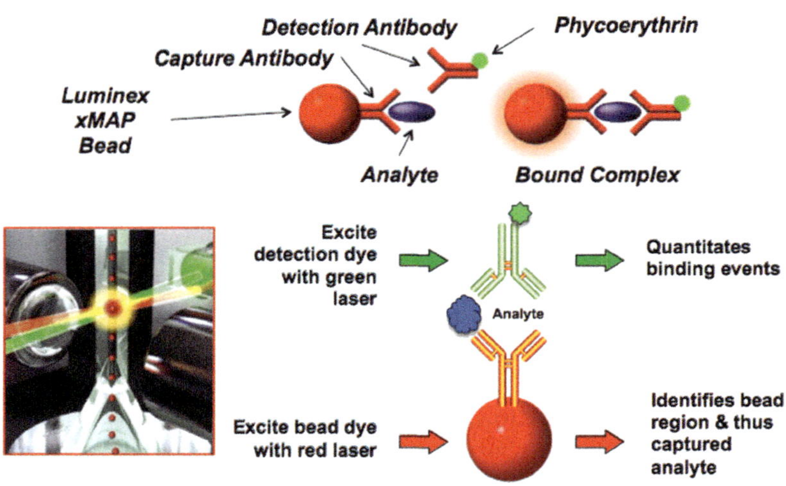

Figure 9.5 Luminex xMAP technology. (*Source:* The Pennsylvania State University/hhd.psu.edu. Accessed February 2023 [94]).

signals, therefore permitting the detection of varied targets [91]. In parasitological studies, in the same reaction, using a small volume, the Luminex assay can detect various targets. The method could be valuable for the investigation of drug resistance alleles, and antigenic variety, and for the detection of parasite infections such as the study conducted on *Cryptosporidium* [90]. *C. parvum* and *C. hominis* cannot be detected with help of antigen finding or serology assays. In both species, DNA-based methods have been effective by using the single nucleotide variance in the ML-2 [92]. Eventually, DNA sequencing is a technique of detection but it is expensive, time-consuming, and laborious. Bandyopadhyay et al., [90] efficiently recognized and differentiate 143 samples of *C. parvum* and *C. hominis* of isolated DNA using this tool by means of oligonucleotide probes precise to the ML-2 regions of each species. This assay is speedy (turnaround time five hours) and less costly than DNA sequencing and PCR. This technique is shown to be more specific and sensitive compared to a DFA test, a technique usually castoff to detect *Cryptosporidium* and *Giardia* species. However, DFA could not distinguish between *C. parvum* and *C. hominis*. In another study, Luminex tools were capable of diagnosing all stages of blood parasites like *P. falciparum*, *P. vivax*, *P. malariae*, and *P. ovale* [92]. Investigation revealed that the Luminex tool can increase the speediness, accuracy, and dependability of other PCR techniques. Furthermore, specimens are controlled at the same time and continue in a 96-well plate from isolated DNA until the analysis of data. The whole procedure is mechanized and thus consistency can be attained. In conclusion, the efficiency and scalability ability of this system and technique give out a pure benefit for extensive investigations over work-intensive microscopy (Figures 9.6 and 9.7).

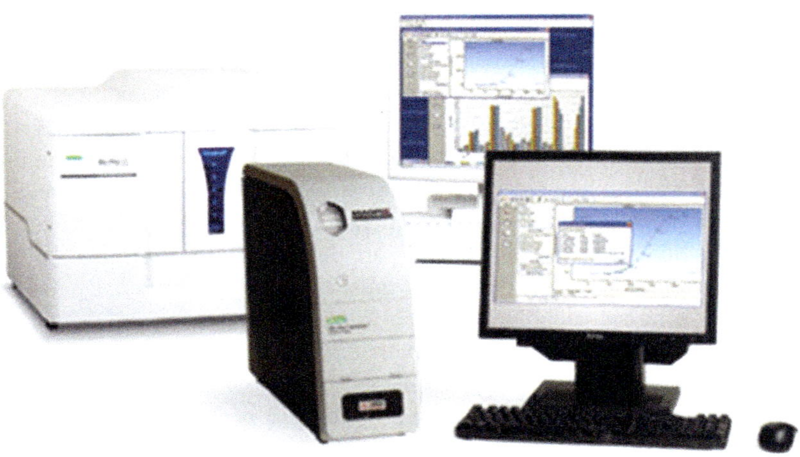

Figure 9.6 Luminex xMAP technology. (*Source:* Bio-Rad Laboratories, Inc. [93]).

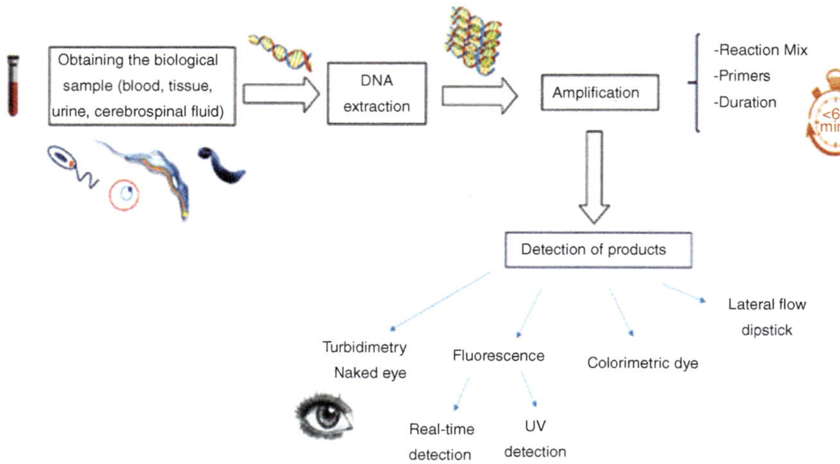

Figure 9.7 LAMP in the diagnosis of leishmaniasis, trypanosomiasis, and, malaria. (*Source:* Adapted from Iseki *et al.*, 2007 [91]).

9.2.4 Proteomics

Proteins are the structural component, key catalysts, the messenger of signals, and act as tissues' molecular motors, and proteomic investigations are apt to offer considerable clinical significance. Proteins are used as biomarkers for cell types, tissues, stages of development, and infection conditions and as possible targets for drug innovation and intercession methods. Proteomic investigation of serum as well as body fluids will provide the next generation of detecting tests for infections or diseases. Current developments in the area of proteomics are imputable mainly due to the overview of the raised area mass spectrometry proficient in screening intricate biological fluids for peptide and distinct protein "biomarkers". The approach of proteomics can recognize proteins in two ways: bottom-up and top-down methods. Biological fluids like proteins are proteolytically broken into small residues that can be sequenced and the subsequent varieties are correlated with those in recognized databases of the peptide. It is the protein equivalent of "shotgun" genomics. Bottom-up strategies are demanding to determine and are unable to recognize altered molecules like those that are glycosylated. Up to now, in the human genome, each open reading frame is supposed to produce at the minimum10 altered proteins; this is the main drawback.

The 2-dimensional gel electrophoresis is the classic top-down approach. Top-down approaches look to recognize proteins and peptides in intricate biological fluids. Through this technique, proteins are determined based on pH as the first dimension and molecular weight as second dimension. This method has low throughput, is labor-intensive, and needs a large number of samples. These drawbacks have led to the quest for better methods. For the expression of protein analysis, some other methods applied are SELDI-TOF MS, MALDI-TOF MS, ICAT, LC-MS-MS, and iTRAQ. The advance of mechanical, high-throughput proteomic tools like MALDI-TOF and SELDI-TOF MS has empowered researchers so that in a short time it is possible to examine concurrently a huge number of samples [95]. The key drawback of SELDI compared with MALDI is that SELDI has a lower resolution and a lower mass accuracy. Furthermore, SELDI is inappropriate for big molecular weight proteins e.g >100 kDa, and on the Protein Chip Array, it is inadequate for the identification of bound proteins.

A number of investigations carried out regarding parasitic infections have concentrated on SELDI. SELDI permits the binding of chemically active Protein Chip surfaces with the sample. Several kinds of Protein Chip arrays exist with varying abilities to fix proteins with various chemicals such as hydrophobic, metallic, anionic, cationic, and biological properties such as enzymes, receptors, and antibodies, thus allowing the direct analysis of proteins from intricate biological samples without the requirement for earlier isolation by 2D gel electrophoresis. The output of SELDI is a range of mass-to-charge ratios with their equivalent comparative intensities. Analyses using SELDI were first used to discover initial predictive biomarkers of carcinoma [96]. In recent times, the SELDI method has been beneficial to investigate serum biomarkers of infectious diseases like SARS [95], African trypanosomiasis [97], fascioliasis [98], and cysticercosis [99]. Such investigations have concentrated on recognizing a distinct conformation of flowing serum proteins that reveal a particular pathophysiological condition, called a "proteomic fingerprint". The actual latent function of proteomic fingerprinting is in its usage as a detection technique for new biomarkers that may then be amalgamated into simple alongside detection based on inexpensive tools like antigen-diagnostic tests (immunologically based) that can be applied in cassette or dipstick set-ups.

9.2.5 Diagnostic Techniques for Ectoparasite Infection

In several studies, the detection of infections caused by ectoparasites is a supplementary direct procedure different to that for endoparasites. Ectoparasites in or on the skin are sometimes large-sized and comparatively motionless, like ticks. Because of their clinical signs or small size or for the reason that of hidden lifestyle a number of species of ectoparasites found on humans on or within the host can be confused with or lead to other situations or infections, e.g, those affected by the medical and economically significant *Sarcoptes* and *Psoroptes* mite species. Direct detection of a parasite based on conventional diagnostic tests e.g., by microscopic investigation of skin scrapings or hair pluck, often lack specificity and sensitivity [100] and are typically reliant on the skill in parasitic detection of the investigator. Furthermore, typical tests cannot identify subclinical infections, which will have a vital part in the perseverance of ectoparasites inside host populaces [101]. Evidently, developed techniques of diagnosis are prerequisites.

9.2.5.1 Immunodiagnostics

Immunodiagnosis of mite epidemics is of veterinary and medical significance. It is often challenging to detect mite infestations, due to their small size, discrete way of life, and the appearance of morbid variations (in a few cases) isolated from the location of infections like scabies mites [102]. The recent and emergent immunodiagnostic assays for mite infection and their benefits and drawbacks are discussed below.

9.2.5.1.1 Sarcoptes Scabies

Microscopic inspection of skin rasps is usually effective technique for sarcoptic mange identification [103] and, in sheep, indirect ELISA is used for identifying mite-specific IgG antibodies. For skin scrapings, there is a noticeable variance among the sensitivities of the two investigative tests, but for rasps, there is an improved link between the two tests in their specificity, while the precision of the tests was likened to that of a testimonial technique for epidemiological studies and assessed by means of the ROC methodology [104]. The augmented sensitivity of the ELISA-based test proposes that it can be used in the detection of the occurrence of sarcoptic mange, evaluate the occurrence, show mange-free position, and define the efficiency of cure or control programs [105].

9.2.5.1.2 Rickettsia prowazekii

In humans, epidemic typhus is caused by *R. prowazekii*. This is an acute recurring disease transferred by the *P. hominis* corpora (body louse) [106]. There are numerous molecular and immunological diagnostic tests for the this infectious parasite and extra sensitive and consistent examinations for the louse can help their identification and control. In this regard, proof that immunoassays for the diagnosis of body louse can be valuable has arisen from investigations in calves, where host antibody reactions have been observed by means of an ELISA which perceives *L. vituli* nymph antigens [107]. The advance of examination of the salivary antigens of ectoparasites of the vector will increase the identification of reappearing ectoparasites and make visible the spreading out of the variety of disease-causing species of the vector. To identify low-level exposure (after 24 hours post infestation) of chickens to the trypanosome vector, IgM antibody to an immunogenic protein from *T. infestans* saliva might be applied [108]. The response of IgM is shorter in duration than the IgG retort and is thus beneficial as a sign of fresh biting action [108, 109]. A parallel test for disclosure to salivary antigens from the pest hemipteran, *C. lectularius*, can benefit from the finding of *C. lectularius* during the detection and cure of domestic ectoparasite swarms.

9.2.6 Molecular Approaches for Improved Diagnoses

9.2.6.1 Scabies

Diagnosis of scabies, is difficult using traditional techniques [110]. For diagnosis of *S. scabiei*, skin rasps from patients suspected to have scabies are tested by means of nested PCR that found a number of patients with definite cases of scabies and showed remaining mite infestation in earlier cured cases [104]. Morphologically, host-specific variants of the mite are vague, therefore it is difficult to precisely distinguish an influx with *S. scabiei* var. hominins from a zoonotic infection with, like *S. scabiei* var. canis by microscopic investigation. The need to identify these differences, in the context of population healthiness and cure systems, is important [111, 112].

9.2.6.2 Diptera

For the detection of ectoparasitic infections, the use of molecular-based approaches has always been useful in the case of blowflies, while current methods have actually been from the fields of medical, veterinary, or forensic science diagnostics. Nowadays for the larvae identification of blowflies, DNA sequence-based investigation is the only molecular-based species identification approach [113]. Using proteomic, instead of genomic techniques, variances among antigenic larval proteins of blowflies can be applied to detect protein markers that will potentially be unified into quick detecting tests like lateral flow devices, decreasing the caseload by means of up-to-date DNA sequencing approaches [114].

9.3 Conclusion

Currently reference and diagnostic laboratories use many testing techniques comprising microscopy, molecular assays, and serological analysis. In order to replace conventional techniques such as microscopy for routine diagnosis, a technique has to be simple, rapid, specific, sensitive, and cheap. A special focus has been given to the development of molecular diagnostic techniques. RT-PCR methodologies are constantly optimized for the identification of several parasites. LAMP has achieved much attention and is viewed as a future molecular tool.

RT-PCR methodologies are constantly optimized for the identification of several parasites. LAMP has achieved much attention and is viewed as a future molecular tool. The current need is for the development of multiplex real-time PCR protocols. These methods can help in the analysis of mixed infections. In addition to being highly specific and sensitive, molecular assays appear to be favorable alternatives for direct microscopy. Even though the high price is a limiting factor for the use of molecular methods, these tests are progressively being used in treatment monitoring, clinical diagnosis, and epidemiological studies of parasitic diseases affecting population worldwide. In the affected populations, the use of these techniques provides comprehensive knowledge of morphology, behavior, and genetic characteristics of parasitic disease.

References

1 Northrop-Clewes, C.A. and Shaw, C. (2000). Parasites. *British Medical Bulletin* 56 (1): 193–208.
2 Fearon, M.A., Scalia, V., Huang, M. et al. (2013). A case of vertical transmission of Chagas disease contracted via blood transfusion in Canada. *Canadian Journal of Infectious Diseases and Medical Microbiology* 24 (1): 32–34.
3 Meyer, A.C. and Birbeck, G.L. (2007). Parasitic infections. In *Neurobiology of Disease* (ed. S. Gilman), 453–472. Elsevier.
4 Garcia, L.S. (2009). Classification and nomenclature of human parasites. In: *Feigin and Cherry's Textbook of Pediatric Infectious Diseases*, (eds S.H. Kaplan, R.D. Feign, G.J. Demmler-Harrison et al.), 2861–2867. Saunders, Elsevier.
5 Assafa, D., Kibru, E., Nagesh, S. et al. (2006). Medical parasitology. *Ethiopia Public Health Training Initiative.*
6 Wakelin, D. (1996). Helminths: pathogenesis and defenses. *In Medical Microbiology*, 4th edition (ed S. Baron), University of Texas Medical Branch at Galveston.
7 Bass, C. and Field, L.M. (2011). Gene amplification and insecticide resistance. *Pest Management Science* 67 (8): 886–890.
8 McNair, C.M. (2015). Ectoparasites of medical and veterinary importance: drug resistance and the need for alternative control methods. *Journal of Pharmacy and Pharmacology* 67 (3): 351–363.
9 Colley, D.G. (2000). Parasitic diseases: opportunities and challenges in the 21st century. *Memórias Do Instituto Oswaldo Cruz* 95: 79–87.
10 Muldrew, K.L. (2009). Molecular diagnostics of infectious diseases. *Current Opinion in Pediatrics* 21 (1): 102–111.
11 Parida, M., Sannarangaiah, S., Dash, P.K. et al. (2008). Loop mediated isothermal amplification (LAMP): a new generation of innovative gene amplification technique; perspectives in clinical diagnosis of infectious diseases. *Reviews in Medical Virology* 18 (6): 407–421.
12 Notomi, T., Okayama, H., Masubuchi, H. et al. (2000). Loop-mediated isothermal amplification of DNA. *Nucleic Acids Research* 28 (12): e63.
13 Shokoples, S.E., Ndao, M., Kowalewska-Grochowska, K., and Yanow, S.K. (2009). Multiplexed real-time PCR assay for discrimination of *Plasmodium* species with improved sensitivity for mixed infections. *Journal of Clinical Microbiology* 47 (4): 975–980.

14 Hawkes, M., Conroy, A.L., Opoka, R.O. et al. (2014). Use of a three-band HRP2/pLDH combination rapid diagnostic test increases diagnostic specificity for *falciparum* malaria in Ugandan children. *Malaria Journal* 13 (1): 1–6.

15 Houzé, S., Boutron, I., Marmorat, A. et al. (2013). Performance of rapid diagnostic tests for imported malaria in clinical practice: results of a national multicenter study. *PloS One* 8 (9): e75486.

16 Aydin-Schmidt, B., Mubi, M., Morris, U. et al. (2013). Usefulness of *Plasmodium falciparum*-specific rapid diagnostic tests for assessment of parasite clearance and detection of recurrent infections after artemisinin-based combination therapy. *Malaria Journal* 12 (1): 1–1.

17 Steel, C., Golden, A., Kubofcik, J. et al. (2013). Rapid *Wuchereria bancrofti*-specific antigen Wb123-based IgG4 immunoassays as tools for surveillance following mass drug administration programs on lymphatic filariasis. *Clinical and Vaccine Immunology* 20 (8): 1155–1161.

18 Johnston, S.P., Ballard, M.M., Beach, M.J. et al. (2003). Evaluation of three commercial assays for detection of *Giardia* and *Cryptosporidium* organisms in fecal specimens. *Journal of Clinical Microbiology* 41 (2): 623–626.

19 Fotedar, R., Stark, D., Beebe, N. et al. (2007). Laboratory diagnostic techniques for *Entamoeba* species. *Clinical Microbiology Reviews* 20 (3): 511–532.

20 Garcia, L.S., Shimizu, R.Y., and Bernard, C.N. (2000). Detection of *Giardia lamblia*, *Entamoeba histolytica*/*Entamoeba dispar*, and *Cryptosporidium parvum* antigens in human fecal specimens using the triage parasite panel enzyme immunoassay. *Journal of Clinical Microbiology* 38 (9): 3337–3340.

21 Sharp, S.E., Suarez, C.A., Duran, Y., and Poppiti, R.J. (2001). Evaluation of the triage micro parasite panel for detection of *Giardia lamblia*, *Entamoeba histolytica*/*Entamoeba dispar*, and *Cryptosporidium parvum* in patient stool specimens. *Journal of Clinical Microbiology* 39 (1): 332–334.

22 Singh, A., Houpt, E., and Petri, W.A. (2009). Rapid diagnosis of intestinal parasitic protozoa, with a focus on *Entamoeba histolytica*. *Interdisciplinary Perspectives on Infectious Diseases* 2009.

23 Ndao, M. (2009). Diagnosis of parasitic diseases: old and new approaches. *Interdisciplinary Perspectives on Infectious Diseases* 2009.

24 Muray, P.R., Rosenthal, K.S., and Pfaller, M.A. (2020). *Medical Microbiology*, 7e. Elsevier.

25 Brooks, G.F., Carroll, K.C., Butel, J.S. et al. (2013). *Jawetz, Melnick, & Adelberg's Medical Microbiology*, 26e. New York: Lange Medical Books, McGraw Hill.

26 Engelkirk, P.G. and Duben-Engelkirk, J.L. (2008). *Laboratory Diagnosis of Infectious Diseases: Essentials of Diagnostic Microbiology*. Lippincott Williams & Wilkins.

27 Visvesvara, G.S., Moura, H., and Schuster, F.L. (2007). Pathogenic and opportunistic free-living amoebae: *Acanthamoeba* spp. *Balamuthia mandrillaris*,

Naegleria fowleri, and Sappinia diploidea. FEMS Immunology & Medical Microbiology. 50 (1): 1–26.

28 Garber, G.E. (2005). The laboratory diagnosis of *Trichomonas vaginalis. Can J Infect Dis Med Microbiol* 16: 35e8.

29 Schuster, F.L. and Ramirez-Avila, L. (2008). Current world status of *Balantidium coli. Clinical Microbiology Reviews* 21 (4): 626–638.

30 Marciano-Cabral, F. and Cabral, G. (2003). *Acanthamoeba* spp. as agents of disease in humans. *Clinical Microbiology Reviews* 16 (2): 273–307.

31 El-Nahas, H.A., Salem, D.A., El-Henawy, A.A. et al. (2013). *Giardia* diagnostic methods in human fecal samples: a comparative study. *Cytometry Part B: Clinical Cytometry* 84 (1): 44–49.

32 Winn, W.C., Jr, Allen, S., Janda, W. et al. (2006). *Koneman's Color Atlas and Textbook of Diagnostic Microbiology*, 6e. Lippincott Philadelphia, PA: Williams & Wilkins.

33 Garcia, L.S. (1993). Macroscopic and microscopic examination of fecal specimens. In: *Diagnostic Medical Parasitology*, 501–535. Wiley.

34 Goldman, E. and Green, L.H. (2008). *Practical Handbook of Microbiology*, 2e. Boc6 Raton, FL: CRC Press.

35 Woods, G.L. and Walker, D.H. (1996). Detection of infection or infectious agents by use of cytologic and histologic stains. *Clinical Microbiology Reviews* 9 (3): 382–404.

36 Fouque, E., Trouilhé, M.C., Thomas, V. et al. (2012). Cellular, biochemical, and molecular changes during encystment of free-living amoebae. *Eukaryotic Cell* 11 (4): 382–387.

37 Ricciardi, A. and Ndao, M. (2015). Diagnosis of parasitic infections: what's going on? *Journal of Biomolecular Screening* 20 (1): 6–21.

38 Phuong, M.S., Lau, R., Ralevski, F., and Boggild, A.K. (2016). Parasitological correlates of *Plasmodium ovale curtisi* and *Plasmodium ovale wallikeri* infection. *Malaria Journal* 15 (1): 1–1.

39 Blasco-Costa, I., Cutmore, S.C., Miller, T.L., and Nolan, M.J. (2016). Molecular approaches to trematode systematics: 'best practice' and implications for future study. *Systematic Parasitology* 93 (3): 295–306.

40 Es-Sette, N., Ajaoud, M., Laamrani-Idrissi, A. et al. (2014). Molecular detection and identification of *Leishmania* infection in naturally infected sand flies in a focus of cutaneous leishmaniasis in northern Morocco. *Parasites & Vectors* 7 (1): 1–8.

41 Djohan, V., Kaba, D., Rayaissé, J.B. et al. (2015). Detection and identification of pathogenic trypanosome species in tsetse flies along the Comoé River in Côte d'Ivoire. *Parasite* 22.

42 Maia, C., Ferreira, A., Nunes, M. et al. (2014). Molecular detection of bacterial and parasitic pathogens in hard ticks from Portugal. *Ticks and Tick-borne Diseases* 5 (4): 409–414.

43 Lempereur, L., Beck, R., Fonseca, I. et al. (2017). Guidelines for the detection of *Babesia* and *Theileria* parasites. *Vector-Borne and Zoonotic Diseases* 17 (1): 51–65.
44 Weerakoon, K.G. and McManus, D.P. (2016). Cell-free DNA as a diagnostic tool for human parasitic infections. *Trends in Parasitology* 32 (5): 378–391.
45 Zimmerman, P.A., Mehlotra, R.K., Kasehagen, L.J., and Kazura, J.W. (2004). Why do we need to know more about mixed *Plasmodium* species infections in humans? *Trends in Parasitology* 20 (9): 440–447.
46 Riddle, M.S., DuPont, H.L., and Connor, B.A. (2016). ACG clinical guideline: diagnosis, treatment, and prevention of acute diarrheal infections in adults. *Official Journal of the American College of Gastroenterology, ACG* 111 (5): 602–622.
47 Claas, E.C., Burnham, C.A., Mazzulli, T. et al. (2013). Performance of the xTAG® gastrointestinal pathogen panel, a multiplex molecular assay for simultaneous detection of bacterial, viral, and parasitic causes of infectious gastroenteritis. *Journal of Microbiology and Biotechnology* 23 (7): 1041–1045.
48 Amar, C.F., Dear, P.H., and McLauchlin, J. (2004). Detection and identification by real time PCR/RFLP analyses of *Cryptosporidium* species from human faeces. *Letters in Applied Microbiology* 38 (3): 217–222.
49 Abanyie, F., Harvey, R.R., Harris, J.R. et al. (2015). 2013 multistate outbreaks of *Cyclospora cayetanensis* infections associated with fresh produce: focus on the Texas investigations. *Epidemiology & Infection* 143 (16): 3451–3458.
50 Wilson, M.L. (2013). Laboratory diagnosis of malaria: conventional and rapid diagnostic methods. *Archives of Pathology and Laboratory Medicine* 137 (6): 805–811.
51 Deng, H. and Alan, T. (2023). Microfluidic devices for biosensing. *Encyclopedia of Sensors and Biosensors*, 3: 474–488.
52 Clark, D.P. and Pazdernik, N. (2015). *Biotechnology*. *Newnes*.
53 Srividya, G., Kulshrestha, A., Singh, R., and Salotra, P. (2012). Diagnosis of visceral leishmaniasis: developments over the last decade. *Parasitol Res* 110: 1065e78.
54 Carter, W.J., Yan, Z., Cassai, N.D., and Sidhu, G.S. (2003). Detection of extracellular forms of *Babesia* in the blood by electron microscopy: a diagnostic method for differentiation from *Plasmodium falciparum*. *Ultrastructural Pathology* 27 (4): 211–216.
55 Diez, M., Favaloro, L., Bertolotti, A. et al. (2007). Usefulness of PCR strategies for early diagnosis of Chagas' disease reactivation and treatment follow-up in heart transplantation. *American Journal of Transplantation* 7 (6): 1633–1640.
56 Hillyer, G.V., de Galanes, M.S., Rodriguez-Perez, J. et al. (1992). Use of the Falcon™ assay screening test-enzyme-linked immunosorbent assay (FAST-ELISA) and the enzyme-linked immunoelectrotransfer blot (EITB) to determine the prevalence of human fascioliasis in the Bolivian altiplano. *The American Journal of Tropical Medicine and Hygiene* 46 (5): 603–609.

57 Klumpp-Thomas, C., Kalish, H., Drew, M. et al. (2021). Standardization of ELISA protocols for serosurveys of the SARS-CoV-2 pandemic using clinical and at-home blood sampling. *Nature Communications* 12 (1): 1–13.
58 Hancock, K. and Tsang, V.C. (1986). Development and optimization of the FAST-ELISA for detecting antibodies to *Schistosoma mansoni*. *Journal of Immunological Methods* 92 (2): 167–176.
59 Campbell, G.H., Aley, S.B., Ballou, W.R. et al. (1987). Use of synthetic and recombinant peptides in the study of host-parasite interactions in the malaria. NavalMedical Research Institute, Bethesda, MD.
60 Maddison, S.E. (1987). The present status of serodiagnosis and seroepidemiology of schistosomiasis. *Diagnostic Microbiology and Infectious Disease* 7 (2): 93–105.
61 Ko, R.C. and Ng, T.F. (1998). Evaluation of excretory/secretory products of larval *Taenia solium* as diagnostic antigens for porcine and human cysticercosis. *Journal of Helminthology* 72 (2): 147–154.
62 World Health Organization. (2011). *Good practices for selecting and procuring* rapid diagnostic tests for malaria.
63 Murray, C.K., Gasser, J.R.A., Magill, A.J., and Miller, R.S. (2008). Update on rapid diagnostic testing for malaria. *Clinical Microbiology Reviews* 21 (1): 97–110.
64 Drakeley, C. and Reyburn, H. (2009). Out with the old, in with the new: the utility of rapid diagnostic tests for malaria diagnosis in Africa. *Transactions of the Royal Society of Tropical Medicine and Hygiene* 103 (4): 333–337.
65 Ramanathan, R., Burbelo, P.D., Groot, S. et al. (2008). A luciferase immunoprecipitation systems assay enhances the sensitivity and specificity of diagnosis of *Strongyloides stercoralis* infection. *The Journal of Infectious Diseases* 198 (3): 444–451.
66 Burbelo, P.D., Ramanathan, R., Klion, A.D. et al. (2008). Rapid, novel, specific, high-throughput assay for diagnosis of *Loa loa* infection. *Journal of Clinical Microbiology* 46 (7): 2298–2304.
67 Vexenat, A.D., Santana, J.M., and Teixeira, A.R. (1996). Cross-reactivity of antibodies in human infections by the kinetoplastid protozoa *Trypanosoma cruzi*, *Leishmania chagasi* and *Leishmania* (*Viannia*) *braziliensis*. *Revista Do Instituto de Medicina Tropical de São Paulo* 38 (3): 177–185.
68 Singh, B. (1997). Molecular methods for diagnosis and epidemiological studies of parasitic infections. *International Journal for Parasitology* 27 (10): 1135–1145.
69 Aye, K.M., Nagayasu, E., Baba, M. et al. (2018). Evaluation of LIPS (luciferase immunoprecipitation system) for serodiagnosis of toxoplasmosis. *Journal of Immunological Methods* 462: 91–100.
70 Moro, P. and Schantz, P.M. (2009). Echinococcosis: a review. *International Journal of Infectious Diseases* 13 (2): 125–133.
71 Papadopoulos, M.C., Abel, P.M., Agranoff, D. et al. (2004). A novel and accurate diagnostic test for human African trypanosomiasis. *The Lancet* 363 (9418): 1358–1363.

72 Mens, P., Spieker, N., Omar, S. et al. (2007). Is molecular biology the best alternative for diagnosis of malaria to microscopy? A comparison between microscopy, antigen detection and molecular tests in rural Kenya and urban Tanzania. *Tropical Medicine & International Health* 12 (2): 238–244.

73 Zarlenga, D.S. and Higgins, J. (2001). PCR as a diagnostic and quantitative technique in veterinary parasitology. *Veterinary Parasitology* 101 (3–4): 215–230.

74 Gasser, R.B. (2006). Molecular tools—advances, opportunities and prospects. *Veterinary Parasitology* 136 (2): 69–89.

75 Farcas, G.A., Soeller, R., Zhong, K. et al. (2006). Real-time polymerase chain reaction assay for the rapid detection and characterization of chloroquine-resistant *Plasmodium falciparum* malaria in returned travelers. *Clinical Infectious Diseases* 42 (5): 622–627.

76 Sales, K.G., Miranda, D.E., da Silva, F.J. et al. (2020). Evaluation of different storage times and preservation methods on phlebotomine sand fly DNA concentration and purity. *Parasites & Vectors* 13 (1): 1–6.

77 Murray, T.S. and Cappello, M. (2008). The molecular diagnosis of parasitic diseases. *The Pediatric Infectious Disease Journal* 27 (2): 163–164.

78 Bretagne, S. and Costa, J.M. (2006). Towards a nucleic acid-based diagnosis in clinical parasitology and mycology. *Clinica Chimica Acta* 363 (1–2): 221–228.

79 Parida, M., Posadas, G., Inoue, S. et al. (2004). Real-time reverse transcription loop-mediated isothermal amplification for rapid detection of West Nile virus. *Journal of Clinical Microbiology* 42 (1): 257–263.

80 Poon, L.L., Wong, B.W., Chan, K.H. et al. (2005). Evaluation of real-time reverse transcriptase PCR and real-time loop-mediated amplification assays for severe acute respiratory syndrome coronavirus detection. *Journal of Clinical Microbiology* 43 (7): 3457–3459.

81 Liang, S.Y., Chan, Y.H., Hsia, K.T. et al. (2009). Development of loop-mediated isothermal amplification assay for detection of *Entamoeba histolytica*. *Journal of Clinical Microbiology* 47 (6): 1892–1895.

82 Njiru, Z.K., Mikosza, A.S., Matovu, E. et al. (2008). African trypanosomiasis: sensitive and rapid detection of the sub-genus Trypanozoon by loop-mediated isothermal amplification (LAMP) of parasite DNA. *International Journal for Parasitology* 38 (5): 589–599.

83 Nkouawa, A., Sako, Y., Nakao, M. et al. (2009). Loop-mediated isothermal amplification method for differentiation and rapid detection of *Taenia* species. *Journal of Clinical Microbiology* 47 (1): 168–174.

84 Poon, L.L., Wong, B.W., Ma, E.H. et al. (2006). Sensitive and inexpensive molecular test for *falciparum* malaria: detecting *Plasmodium falciparum* DNA directly from heat-treated blood by loop-mediated isothermal amplification. *Clinical Chemistry* 52 (2): 303–306.

85 Bakheit, M.A., Torra, D., Palomino, L.A. et al. (2008). Sensitive and specific detection of Cryptosporidium species in PCR-negative samples by loop-mediated isothermal DNA amplification and confirmation of generated LAMP products by sequencing. *Veterinary Parasitology* 158 (1–2): 11–22.

86 Alhassan, A., Thekisoe, O.M., Yokoyama, N. et al. (2007). Development of loop-mediated isothermal amplification (LAMP) method for diagnosis of equine piroplasmosis. *Veterinary Parasitology* 143 (2): 155–160.

87 Guan, G., Chauvin, A., Luo, J. et al. (2008). The development and evaluation of a loop-mediated isothermal amplification (LAMP) method for detection of *Babesia* spp. *Infective to Sheep and Goats in China. Experimental Parasitology.* 120 (1): 39–44.

88 Aonuma, H., Suzuki, M., Iseki, H. et al. (2008). Rapid identification of *Plasmodium*-carrying mosquitoes using loop-mediated isothermal amplification. *Biochemical and Biophysical Research Communications* 376 (4): 671–676.

89 Aonuma, H., Yoshimura, A., Perera, N. et al. (2009). Loop-mediated isothermal amplification applied to filarial parasites detection in the mosquito vectors: *Dirofilaria immitis* as a study model. *Parasites & Vectors* 2 (1): 1–7.

90 Mori, Y., Nagamine, K., Tomita, N., and Notomi, T. (2001). Detection of loop-mediated isothermal amplification reaction by turbidity derived from magnesium pyrophosphate formation. *Biochemical and Biophysical Research Communications* 289 (1): 150–154.

91 Iseki, H., Alhassan, A., Ohta, N. et al. (2007). Development of a multiplex loop-mediated isothermal amplification (mLAMP) method for the simultaneous detection of bovine Babesia parasites. *Journal of Microbiological Methods* 71 (3): 281–287.

92 Dunbar, S.A. (2006). Applications of Luminex® xMAP™ technology for rapid, high-throughput multiplexed nucleic acid detection. *Clinica Chimica Acta* 363 (1–2): 71–82.

93 Bio-Rad Laboratories Inc. bio-rad.com. Accessed March 2023.

94 The Pennsylvania State University. hhd.psu.edu. Accessed February 2023.

95 Bandyopadhyay, K., Kellar, K.L., Moura, I. et al. (2007). Rapid microsphere assay for identification of *Cryptosporidium hominis* and *Cryptosporidium parvum* in stool and environmental samples. *Journal of Clinical Microbiology* 45 (9): 2835–2840.

96 McNAMARA, D.T., Kasehagen, L.J., Grimberg, B.T. et al. (2006). Diagnosing infection levels of four human malaria parasite species by a polymerase chain reaction/ligase detection reaction fluorescent microsphere-based assay. *The American Journal of Tropical Medicine and Hygiene* 74 (3): 413.

97 Luminex™ bead-based immunoassays drive immunoassays towards higher-content biomarker discovery By Behind The Bench (2016). Staff05.19.

98 Xiao, Z., Prieto, D., Conrads, T.P. et al. (2005). Proteomic patterns: their potential for disease diagnosis. *Molecular and Cellular Endocrinology* 230 (1–2): 95–106.

99 Pang, R.T., Poon, T.C., Chan, K.A. et al. (2006). Serum proteomic fingerprints of adult patients with severe acute respiratory syndrome. *Clinical Chemistry* 52 (3): 421–429.

100 Rioux, M.C., Carmona, C., Acosta, D. et al. (2008). Discovery and validation of serum biomarkers expressed over the first twelve weeks of Fasciola hepatica infection in sheep. *International Journal for Parasitology* 38 (1): 123–136.

101 Deckers, N., Dorny, P., Kanobana, K. et al. (2008). Use of Protein Chip technology for identifying biomarkers of parasitic diseases: the example of porcine cysticercosis (*Taenia solium*). *Experimental Parasitology* 120 (4): 320–329.

102 Swets, J.A. (1988). Measuring the accuracy of diagnostic systems. *Science* 240 (4857): 1285–1293.

103 Van den Broek, A. and Huntley, J. (2003). Sheep scab: the disease, pathogenesis and control. *The Journal of Comparative Pathology* 128: 79e91.

104 Walton, S.F. and Currie, B.J. (2007). Problems in diagnosing scabies, a global disease in human and animal populations. *Clinical Microbiology Reviews* 20 (2): 268–279.

105 Hengge, U.R., Currie, B.J., Jäger, G. et al. (2006). Scabies: a ubiquitous neglected skin disease. *The Lancet Infectious Diseases* 6 (12): 769–779.

106 Rodríguez-Cadenas, F., Carbajal-González, M.T., Fregeneda-Grandes, J.M. et al. (2010). Development and evaluation of an antibody ELISA for sarcoptic mange in sheep and a comparison with the skin-scraping method. *Preventive Veterinary Medicine* 96 (1–2): 82–92.

107 Raoult, D. and Roux, V. (1999). The body louse as a vector of reemerging human diseases. *Clinical Infectious Diseases* 29 (4): 888–911.

108 Colwell, D.D. and Himsl-Rayner, C. (2002). *Linognathus vituli* (Anoplura: linognathidae): population growth, dispersal and development of humoral immune responses in naıve calves following induced infestations. *Veterinary Parasitology* 108 (3): 237–246.

109 Dias, J.C. (2007). Southern cone initiative for the elimination of domestic populations of *Triatoma infestans* and the interruption of transfusion Chagas disease: historical aspects, present situation, and perspectives. *Memórias Do Instituto Oswaldo Cruz* 102: 11–18.

110 Schwarz, A., Medrano-Mercado, N., Billingsley, P.F. et al. (2010). IgM-antibody responses of chickens to salivary antigens of *Triatoma infestans* as early biomarkers for low-level infestation of triatomines. *International Journal for Parasitology* 40 (11): 1295–1302.

111 Fukuyama, S., Nishimura, T., Yotsumoto, H. et al. (2010). Diagnostic usefulness of a nested polymerase chain reaction assay for detecting *Sarcoptes scabiei* DNA in skin scrapings from clinically suspected scabies. *British Journal of Dermatology* 163 (4): 892–894.

112 Currie, B.J., Maguire, G.P., and Wood, Y.K. (1995). Ivermectin and crusted (Norwegian) scabies. *Medical Journal of Australia* 163 (10): 559–560.

113 Walton, S.F., Choy, J.L., Bonson, A. et al. (1999). Genetically distinct dog-derived and human-derived *Sarcoptes scabiei* in scabies-endemic communities in northern Australia. *The American Journal of Tropical Medicine and Hygiene* 61 (4): 542–547.

114 Sperling, F.A., Anderson, G.S., and Hickey, D.A. (1994). A DNA-based approach to the identification of insect species used for postmorten interval estimation. *Journal of Forensic Science* 39 (2): 418–427.

10

Drugs for the Control of Parasitic Diseases

Current Status and Case Studies

Pratichi Singh[1], Swetanshu[1], Shikha Yadav[2,], Adeline Lum Nde[3], and Vijay Jyoti Kumar[4]*

[1] *Department of Biosciences, School of Basic and Applied Sciences, Galgotias University, Greater Noida, Uttar Pradesh, India*
[2] *Department of Pharmacy, School of Medical and Allied Sciences, Galgotias University, Greater Noida, Uttar Pradesh, India*
[3] *Department of Pharmacology, School of Clinical Medicine, Faculty of Health Sciences, University of the Free State, Bloemfontein, South Africa*
[4] *Department of Pharmaceutical Sciences, H. N. B. Garhwal University, Srinagar Garhwal, Uttarakhand, India*
* *Corresponding author*

10.1 Introduction

Parasites are generally microorganisms and include both animals and plants. Organisms that fall under this category have some sort of association with other organisms. The larger organism called the host harbors these parasites. In general, the host is the reason behind the survival of parasites as they depend on them for their basic needs such as nutrients, shelter, protection, etc. This relationship is called parasitism. Generally, the host gets harmed by the action of the parasite but in a few cases, the interaction is neutral. It is possible to have a neutral relationship but in any of the scenarios it is not beneficial for the host. The host often dies from by-product secretion [1].

Parasites are subdivided into two categories: ectoparasites, which are found on the outside of the host, and endoparasites which are inside the host. A majority of ectoparasites can be seen by the naked eye whereas endoparasites are invisible. Endoparasites usually dwell in the bloodstream, liver, muscle, digestive system, etc. Some can also

Parasitic Infections: Immune Responses and Therapeutics, First Edition.
Edited by Abhay Prakash Mishra and Manisha Nigam.

travel to the brain. Protozoa, arthropods, and helminths are the most studied parasites associated with infections in humans [1]. Parasite-associated diseases cause millions of deaths worldwide. Transmission occurs in numerous ways like through zoonotic (animal), and blood transmissions, food, water, insect vectors, etc.

In this chapter, we discuss a few parasitic diseases like malaria, babesiosis, leishmaniasis, lymphatic filariasis, and schistosomiasis. Also, the details of drugs commonly used in the treatment of these respective diseases, along with case studies, will be discussed.

10.2 Malaria

Being a member of the Phylum Apicomplexa, *Plasmodium* is categorized into five different species: *Plasmodium vivax, Plasmodium falciparum, Plasmodium malariae, Plasmodium ovale*, and *Plasmodium knowlesi* [2]. Among these five species, only two *P. vivax* and *P. falciparum* are of interest to humans as they are the most pathogenic to humans. *Plasmodium falciparum* is the deadliest and is also the dominant species in Africa, Dominican Republic, Haiti, New Guinea whereas, *P. vivax* is the dominant species in the islands of the West Pacific and America [3] However, in eastern Asia and the Indian subcontinent, they are both present [4]. The female *Anopheles* mosquito is a vector for this parasite [5]. The pathogenesis of malaria includes changes like the increase in the rigidity of the RBC membrane, production of hemozoin, amplified splenic clearance, production of proteins like KAHRP and PfEMP-1, etc., activation of T-cell (CD4+), and the presence of Th1 cytokines like INF-gamma and TNF-alpha.

In 2019, about 229 million cases were reported worldwide leading to the death of around 409,000 people. The annual malaria report of WHO published in 2020 shows that between the years 2000–2019, a total of 1.5 billion cases were reported worldwide causing the death of 7.6 million people.

There are several drugs such as tafenoquine, artesunate, etc. listed and deposited in drug banks that are used in the treatment of malaria.

10.2.1 Tafenoquine

Tafenoquine ($C_{24}H_{28}F_3N_3O_3$) is a type of 8-aminoquinoline and has a broad range of activities [6].

IUPAC Name: 4-*N*-[2,6-dimethoxy-4-methyl-5-[3-(trifluoromethyl) phenoxy] quinolin-8-yl] pentane-1,4-diamine.

Tafenoquine (Figure 10.1a) is used for inhibiting the multiple-level life stage of many *Plasmodium* species. The prolonged half-life of tafenoquine allows its single-dose use in malaria treatment. The reoccurrence of malaria after a week of treatment to years after the first encounter with *Plasmodium vivax* is because the sporozoites enter into a dormant stage in the liver. The parasite is called a hypnozoite at this stage. This hypnozoite remains in the liver for a long period and works as the reservoir which ensures the reoccurrence of the disease. So, for the complete eradication of parasites, elimination of this reservoir is an essential task [6, 7]. Primaquine was previously used for reservoir removal before the discovery of tafenoquine [8]. The only difference between these drugs is that tafenoquine has a 5-phenoxy group and is better than primaquine. Primaquine requires approximately 14 days for treatment while, tafenoquine is a single dose drug [9].

10.2.2 Artesunate

IUPACName:4-Oxo-4-(((3R,5aS,6R,8aS,9R,10S,12R,12aR)-3,6,9-trimethyldecahydro-12H-3,12-epoxy [1,2] dioxepino [4,3-i] isochromen-10-yl) oxy) butanoic acid.

On May 26, 2020, artesunate ($C_{19}H_{28}O_8$) got approved by FDA. It is now used to treat the onset of severe malaria cases. WHO also supports the use of artesunate as the primary drug in the treatment of severe malaria. Artesunate (Figure 10.1b) is a derivative of a compound called artemisinin which was developed to overcome the low hydrophilicity of artemisinin. Artesunate has better water solubility. The normal functioning of *Plasmodium* is inhibited by free radical generation, which occurs as a result of the action of artesunate on dihydroartemisinin (DHA). The free radical act as an inhibitory substance during the erythrocytic stage of the *Plasmodium* lifecycle causing the blockage in nucleic acid and protein synthesis [10]. Sometimes parasitic protein is alkylated due to the free radical reaction. Due to the shorter half-life of artesunate, this metabolism process does not last long, hence enlisting it as a moderate level drug in the therapeutic index. Hypersensitivity and haemolytic anemia are some of the side effects associated with treatment with this drug.

10.2.3 Mefloquine

IUPAC NAME: [2,8-bis (trifluoromethyl) quinolin-4-yl]–piperidin-2-yl methanol

Malaria is also treated with mefloquine ($C_{17}H_{16}F_6N_2O$), which is a derivative of quinoline. WHO approved its use in the treatment of mild and moderate cases of malaria associated with *P. falciparum* and *P. vivax*. It is the most effective drug against chloroquine-resistant *P. falciparum* (currently not preferred to be used in treatment due to drug-resistance). Mefloquine (Figure 10.1c) act on sporozoites in the bloodstream and eliminates them. The mode of action of this drug is not understood. Some researchers suggest that the drug targets the 80s ribosomal component of *P. falciparum* for its elimination. The enzyme called CYP3A4 is responsible for the metabolism of mefloquine in the liver. In adults, they have a half-life of around 21 days.

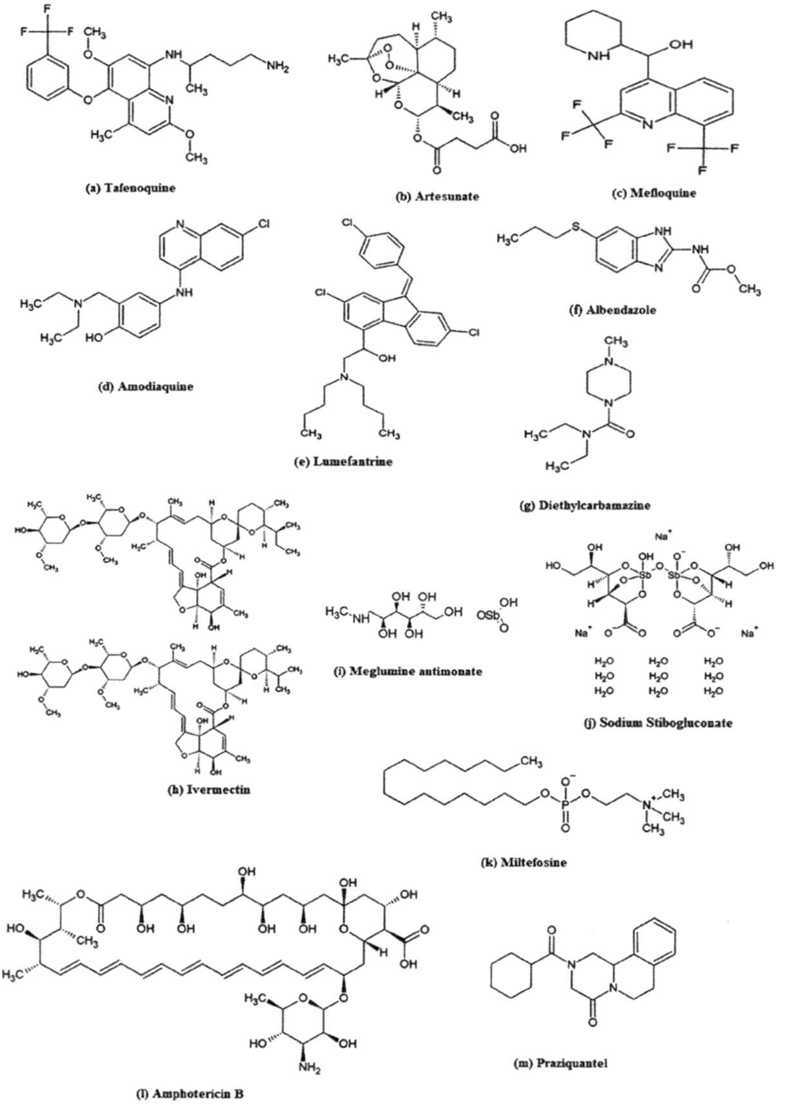

Figure 10.1 Chemical structure of drugs used in the treatment of malaria (a–e), lymphatic filariasis, (f–h) leishmaniasis, (i–l) schistosomiasis (m) are presented.

10.2.4 Amodiaquine

IUPAC NAME: 4-[(7-chloroquinolin-4-yl) amino]-2-(diethylaminomethyl)phenol

Amodiaquine (Figure 10.1d) is a 4-aminoquinoline derivative. It was first synthesized in 1948 [11]. It is used to treat patients with malaria caused by

P. falciparum, inclusive of those suffering from chloroquine resistance [12]. The mode of action of this drug is still not understood, but some findings suggest that the polymerization of haem gets inhibited resulting in free haem accumulation in the body which acts as a toxic matter for parasites [13]. Since amodiaquine is found to be linked with acute hepatitis it is only allowed to be used in the treatment of malaria, not in its prophylaxis.

10.2.5 Lumefantrine

IUPAC NAME: 2-(dibutylamino)-1-[(9*Z*)-2,7-dichloro-9-[(4-chlorophenyl) methylidene] fluoren-4-yl] ethanol.

Lumefantrine (Figure 10.1e) also known as benflumetol, was synthesized in 1976. Lumefantrine in combination with artemether is used in the treatment of acute uncomplicated malaria which results from *P. falciparum* accumulation in the body. It is also effective for chloroquine-resistant strains. This drug is effective in the erythrocytic stage as it is a type of schizonticidal drug. The exact working mechanism is unknown but some studies suggest that it inhibits the synthesis of protein and nucleic acid by the formation of beta-hematin [13].

10.3 RTS, S/AS01

After seven decades, WHO finally approved the first-ever malaria vaccine on October 6, 2021. RTS, S/AS01 commonly known as mosquirix is being developed by GlaxoSmithKline (GSK) Biologicals [14]. The mosquirix is delivered intramuscularly via lyophilized injection. Four doses of the vaccine are being administered to children in-between the ages of five months to 15–18 months [15]. The efficacy of the vaccine in phase three clinical trials is as follows: children of age between 5 and 17 months showed 36.3% in case of clinical malaria, 32.2% in case of severe malaria, and 40% in malaria hospitalization, and infants of 6–12 weeks showed 25.9% in case of severe malaria, 17.3% in case of severe malaria and 18% in malaria hospitalization [16].

Case Study

More or less every country in the world is affected by malaria. But we cannot ignore the fact that malaria is often found to be more widespread in developing countries while it has a lesser impact on developed nations.

In the case of Bangladesh, around 152,700 malaria cases were reported in 1995 [17] which increased to 216 million in 2010 causing the death of around 655,000 people [18]. The frequency of this disease in Bangladesh is at a moderate level but is maintained by to the periodic migration of people from countries where it is an epidemic. Chloroquine is one of the primary drugs used for treatment, but will soon be replaced due to chloroquine resistance in the species. In developing countries like Bangladesh, it has been observed that malnutrition causes the failure to thrive, which is responsible for the illness related to malaria [19]. The geographical distribution, the housing arrangement, people's behavior, and living habits also play an important role in malarial transmission [20, 21]. To understand the dynamics of transmission statistical modeling has been developed. The Susceptible-Exposed-Infectious-Removed (SEIR) model was developed to calculate the number of infected people at a certain time. This can also be utilized to obtain the number of required health workers to tackle malaria. The data gathered by using the SEIR model helped draft an effective policy to regulate the health of individuals as defined by various international bodies. This also arranges in order of importance the areas where the awareness program is required [22]. The National Malaria Control Program was launched in 2014 to roll out new strategies to stop the exponential transmission of malaria by taking into mind risk factors. Of the total of 64 districts of Bangladesh, 13 are hotspots for malaria. The common thing in these districts is that 12 districts are on the international border with India and Myanmar. These are also heavily densified districts as out of a total of 13.25 million people 9.95 million are the resident of these districts only [23].

10.4 Lymphatic Filariasis

Globally, lymphatic filariasis is caused by a microscopic worm with a thread-like appearance. When the worm reaches its adult form it travels to the lymphatic system. This parasite causes lymphedema, hydrocele (only in men), and elephantiasis, and may also lead to permanent disability in patients. Lymphatic filariasis is a disease that affects roughly 81 nations in North and South America, Africa, and the Indo-Pacific region. Approximately 1.3 billion people are infected by the nematode parasite, which is mosquito-borne thus leading to an increase in the infectivity rate [24]. It has been estimated that around 120 million people host this parasite, out of which *Wuchereria bancrofti* is leading with 91% infection while *Brugia timori* and *Brugia malayai* collectively have 9%. The vector of these parasites are mosquitoes (*Aedes*, *Anopheles*, and *Mansonia*).

10.4.1 Albendazole

IUPAC NAME: methyl *N*-(6-propylsulfanyl-1*H*-benzimidazol-2-yl) carbamate.

Albendazole ($C_{12}H_{15}N_3O_2S$) is a synthetically made parasiticidal which is a derivative of benzimidazole. Tubulin polymerization gets disrupted leading to imperfect microtubules by the action of albendazole (Figure 10.1f). These imperfect microtubules slow down the cell division and embryogenesis of the parasite [25].

10.4.2 Diethylcarbamazine

IUPAC NAME: N, N-diethyl-4-methyl piperazine-1-carboxamide

Diethylcarbamazine ($C_{10}H_{21}N_3O$) is a type of anthelmintic drug used for the treatment of infections caused by *Wuchereria brancrofti* [26]. Diethylcarbamazine (Figure 10.1g) is an organic compound synthesized synthetically and is specific for numerous parasites present in nature. Studies have also revealed that the diethylcarbamazine mechanism of action against *Brugia malayi* involves the stimulation of the cyclooxygenase pathway and nitric oxide synthase for phagocytosis. The arachidonic acid metabolic pathway and 5-lipoxygenase pathway are also involved in the working mechanism of this drug. The half-life of this drug is approximately eight hours [26]. Fever, skin rash, itching in the eye, and swelling in the face are some of the commonly occurring side effects of using this drug.

10.4.3 Ivermectin

IUPACNAME:-(1R,4S,5'S,6R,6′R,8R,10E,12S,13S,14E,16E,20R,21R,24S)-6′-[(2S)-butan-2-yl]-21,24-dihydroxy-12-[(2R,4S,5S,6S)-5-[(2S,4S,5S,6S)-5-hydroxy-4-methoxy-6-methyloxan-2-yl]oxy-4-methoxy-6-methyloxan-2-yl]oxy-5′,11,13,22-tetramethylspiro[3,7,19-trioxatetracyclo[15.6.1.14,8.020,24]pentacosa-10,14,16,22-tetraene-6,2′-oxane]-2-one.

Ivermectin ($C_{47}H_{72}O_{14}$) is a medicine that has a semi-synthetic origin having pentacyclic sixteen-membered lactones and is used as an anti-parasitic drug. Ivermectin (Figure 10.1h) is an effective anti-parasitic drug isolated from the soil bacterium, *Streptomyces avermitilis*. It is a product obtained after mixing two avermectins [27]. Apart from its use in the treatment of worms, it is also used in treating diseases like onchocerciasis and infections of head lice. Systemic ivermectin therapy is widely used at the international level for the treatment of numerous tropical diseases like *Loa loa* infection, and migraine caused by cutaneous larvae.

In invertebrate, the binding of glutamate-gated chloride ion channel with ivermectin occurs in nerve and muscle cells that are of very high affinity and selective in nature. This kind of binding induces hyperpolarization in the targeted cell through increased permeability of the cell membrane to chloride ions, which paralyzes the parasite leading it to its death [28]. Moreover, the interaction of ivermectin with various types of ligand-gated chloride channels is gated by gamma-aminobutyric acid. Some studies also showed that the intrauterine development of microfilariae is inhibited by ivermectin action. Nausea, dizziness, enema, asthenia, vomiting, headache, and diarrhea are some of the common side effects of this drug.

10.4.4 Triple Drug

Based on WHO recommendations, ivermectin is administered in different combinations; as DA (combination of diethylcarbamazine with albendazole) and IA (combination of albendazole with ivermectin). These three drugs were once used under the name, IDA. In different clinical trials, the drug IDA shows its effectiveness in the elimination of *Wuchereria brancrofti* microfilariae [29].

Case Study

The global programme to eliminate lymphatic filariasis (GPELF) was launched in 2000. Lymphatic filariasis is endemic to about 75 countries. Mass drug administration (MDA) is conducted in these countries to eliminate the disease [30]. The GPELF programme is still effectively operational in many eastern and western areas of Africa. Some African countries are also administrating the approaches of target eradication [31]. The programme is still in its early stage or is not showing as satisfactory a result as shown in another region because in Central Africa there is a complex distribution of disease due to geographical situations. The existence of liaisons in Central Africa affects the employment of MDA with albendazole and ivermectin. The weak financial background of countries in Central Africa is also a reason for not employing programmes on a large scale. The survey was done by those who disclosed a high infection rate. "Death to Onchocerciasis and Lymphatic Filariasis" (DOLF) was launched having the aim to improvise the already going drug delivery programme MDA. This gives assess to many patients of lymphatic filariasis to proper medication of ivermectin and albendazole either solely or in combination as required. Results of studies carried out in 40 villages in the Republic of Congo is tabulated below (Tables 10.1–10.3) [30].

Table 10.1 Parasitological results of the patients suffering with lymphatic filariasis are presented.

				W. bancrofti microfilariae count	
Age (years)	**No. of Patients**	**ICT-Positive (%)**	**Mf Positive (%)**	**Geometric Mean**	**Median**
5–10	203	1.5	0.5	111.5	111.5
11–15	110	12.7	8.2	24.8	22.0
16–20	57	35.1	10.5	3.8	3.0
21–30	113	25.7	5.3	13.5	16.3
31–40	107	25.2	5.6	6.7	9.8
41–50	96	25.0	7.3	25.3	23.0
>50	88	19.3	6.8	16.4	18.0
Total	774	17.1	5.3	14.0	15

Source: adapted from [30].

Table 10.2 *Wuchereria bancrofti* prevalence and intensity of infection in males.

FOR MALES					
				W. bancrofti microfilariae count	
Age (years)	No. of Patients	ICT-Positive (%)	Mf Positive (%)	Geometric mean	Median
5–10	94	2.1	1.1	111.5	111.5
11–15	52	15.4	11.5	20.	17
16–20	32	43.8	15.6	5.0	4.5
21–30	45	33.3	8.0	19.6	16.3
31–40	54	33.3	11.1	6.7	9.8
41–50	43	32.6	11.6	29.4	28.5
>50	34	29.4	8.8	8.4	11.5
Total	35	22.9	8.5	13.2	13.5

Source: adapted from [30].

Table 10.3 *Wuchereria bancrofti* prevalence and intensity of infection in females.

FOR FEMALES					
				W. bancrofti microfilariae count	
Age	No. of Patients	ICT-Positive (%)	Mf Positive (%)	Geometric Mean	Median
5–10	109	0.9	0	0	0
11–15	58	10.3	5.2	37.9	33.0
16–20	25	24	4.0	1.0	1.0
21–30	68	20.6	2.9	6.5	21.5
31–40	53	17.0	0	0	0
41–50	53	18.9	3.8	17.3	18.0
>50	54	13.0	5.6	31.9	24.5
Total	420	12.6	2.6	16.3	23.0

Source: adapted from [30].

10.5 Leishmaniasis

Leishmania is an intracellular parasite that invades phagocytic host cells [32]. The phlebotomine sand fly is the carrier of this parasite, and it gets transmitted either zoonotically or anthropologically. There are around 30 vector species that transmit this parasite [33]. According to the report published by WHO this disease is endemic in around 98 countries where around one billion people are vulnerable to this parasite. This disease is classified into three forms: visceral leishmaniasis (VL) commonly called kala-azar, is the most severe form; cutaneous leishmaniasis (CL) is most common, and mucocutaneous leishmaniasis (MCL) is a form that causes the greatest number of disabilities in patients [34]. *Leishmania tropica* is a major cause of CL in Central Asia and Middle East countries while *Leishmania mexicana* and *Leishmania braziliensis* are the main cause in America. *Leishmania chagasi, Leishmania infantum, Leishmania donovani* are the causative agent of VL [35]. Suppressed immunity, hepatosplenomegaly, anaemia and fever are the clinical manifestation of leishmania. Without any treatment, most patients who suffer from VL die. Few drugs are available and due to high costs many are unable to afford them. The side-effect and high toxicity of these drugs must also be considered a factor of their limited use. Some studies suggest that the reoccurrence of infection also occurs due to the development of drug resistance [36]. In some studies, the very prolonged treatment, which is required for complete eradication of infection and the infection-causing parasite from the body, is considered one of the reasons the parasite develops resistance to drugs that are being used [35, 36]. A good and effective vaccine is therefore required for protection from this infection.

10.5.1 Meglumine Antimoniate

IUPAC NAME: Hydroxy-dioxostiborane; (2*R*,3*R*,4*R*,5*S*)-6-methyl amino hexane -1,2,3,4,5 pentol

Meglumine antimonate ($C_7H_{18}NO_8Sb$) drug is exclusively used for the treatment of leishmaniasis [37]. This drug is applied on VL and CL (Figure 10.1i). It is an intramuscular drug generally administered at the site of infection [37]. The mechanism of action is not known. The common side effects are nausea, cough, appetite, abdominal pain tiredness, irregular heartbeat, muscular pain, and kidney, and liver damage [37].

10.5.2 Sodium Stibogluconate

IUPAC NAME: trisodium (3R,4S,5R)-1-{[(3R,4S,5R)-3-carboxylato-5-[(1R)-1,2-dihydroxyethyl]-1-oxido-2,6,7-trioxa-1-stibabicyclo[2.2.1]heptan-1-yl]oxy}-5-[(1R)-1,2-dihydroxyethyl]-1-hydroxy-2,6,7-trioxa-1-stibabicyclo heptane-3-carboxylate nonahydrate.

Sodium stibogluconate ($C_{12}H_{35}Na_3O_{26}Sb_2$) is a derivative of dicarboxylic acid. It is from the class of medicine called pentavalent antimonial. It has a molecular framework of aliphatic hetero-polycyclic compounds. Sodium stibogluconate (Figure 10.1j) is available for administration in the form of an injection. It is used in the treatment of leishmaniasis with the trade name pentostam. Some studies suggest it is used as an anti-tumor agent. MCL and VL are the forms of leishmaniasis treated by this drug. Several studies ongoing to establish the exact mode of action of this drug. Available data indicates a remarkable decrease of about 50% in parasitic DNA and RNA . Some theories suggest that decreasing the level of ATP and GTP will affect the synthesis of a certain macromolecule that is required for parasitic metabolic activities [38]. The only known mechanism of action of this drug is that it obstructs the DNA topoisomerase I which causes downregulation of DNA replication and DNA transcription. Major side effects include severe diarrhea, nausea, and vomiting. In some cases, hepatitis and nephritis have also been noted.

10.5.3 Miltefosine

IUPAC NAME: hexadecyl 2-(trimethylazaniumyl) ethyl phosphate

Miltefosine ($C_{21}H_{46}NO_4P$) is a drug with very diversified antagonistic action. It has anti-microbial and anti-cancerous activities. The route of administration of this drug is oral. Miltefosine (Figure 10.1k) is the only drug that is used to treat all three forms of leishmaniasis i.e., CL, VL, MCL [39]. Normally this drug is prescribed to patients who are above 12 years old. Among all species of leishmania, *Leishmania donovani* is the most sensitive to this drug. The drug affects the signaling pathway of the cell, which is dependent on lipids. Moreover, its role in apoptosis has also been identified. Within the mitochondria, the drug hinders the action of cytochrome-c oxidase which eventually leads to the dysfunction of the mitochondria. In the gastrointestinal tract, the drug gets absorbed and distributed to the kidney, spleen intestinal mucosa, and liver [40].

The target of miltefosine is P-glycoprotein 1. The drug is broken down into phospholipase D in order to release choline as a metabolite. All the metabolites which are produced are endogenous and are used in the production of the cell membrane, acetylcholine, and long-chain fatty acids. The complete elimination of this drug takes a very long time as it has a very slow process and prolonged elimination half-life [41].

10.5.4 Amphotericin B

IUPAC NAME: (1R,3S,5R,6R,9R,11R,15S,16R,17R,18S,19E,21E,23E,25E,27E,29E,31E,33R,35S,36R,37S)-33-[(2R,3S,4S,5S,6R)-4-amino-3,5-dihydroxy-6-methyloxan-2-yl] oxy-1,3,5,6,9,11,17,37-octahydroxy-15,16,18-trimethyl-13-oxo-14,39-dioxabicyclo [33.3.1] nonatriaconta-19,21,23,25,27,29,31-heptaene-36-carboxylic acid.

Amphotericin B ($C_{47}H_{73}NO_{17}$) is a drug that has diverse uses. Apart from its use in the treatment of leishmaniasis, it can also be used in the treatment of fungal infections, cryptococcal meningitis, etc. It has a very high antagonistic activity against a number of fungal species. Amphotericin B (Figure 10.1l) is ineffective on bacteria and viruses. The drug is administered in the form of intravenous injection. It has very high protein binding activity of more than 90% with plasma protein. Its metabolism is exclusively renal. The eliminatory half-life is of around 15 days. This drug binds to ergosterol, which causes damage to membrane integrity leading to the death of the cell [42].

Case Study

Patient 1

A man of age 77 years who lived in New York for around 50 years and not travelled to any country which is enlisted as an endemic region. He had never been treated for leishmaniasis nor any kind of cutaneous lesions before [43]. In 2014, after otolaryngology evaluation, a large lobule causing swelling in the nasal area obstructing the passage of air was found which had originated from the septum. The result of the CT scan confirms the formation of oedema and bowing of the nasal septum. Upon the PCR, the presence of *Leishmania tropica* was confirmed. Then he was transferred to the Weill Cornel Medical College for treatment. Although he was tested negative for HIV and was not immunocompromized, nor had a cutaneous scar, his age and cardiac disease along with chronic renal problems became a cause for concern because treatment with liposomal amphotericin B and sodium stibogluconate may cause an adverse reaction [43]. So, to avoid any risky situation, he was given miltefosine following the dose already stated and approved by FDA. After two weeks of treatment, the symptoms were gradually disappearing which was a good sign for continuing the treatment. In about two months, the symptoms disappeared. When the otolaryngologic examination was done in week nine the result was very satisfying as it showed no signs of the infection. The patient was followed up for six months till he was asymptomatic [43].

Patient 2

This was the case of a 34 years old woman. She had a suppressed immunity and was already dealing with weight loss and malabsorption. She had a complicated set of diseases like antiphospholipid syndrome, systemic lupus erythematosus (SLE), lupus nephritis, and myocardial pericarditis which reoccurred in nature. For this reason, she was treated with various kinds of immunosuppressive techniques and drugs. The presence of atrophy in duodenal mucosa was identified as a result of esophagogastroduodenoscopy. For disease staging and characterization of disease, capsule endoscopy of the small intestine was done. The results of this endoscopy revealed that there were sparse hyperaemic foci, enlarged and flattened oedematous villi in the intestine specifically in the ileum. One after another continuous histopathological examination of the patient revealed that she had the *Leishmania* protozoan, in the amastigotes stage of its life cycle. The analysis of the parasite revealed that it was visceral leishmaniasis. VL is a parasite popularly known for its life-threatening activities. Some symptoms that arise due to VL are fever, pancytopenia, hepatomegaly, and splenomegaly. The patient had resistance to Amphotericin B so pentamidine was administered to her in form of an intravenous injection.

10.6 Schistosomiasis

Schistosomiasis is both an acute and chronic type of parasitic disease, which is caused by blood flukes. The worm falls under genus *Schistosoma* and is the causative agent. According to the data provided by WHO in 2019, around 236.6 million people were subjected to preventive treatment. The active case data of schistosomiasis can be drawn out from 78 countries, whereas it is endemic in 51 countries. The cases are higher in the tropical or subtropical regions, where people don't have access to safe and contamination-free drinking water [44].

There are mainly two types of schistosomiasis namely intestinal schistosomiasis and urogenital schistosomiasis. Species like *Schistosoma mansoni, Schistosoma mekongi, Schistosoma japonicum*, and *Schistosoma guineensis* are responsible for intestinal schistosomiasis in the Middle East, Africa, Latin America, China, Philippines, Indonesia, etc. while *Schistosoma haematobium* is responsible for causing urogenital schistosomiasis in Africa and the Middle East. In women, urogenital schistosomiasis is considered a potential risk factor for HIV infection. Freshwater snails are the vector for this parasite. Intestinal schistosomiasis shows symptoms like diarrhea, abdominal pain, and haematochezia. Hypertension in the blood vessels that lie in the abdomen, and enlargement of the spleen and liver may occur in advance cases. While in the case of urogenital schistosomiasis, symptoms are haematuria, and bladder fibrosis; damage in the kidney may occur in the advanced stage and it may also cause infertility.

10.6.1 Praziquantel

IUPAC NAME: 2-(cyclo hexane carbonyl)-3,6,7,11b-tetrahydro-1H-pyrazino[2,1-a] isoquinolin-4-one.

Praziquantel ($C_{19}H_{24}N_2O_2$) is an anti-parasitic drug that is used in the treatment of parasitic worms which cause schistosomiasis. Praziquantel (Figure 10.1m) is effective against all species of parasites which cause schistosomiasis. This drug destroys the schistosomes by regulating cell membrane permeability and disintegrating the skin of the schistosomes [45]. The effect of this drug is remarkable in adult worms. Some of the studies show that the invasion calcium plays an important regulatory role here. Some secondary effects observed are increase in the release of lactate and a decrease in glucose uptake and glucagon level. This drug paralyzes its target by increasing calcium ion uptake inside the schistosomes. Some morphological changes can also be seen on the drug's target, which occur as the parasite's outer surface becomes more exposed to schistosome antigens. The complete mechanism of action of this drug is a matter of ongoing research. The drawback of this drug is that it is less effective on young parasites.

Case Study

In the village of Madagascar, there was the case of a 41-year-old woman. She became pregnant five times but, unfortunately, she is a mother of only four children as she had one miscarriage. She is an 18-year-old farmer who became sexually active at the age of 18. She gave birth to her first child at 20. She is against the use of contraceptives. Her HIV status was unknown. In 2015, there was an ongoing campaign in which she was first diagnosed with cervical cancer. The diagnosis was done by the detection of the human papilloma virus (HPV) as instructed by the WHO [46]. She got a positive HPV report, then she was called to France at the medical centre at Chirurgical for the visual inspection of Lugol's iodine (VIIL), visual inspection acetic acid (VIA), and colposcopy [46–48]. The digital image of her cervix was captured during the gynaecologic examination. Lesions in the ectocervical were not disclosed by visual inspection. Even the histologic assessment also did not reveal any lesions. Based on the test results in 2016, she was declared HIV positive. When the biopsy was done, the presence of *Schistosoma haematobium* eggs was found, but she was asymptomatic. She was administered and treated with praziquantel.

10.7 Babesiosis

The microscopic protozoan parasite belonging to the genus *Babesia* is a major cause of human babesiosis. It is an intraerythrocytic parasite that is carried by ticks. The infection occurs in red blood corpuscles (RBC) [49]. Hundreds of *Babesia* species have been identified, but only a few have the potential to cause an infection in humans. *Babesia duncani, Babesia microti, Babesia venatorum, Babesia crassa*, and *Babesia divergens* are some species that are related to human babesiosis [50]. In United States of America (USA), *Babesia microti* is a causative parasite, while in Europe, *Babesia divergens* is the main species responsible for infection. In China, *Babesia venatorum* and *Babesia microti*, both are responsible for the widespread infection. Some studies on the distribution pattern suggest that *Babesia venatorum* and *Babesia microti* are dominant when compared to *Babesia divergens* [51]. Since 2011, human babesiosis has been considered an alarming disease in the USA. Most infections caused by these parasites are placed under the category of asymptomatic diseases. It may act as a trigger for several other conditions which may be considered lethal to people with suppressed immunity or one who is splenic. Elderly people also share the same amount of risk [52]. The incubation period lies between one to nine weeks. The clinical feature is somewhat similar to malaria displaying several signs and symptoms like malaise, fever, sweats, headache, sore throat, dark urine, splenomegaly, jaundice, hepatomegaly, etc. [53]. Usually the parasite gets transmitted by a tick, but sometimes the transfusion of blood may also be a cause of transmission. Transmission rarely occurs by organ transplantation or by perinatal transmission [54–56]. Neither vaccine nor prophylaxis is available for babesiosis [57].

10.8 Treatment

The presence of the *Babesia divergens* is interlinked with immunosuppressed patients, hence its treatment is challenging. The drug called diminazene aceturate ($C_{22}H_{29}N_9O_6$) was observed to be effective in the case of animal babesiosis, but does not provide any promising data in human trials where it was found to be fatal in a few cases. When administered to patients with *Babesia microti* infection, it caused damage to the neurological system [58]. Trimethoprim-sulfamethoxazole ($C_{24}H_{29}N_7O_6S$) in corporation with pentamidine ($C_{19}H_{24}N_4O_2$) provide satisfactory results in human babesiosis which is caused by *Babesia divergens* [59]. The use of clindamycin ($C_{18}H_{33}ClN_2O_5S$) along with quinine provide satisfactory results in various cases of *Babesia divergens*, without any side effects [60]. Studies suggest atovaquone ($C_{22}H_{19}ClO_3$) as an effective drug in the treatment of animal babesiosis for both causative agents, *Babesia divergens* and *Babesia bovis*. A trial was done to evaluate its effectiveness in human babesiosis which also provided satisfying results [60]. Pentamidine isethionate ($C_{19}H_{24}N_4O_2.2C_2H_6O_4S$) has very low success in the treatment of *Babesia microti*, but the overall treatment of *Babesia microti* is very promising. Azithromycin ($C_{38}H_{76}N_2O_{14}$) and Atovaquone ($C_{22}H_{19}ClO_3$) both can be used to treat *Babesia microti*. Figure 10.2 shows the chemical structure of all the drugs used in treating babesia species.

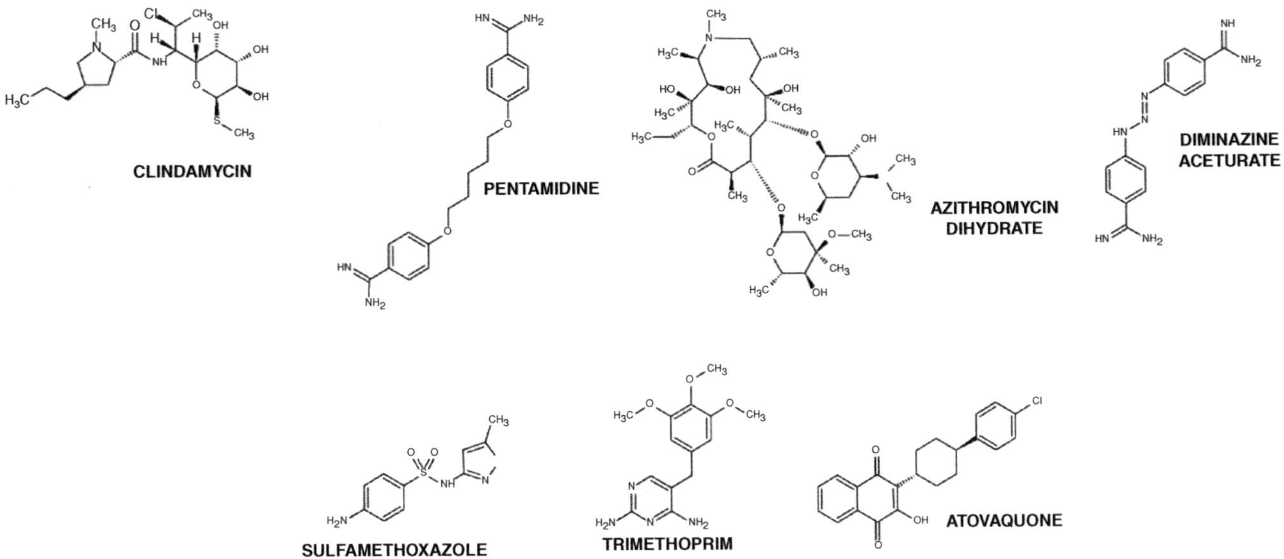

Figure 10.2 Chemical structure of drugs treating babesiosis.

Case Study

In the Wenzhou Medical University of China, an interesting case was identified in September 2019 of a man diagnosed with unusual symptoms including loss of appetite, jaundice, severe asthenia, and greenish-brown urine [61]. His medical history suggested that a few months before he had repeated blood transfusions, vacuum sealing drainage, plasma infusion, and debridement, during the treatment of a wound in his upper limb [50]. He was first treated with cefuroxime sodium, imipenem, and sulperazon. The drugs displayed no promising effect. His body still maintained a high temperature and the jaundice intensified as the days went by. Moreover, the patient neither showed any traveling history to a babesiosis endemic area nor was he having a wound from a tick bite.

At the time of hospital admission, his heartbeat was 144/minute but surprisingly his rate of respiration and blood pressure was normal and his body temperature was measured to be 101.66°F. He was at a normal consciousness level. The report of his blood examination depicts that several biochemical indicators were on the high side. Direct bilirubin (371 μmol/L), aspartate aminotransferase (89 U/L), total bilirubin (407 μmol/L), and lactate dehydrogenase (2421 U/L), but the level of haptoglobin was on the lower side. The routine blood examination illustrated a decrease in erythrocyte level (2.00 × 1012/L) along with haemoglobin (57 g/L) but the level of platelet and leukocyte was normal by 125 × 10^9/L and 4.91 × 10^9/L respectively [61]. Other reports confirmed that he had a raised level of C-reactive protein and procalcitonin. He had a negative blood-based bacterial culture report. The ultrasound report showed hepatosplenomegaly. Later on, when the peripheral blood smears and bone marrow aspiration were done, the parasite was found to be present. The parasite was isolated from the cells and sent to Beijing Genomics Institute for genomic analysis. The result from that lab suggested that the parasite was *Babesia microti*. After the validation of the parasite, the drug used for the treatment of babesiosis was administered. The combination of the drug atovaquone and azithromycin gave promising results [62]. After three weeks of treatment, the patient showed increased haemoglobin levels and other associated factors.

10.9 Conclusion and Future Perspectives

In this chapter, we discussed some of the parasitic diseases which affect a large group of people. In most of the cases, patients are unknowingly affected by endoparasites which develop into different diseases [63]. Here our main focus was on parasitic diseases like malaria, leishmaniasis, lymphatic filariasis, schistosomiasis, and babesiosis. We elaborated the mechanism of action of various kinds of drugs used in the treatment of parasitic diseases. Malaria can be treated by administrating drugs such as tafenoquine, amodiaquine, mefloquine, and artesunate. Recently, researchers have developed a vaccine RTS, S/AS01 for the treatment of malaria. Lymphatic filariasis are treated with albendazole, diethylcarbamazine, and ivermectin. Similarly, drugs like meglumine antimoniate, sodium stibogluconate, miltefosine, and amphotericin B are used for the treatment of leishmaniasis. Babesiosis is treated with multiple drug combinations like diminazene aceturate, trimethoprim-sulfamethoxazole, clindamycin, atovaquone, pentamidine isethionate, and azithromycin, while schistosomiasis is treated with praziquantel.

Moreover, we examined some case studies of each reported disease with the aim of understanding the mode of action, symptoms of disease, and other physiochemical changes occurring in the body during diagnosis [64]. To conclude, a remarkable improvement is observed in patients upon the administration of these drugs. Although, these parasitic diseases are a serious concern, by applying some precautions, like having a hygienic lifestyle, is an effective way of preventing the entry of these parasites into the body.

References

1 Serres, M. (2013). *The Parasite*. University of Minnesota Press.

2 Cox, F.E. (2010). History of the discovery of the malaria parasites and their vectors. *Parasites & Vectors* 3 (1): 1–9.

3 Guerra, C.A., Gikandi, P.W., Tatem, A.J. et al. (2008). The limits and intensity of *Plasmodium falciparum* transmission: implications for malaria control and elimination worldwide. *PLoS Medicine* 5 (2): e38.

4 White, N.J., Pukrittayakamee, S., Hien, T.T. et al. (2014). Erratum: malaria. *The Lancet* 383: 723–735.

5 Theel, E.S. and Pritt, B.S. (2016). Parasites. *Microbiology Spectrum* 4 (4).

6 Lu, K.Y. and Derbyshire, E.R. (2020). Tafenoquine: a step toward malaria elimination: published as part of the biochemistry series "biochemistry to bedside". *Biochemistry* 59 (8): 911–920.

7 White, M.T., Walker, P., Karl, S. et al. (2018). Mathematical modelling of the impact of expanding levels of malaria control interventions on *Plasmodium vivax*. *Nature Communications* 9 (1).

8 Llanos-Cuentas, A., Lacerda, M.V., Hien, T.T. et al. (2019). Tafenoquine versus primaquine to prevent relapse of *Plasmodium vivax* malaria. *New England Journal of Medicine* 380 (3): 229–241.

9 Ashley, E.A. and Phyo, A.P. (2018). Drugs in development for malaria. *Drugs* 78 (9): 861–879.

10 Lisewski, A.M., Quiros, J.P., Ng, C.L. et al. (2014). Supergenomic network compression and the discovery of EXP1 as a glutathione transferase inhibited by artesunate. *Cell* 158 (4): 916–928.

11 Berliner, R.W., Earle, D.P., Taggart, J.V. et al. (1948). Studies on the chemotherapy of the human malarias. VI. The physiological disposition, antimalarial activity, and toxicity of several derivatives of 4-aminoquinoline. *The Journal of Clinical Investigation* 27 (3): 98–107.

12 Bompart, F., Kiechel, J.R., Sebbag, R., and Pecoul, B. (2011). Innovative public-private partnerships to maximize the delivery of anti-malarial medicines: lessons learned from the ASAQ Winthrop experience. *Malaria Journal* 10 (1): 1–9.

13 Combrinck, J.M., Mabotha, T.E., Ncokazi, K.K. et al. (2013). Insights into the role of heme in the mechanism of action of antimalarials. *ACS Chemical Biology* 8 (1): 133–137.

14 Sahoo, J.P., Mishra, P., Sahoo, S., and Samal, K.C. (2021). Historic moment: WHO approves the world's first malaria vaccine "mosquirix or RTS, S/AS01". *Biotica Research Today* 3 (10): 892–896.

15 Laurens, M.B.R.T.S. (2020). S/AS01 vaccine (Mosquirix™): an overview. *Human Vaccines & Immunotherapeutics* 16 (3): 480–489.

16 Vekemans, J., Schellenberg, D., Benns, S. et al. (2021). Meeting report: WHO consultation on malaria vaccine development. *Geneva, Vaccine* 39 (22): 2907–2916.

17 Nahar, N., Afroza, S., and Hossain, M. (1998). Incidence of low birth weight in three selected communities of Bangladesh. *Bangladesh Medical Research Council Bulletin* 24 (2): 49–54.

18 Basu, S. and Sahi, P.K. (2017). Malaria: an update. *The Indian Journal of Pediatrics* 84 (7): 521–528.

19 Rahman, A., Chowdhury, S., Karim, A., and Ahmed, S. (2008). Factors associated with nutritional status of children in Bangladesh: a multivariate analysis. *Demography India* 37 (1): 95–109.

20 Binns, C. and Low, W.Y. (2015). Malaria continues as a major public health problem. *Asia Pacific Journal of Public Health* 27 (3): 261–262.

21 Wang, L., Liao, Y., Yang, L. et al. (2016). Emergency response to and preparedness for extreme weather events and environmental changes in China. *Asia Pacific Journal of Public Health* 28 (2_suppl): 59S–66S.

22 Bacaër, N., Ouifki, R., Pretorius, C. et al. (2008). Modeling the joint epidemics of TB and HIV in a South African township. *Journal of Mathematical Biology* 57 (4): 557–593.

23 Rahman, A. and Kuddus, M.A. (2020). Cost-effective modeling of the transmission dynamics of malaria: a case study in Bangladesh. *Communications in Statistics: Case Studies, Data Analysis and Applications* 6 (2): 270–286.

24 Fischer, P., Supali, T., and Maizels, R.M. (2004). Lymphatic filariasis and *Brugia timori*: prospects for elimination. *Trends in Parasitology* 20 (8): 351–355.

25 Molina, A.J., Merino, G., Prieto, J.G. et al. (2007). Absorption and metabolism of albendazole after intestinal ischemia/reperfusion. *European Journal of Pharmaceutical Sciences* 31 (1): 16–24.

26 Chen, X., Ji, Z.L., and Chen, Y.Z. (2002). TTD: therapeutic target database. *Nucleic Acids Research* 30 (1): 412–415.

27 El-Saber Batiha, G., Alqahtani, A., Ilesanmi, O.B. et al. (2020). Avermectin derivatives, pharmacokinetics, therapeutic and toxic dosages, mechanism of action, and their biological effects. *Pharmaceuticals* 13 (8): 196.

28 Alam, S., Kamal, T.B., Sarker, M.M. et al. (2021). Therapeutic effectiveness and safety of repurposing drugs for the treatment of COVID-19: position standing in 2021. *Frontiers in Pharmacology* 12.

29 Mathachan, S.R., Sardana, K., and Khurana, A. (2021). Current use of ivermectin in dermatology, tropical medicine, and COVID-19: an update on pharmacology, uses, proven and varied proposed mechanistic action. *Indian Dermatology Online Journal* 12 (4): 500.

30 Chesnais, C.B., Missamou, F., Pion, S.D. et al. (2014). A case study of risk factors for lymphatic filariasis in the Republic of Congo. *Parasites & Vectors* 7 (1): 1–2.

31 Sodahlon, Y.K., Dorkenoo, A.M., Morgah, K. et al. (2013). A success story: Togo is moving toward becoming the first sub-Saharan African nation to eliminate lymphatic filariasis through mass drug administration and countrywide morbidity alleviation. *PLoS Neglected Tropical Diseases* 7 (4): e2080.

32 Cabezas, Y., Legentil, L., Robert-Gangneux, F. et al. (2015). Leishmania cell wall as a potent target for antiparasitic drugs. A focus on the glycoconjugates. *Organic & Biomolecular Chemistry* 13 (31): 8393–8404.

33 De Freitas, E.O., Leoratti, F.M., Freire-de-lima, C.G. et al. (2016). The contribution of immune evasive mechanisms to parasite persistence in visceral leishmaniasis. *Frontiers in Immunology* 7: 153.

34 Pearson, R.D. and de Queiroz Sousa, A. (1996). Clinical spectrum of leishmaniasis. *Clinical Infectious Diseases* 22: 1–13.

35 Savoia, D. (2015). Recent updates and perspectives on leishmaniasis. *The Journal of Infection in Developing Countries* 9 (06): 588–596.

36 Freitas-Junior, L.H., Chatelain, E., Kim, H.A., and Siqueira-Neto, J.L. (2012). Visceral leishmaniasis treatment: what do we have, what do we need, and how to deliver it? *International Journal for Parasitology: Drugs and Drug Resistance* 2: 11–19.

37 Rajasekaran, R. and Chen, Y.P. (2015). Potential therapeutic targets and the role of technology in developing novel antileishmanial drugs. *Drug Discovery Today* 20 (8): 958–968.

38 Walker, J. and Saravia, N.G. (2004). Inhibition of *Leishmania donovani* promastigote DNA topoisomerase I and human monocyte DNA topoisomerases I and II by antimonial drugs and classical antitopoisomerase agents. *Journal of Parasitology* 90 (5): 1155–1162.

39 Monge-Maillo, B. and López-Vélez, R. (2015). Miltefosine for visceral and cutaneous leishmaniasis: drug characteristics and evidence-based treatment recommendations. *Clinical Infectious Diseases* 60 (9): 1398–1404.

40 Dorlo, T.P., van Thiel, P.P., Huitema, A.D. et al. (2008). Pharmacokinetics of miltefosine in old world cutaneous leishmaniasis patients. *Antimicrobial Agents and Chemotherapy* 52 (8): 2855–2860.

41 Laniado-Laborín, R., Cabrales-Vargas, M.N., and Amphotericin, B. (2009). Side effects and toxicity. *Revista Iberoamericana De Micologia* 26 (4): 223–227.

42 Murray, H.W., Eiras, D.P., Kirkman, L.A. et al. (2020). Case report: mucosal leishmaniasis in New York city. *The American Journal of Tropical Medicine and Hygiene* 102 (6): 1319.

43 Marinoni, B., Tontini, G.E., Maggioni, M. et al. (2020). Intestinal leishmaniasis: a rare case of enteropathy. *Endoscopy* 52 (09): E335–6.

44 Doenhoff, M.J., Cioli, D., and Praziquantel:, U.J. (2008). Mechanisms of action, resistance and new derivatives for schistosomiasis. *Current Opinion in Infectious Diseases* 21 (6): 659–667.

45 Randrianasolo, B.S., Jourdan, P.M., Ravoniarimbinina, P. et al. (2015). Gynecological manifestations, histopathological findings, and schistosoma-specific polymerase chain reaction results among women with *Schistosoma haematobium* infection: a cross-sectional study in Madagascar. *The Journal of Infectious Diseases* 212 (2): 275–284.

46 Poggensee, G., Sahebali, S., Van Marck, E. et al. (2001). Diagnosis of genital cervical schistosomiasis: comparison of cytological, histopathological and parasitological examination. *The American Journal of Tropical Medicine and Hygiene* 65 (3): 233–236.

47 Kjetland, E.F., Hove, R.J., Gomo, E. et al. (2009). PCR in vaginal lavage as an indicator of genital *Schistosoma haematobium* infection in rural Zimbabwean women. *American Journal of Tropical Medicine and Hygiene* 81 (6): 1050.

48 Arthropathies, C. and Rheumatology, G. (2021). ACR. What Rheumatologists Need to Know About Lyme Disease and Tick-Borne Illnesses. https://www.rheumatologyadvisor.com/home/topics/lyme-disease/acr-2021-session-rheumatologists-lyme-disease-tick-borne-illnesses.

49 Krause, P.J. (2019). Human babesiosis. *International Journal for Parasitology* 49 (2): 165–174.

50 Madison-Antenucci, S., Kramer, L.D., Gebhardt, L.L., and Kauffman, E. (2020). Emerging tick-borne diseases. *Clinical Microbiology Reviews* 33 (2): e00083–18.

51 Gray, E.B. and Herwaldt, B.L. Babesiosis surveillance-United States, 2011–2015.

52 Butler-Haughton, M. Babesiosis. Workplace Health & Safety. 2020;68(11): 545.

53 Herwaldt, B.L., Linden, J.V., Bosserman, E. et al. (2011). Transfusion-associated babesiosis in the United States: a description of cases. *Annals of Internal Medicine* 155 (8): 509–519.

54 Vannier, E. and Krause, P.J. (2012). Human babesiosis. *New England Journal of Medicine* 366 (25): 2397–2407.

55 Cornett, J.K., Malhotra, A., and Hart, D. (2012). Vertical transmission of babesiosis from a pregnant, splenectomized mother to her neonate. *Infectious Diseases in Clinical Practice* 20 (6): 408–410.

56 Parsh, B. and Whitney, K. (2021). What is babesiosis? *Nursing* 50 (5): 13–14.

57 Ruebush, T.K., 2nd, Rubin, R.H., Wolpow, E.R. et al. (1979). Neurologic complications following the treatment of human *Babesia microti* infection with diminazene aceturate. *The American Journal of Tropical Medicine and Hygiene* 28 (2): 184–189.

58 Francioli, P.B., Keithly, J.S., Jones, T.C. et al. (1981). Response of babesiosis to pentamidine therapy. *Annals of Internal Medicine* 94 (3): 326–330.

59 Smith, R.P., Hunfeld, K.P., and Krause, P.J. (2020). Management strategies for human babesiosis. *Expert Review of Anti-Infective Therapy* 18 (7): 625–636.

60 Chen, Y., Yan, D., and Zhang, Y.C. (2020). Transfusion-associated babesiosis in China: a case report. *Transfusion and Apheresis Science* 59 (6): 102902.

61 Cangelosi, J.J., Sarvat, B., Sarria, J.C. et al. (2008). Transmission of *Babesia microti* by blood transfusion in Texas. *Vox Sanguinis* 95 (4): 331–334.

62 Krause, P.J., Lepore, T., Sikand, V.K. et al. (2000). Atovaquone and azithromycin for the treatment of babesiosis. *New England Journal of Medicine* 343 (20): 1454–1458.

63 WHO Model Formulary (2008). (eds W.W., Stuart, M.C., Kouimtzi, M., and Hill, S.R.) *World Health Organization*. Geneva: Switzerland.

64 Wang, C., Han, L., Stein, G. et al. (2020). Crowdsourcing in health and medical research: a systematic review. *Infectious Diseases of Poverty* 9 (1): 1–9.

11

Opportunities and Challenges in the Development of Antiparasitic Drugs

Maryam Bello-Akinosho[1,], Kayode Olayinka Afolabi[1,2], Harish Chandra[3], Dearikha Karina Mayashinta[4], Yulia Dwi Setia[5], Abhay Prakash Mishra[6], and Carolina Pohl-Albertyn[1]*

[1] *Pathogenic Yeast Research Group, Department of Microbiology and Biochemistry, Faculty of Natural and Agricultural Sciences, University of the Free State, Bloemfontein, Republic of South Africa*
[2] *Molecular Epidemiology and Public Health Research Group, Department of Biological Sciences, Anchor University Lagos, Nigeria*
[3] *Department of Botany and Microbiology, Gurukul Kangri (Deemed to be University), Haridwar, Uttarakhand, India*
[4] *Department of Parasitology, Faculty of Medicine, Universitas Brawijaya, Malang, East Java, Indonesia*
[5] *Department of Parasitology, Department of Infectious Disease, Faculty of Medicine, University of Miyazaki, Japan*
[6] *Department of Pharmacology, Faculty of Health Sciences, University of the Free State, Bloemfontein, Republic of South Africa*
* *Corresponding author*

11.1 Introduction

The need to develop new antiparasitic agents has become more compelling than ever before due to the high burden of parasitic diseases, especially in the tropics where there is high prevalence due to poor sanitation and endemic poverty. Breaking the cycle of parasitic infections in these regions would therefore require a concerted effort and the discovery of new antiparasitic prophylaxis and treatment regimens should be of paramount priority [1]. Parasitic diseases are caused by parasites, which are a group of pathogens that often cause chronic disease in humans and animals, in contrast to bacterial and viral diseases which generally have a rapid onset. Parasitic diseases are more difficult to diagnose, thereby magnifying their economic losses due to decreased productivity. In addition, many parasitic infections are zoonotic, so there is a risk of wider spread. The life cycle of parasites is more complex than bacteria or viruses, they have several stages in their life cycle, each of which has a different sensitivity to antiparasitic drugs. The

Parasitic Infections: Immune Responses and Therapeutics, First Edition.
Edited by Abhay Prakash Mishra and Manisha Nigam.

presence of vectors that aid in the spread of the parasite, as well as the presence of a reservoir host, are also important factors connected to the reoccurrence of parasitic infection [2].

Parasitic diseases caused by helminths and protozoa have remained a burden to health systems globally, especially in tropical countries. The prevalence of these diseases is quite high in areas with inadequate access to water, sanitation, and hygiene. Although the morbidity, mortality, and economic losses caused by parasitic infections are quite significant, the existing antiparasitic drugs are still far from ideal, and have been used for several decades [3, 4]. The market for antiparasitic drugs that are considered less promising is one of the causes of the slow progress being experienced in the discovery of novel drugs for parasitic diseases. Furthermore, the peculiar situation of poverty in the endemic regions precludes any commercial drive for drug development by investors and profit-oriented pharmaceutical companies [5]. The involvement of various sectors consisting of governments in the endemic countries, researchers, the pharmaceutical industry, and non-government organizations are needed to support the discovery and development of antiparasitic drugs. In the future, the existence of new antiparasitic drugs is expected to enhance the quality of life and the welfare of the global community [3].

Despite the various shortcomings, recently, there has been a paradigm shift in the screening for new antiparasitic drugs in which the focus is now more on mechanism-based high throughput screening rather than empirical whole parasite screening. There have been many investments into molecular parasitology and understanding of the basic biology of parasites [6]. From a historical point of view, drugs for treating parasitic infections were discovered by experimental, verifiable screening using intact parasites either in vitro or in animal models. At present, cellular target-based approaches now make possible a quicker and more efficient screening of very large compound libraries for new drug identification and effective optimization [6–8]. Furthermore, research efforts in recent decades on the biology of the host–parasite interactions have revealed many new findings on the pathways of many parasites' biochemical processes. Moreover, significant developments in the field of genomics accompanied by the discovery of many molecular targets have also contributed to the discovery of new drugs [4]. Research efforts have also been supported by financial assistance from various sectors and foundations which make the genome sequencing of several parasites a reality. This chapter focuses on the prospects for the identification of novel antiparasitic compounds from natural products as well as some recent advancements in the field. Along with highlighting the numerous scientific and industry-sponsored collaboration efforts toward the discovery and development of new antiparasitic drugs, it also provides insight into the current stage of antiparasitic drug discovery and the associated challenges in terms of resistance, side effects, and efficacy.

11.2 Opportunities in the Development of Antiparasitic Drugs

11.2.1 Prospects in the Development of Novel Antiparasitic Natural Products

Humans have used natural products, including medicinal plants, to treat ailments and various health disorders for many thousands of years [9]. Interestingly, the majority of the notable drugs listed in the modern pharmacopoeia originated from nature [10]. According to Newman and Cragg [11], 28% out of 1184 registered new drugs between the years 1981 and 2006 were natural products or their potent derivatives. In addition, about 24% of the registered new therapeutic agents had functional groups with pharmacological activity derived from natural products. There is a progressive drive in exploring the wealth of resources found in nature for combating many infectious agents, including parasites, in human history. The problem of the incessant emergence of resistant parasitic strains to conventional drugs, coupled with the debilitating side effects of many synthetic drugs, has further necessitated the progressive combing of natural products for better alternatives. Traditional natural products have been tipped as a good source of synthesizing novel life-saving drugs [9, 12].

Among tropical diseases, which are the major source of morbidity and mortality globally, malaria remains the most widely distributed parasitic infection caused by protozoan *Plasmodium* spp. The economic importance of malaria cannot be overemphasized, with an estimated 228 million cases and 405,000 reported deaths globally in 2018 alone [13, 14]. Malaria has continued to exert a significant burden despite various mitigation measures employed over the years in curbing its menace. For example, the majority of the chemotherapeutic agents that are used for its treatment have become ineffective due to the development of resistance by the parasites [14]. Due to the economic importance of malaria, the infection has become a centre of attention with regard to looking for mitigating measures. Both practitioners of traditional and modern medicines have greatly explored the invaluable gifts of nature in a bid to curb its menace. Hence, malaria chemotherapy has a long history of being based on natural products.

Plant metabolites are the source of the most effective antimalarial agents. The very first and highly successful antimalarial agent was quinine, which was obtained from the bark of the Southern American tree called *Cinchona*, from which the successful synthetic antimalarial chloroquine was developed. Chloroquine has, however, become an ineffective antimalarial agent due to the development of resistance against it by the parasite. Similarly, a Chinese medicinal plant called *Artemisia annua* was traditionally used for the treatment of fever and eventually became the source of artemisinin, an important antimalarial agent [15].

In addition to the many feats such as the successful sequencing of the *P. falciparum* genome and the invention of sophisticated target identification strategies, myriads of novel antimalarial drug candidates are being discovered from natural products [16, 17]. There has been a continuous exploration of natural products as antiplasmodial agents by scientists globally [15]. Antimalarial drugs, developed from natural products, possessing novel modes of action have been on the rise in recent years. [18, 19]. A similar scenario applies to other potential antiparasitic agents derived from natural products [12, 20]. However, the increase in the number of successful novel antiparasitic agents up to the clinical level is still below expectations.

Many novel bioactive natural products from terrestrial plants, rich marine flora, and fauna as well as microorganisms, are being extensively explored as antiplasmodial [15], antitrypanosomal [21], antileishmanial [22], anticryptosporidial [23], and anthelmintic [24] agents; however, there is need for accelerated efforts in testing most of them in greater detail, mainly using animal models and subsequently, in clinical trials [9]. This will help in heralding the "New golden age of natural products drug discovery" as foreseen by Shen [16]. Table 11.1 shows some of the recently discovered potential antiparasitic drugs from various groups of natural products and their derivatives with some details on their level of development.

11.2.2 Prospects in Repositioning or Repurposing Old Drugs

Drug repositioning or repurposing is an alternative way of adopting novel applications for old therapeutic agents beyond their real or initial indication. Generally, prospecting, development, and certification of a new drug for use is a relatively long process that is expensive and time-consuming. It has been estimated that about $2 billion is needed for the development of novel drugs to the point of making them available in the market for use. Also, it has been estimated that the development takes 10–17 years with less than a 10% probability of success in coming up with a new drug compared to 3–12 years for a repositioned drug (Figure 11.1) [46, 47]. Therefore, drug repositioning has huge advantages in terms of cost-effectiveness and timeliness, especially for parasitic diseases for which therapeutic agents are needed urgently. The use of a drug repositioning strategy helps in bypassing many developmental phases required for making new drugs since the pharmacokinetic and safety profiles of the old drug to be repurposed are well known and documented [47].

Historically, drugs have been developed for other uses and then found use as antiparasitic drugs. For example, a team of researchers from the University of North Carolina, who received research funding from the Bill and Melinda Gates Foundation, were working in conjunction with Imtech International to explore antifungal drugs that could effectively serve as drugs for African trypanosomiasis.

Table 11.1 Some recent potential antiparasitic drugs from various groups of natural products.

Category	Drug/Chemical Name	Source	Activity	References
Bioactive Natural	**Products from Plants**			
Triterpenoids	α and β-amyrin	*Eugenia pyriformis* Cambess	Antileishmanial and antitrypanosomal activity against *L. amazonensis* (promastigotes- IC_{50} of 5.98 and 9.38 μg/mL) and *T. cruzi* (epimastigotes- 5.56 and 34.34 μg/mL, and trypomastigotes-LC_{50} of 16.69 and 7.80 μg/mL)	[25]
Sesquiterpene lactone	Tagitinin C	*Tithonia diversifolia* (Hemsl.)	Antitrypanosomal activity against the epimastigote forms of *T. cruzi*, with IC_{50} of 1.15 μM	[26]
Withanolides	Dichloromethane and crude extracts	Dichloromethane extract of *Aureliana fasciculata* var. *fasciculata* and crude extract of *Pipertectoniifolium*	Antitrypanosomal activity against the intracellular forms of *T. cruzi* (Tulahuen strain) with EC_{50} values of 10.7 ± 2.8 μg/mL for *A. fasciculata* var. *fasciculata* and 12.85 ± 1.52 μg/mL for *P.tectoniifolium* Activity against the bloodstream trypomastigotes with EC_{50} 2.2 ± 1.0 μg/mL for *A. fasciculata* var. *fasciculata* and 38.8 ± 2.1 μg/mL for *P.tectoniifolium.*	[27]
Withanolides	Aurelianolides A and B	*Aureliana fasciculata*	Trypanosomicidal effect against *Trypanosoma cruzi* with EC_{50} of 4.6 ± 1.3 μM for compound A and 1.6 ± 0.4 μM for compound B against the intracellular forms of the parasite.	[27]
Amide alkaloid	Piplartine	*Pipertectoniifolium*	Anti-*Trypanosoma cruzi* activity with EC_{50} of 8.1 ± 0.9 μM	[27]

(Continued)

Table 11.1 (Continued)

Category	Drug/Chemical Name	Source	Activity	References
Multiple chemical compounds	*Prosopis strombulifera* aqueous extract (PsAE)	*P. strombulifera* (Lam.) Benth	In vivo antileishmanial activity against promastigotes in a murine model of cutaneous leishmaniasis	[28]
Multiple chemical compounds	Crude extract and fractions of *S. nigrum* L. (Solanaceae) leaf	*Solanum nigrum* L. (Solanaceae)	Antiplasmodial activity against *P. berghei* in mice	[29]
Multiple chemical compounds	Ethanolic leaf and twig extract of *Faurea speciosa*	Faurea speciosaWelw. (Proteaceae)	Antiplasmodial activities in mice with parasite inhibition rates of 66.63%, 56.93%, and 71.70% in the suppressive, prophylactic, and curative tests respectively	[30]
Flavonoids	Leaf and twig extractof *Pappea capensis*	*Pappea capensis* (Eckl. and Zeyh.) leaf	In vitro antiplasmodial activity against *P. falciparum* by more than 80% at 50 μg/mL	[31]
Multiple chemical compounds	Phenolic extract of *Sida acuta* leaf	*Sida acuta*	In vivo antiplasmodial activity on *P. berghei*-infected animals with phenolic extract (600 mg/kg bw) possessing the highest plasmodial suppressing ability of over 64%	[32]
Lignan	Myrislignan	*Myristica fragrans* Houtt	Antitrypanosomal activity against *T. gondii* in vitro and in vivo. Myrislignan showed in vitro inhibition of 50% against the tachyzoite of *T. gondii* at a concentration of 32.41 μg/ml	[24]

Lignan	Licarin-B	Nutmeg	In vitro antitrypanosomal activity against the invasion and proliferation of *T. gondii* based on doses with an EC_{50} of 14.05 ± 3.96 mg/ml. It equally showed in vivo activity by protecting about 90% of mice infected with *T. gondii* from death when treated at 50 mg/kg bw	[33]
Bioactive Natural Products from Marine Sources				
Lactone	Trans-(+)-harzialactone A	Marine-derived fungus; *Paecilomyces* sp. 7A22	Antileishmanial activity against promastigotes (IC_{50} of 5.25 mg mL^{1}) and intracellular amastigotes (IC_{50} of 18.18 mg mL^{1}) forms of *Leishmania amazonensis.*	[34]
Cyclic polyketide-peptide	Janadolide	Marine coast cyanobac-terium; *Okeania* sp.	Antitrypanosomal activity against *T. brucei brucei* trypanosomes (IC_{50} of 47 nmol/L)	[35]
Lipopeptide	Ikoamide	Marine cyano-bacterium; *Okeania* sp.	Antimalarial activity against *P. falciparium* late trophozoites and schizonts (IC_{50} of 0.14 µmol/L)	[36]
Diterpene	Atomaric acid and its methyl ester derivative	*Stypopodium zonale* (brown algae)	Antileishmanial activity against *L. amazonensis* intracellular amastigotes (IC_{50} of 20.2 and 22.9 µmol/L)	[37]
Sterol	Kaimanol and Saringosterol	Marine sponge; *Xestospongia* sp.	Antimalarial activity against the blood stage of *P. falciparum* 3D7 strains (IC_{50} of 359 and 0.250 nmol/L)	[38]

(Continued)

Table 11.1 (Continued)

Category	Drug/Chemical Name	Source	Activity	References
Alkaloid	Paenidigyamycin A	Mangrove rhizospheresoil bacterium; *Paenibacillus* sp. DE2SH	Potent antiparasitic imidazole against the promastigotes of *Leishmania major* with IC_{50} of 0.75 μmol/L, and *Leishmania donovani* with IC_{50} of 7.02 μmol/L Antiplasmodial activity against the late trophozoites and schizonts of *P. falciparum* strain 3D7 with IC_{50} of 9.08 μmol/L. Antitrypanosomal activity against the intracellular amastigotes *of T. brucei brucei* with IC_{50} of 0.78 μmol/L.	[39]
Estradiol analog	Estradiol benzoate	Marine natural products	In vitro anti-trypanosomal activity against the tachyzoite of *T. gondii* with over 70% inhibition of growth and IC_{50} values of 4.41 ± 0.94.	[40]
Octyl gallate	3,4,5-trihydroxybenzoate	Marine natural products	In vitro anti-trypanosomal activity against the tachyzoite of *T. gondii* activity with over 70% inhibition of growth and IC_{50} values of 5.66 ± 0.35 μM.	[40]
Polyene macrolide antibiotic	Actinomycins, X2, and D	Mangrove sediment bacterium; *Streptomyces smyrnaeus* Strain UKAQ_23	In vitro and in-silico antileishmanial activity against promastigotes and amastigotes of *Leishmania major* (EC_{50} at 2.10 ± 0.10 μg/mL and 0.10 ± 0.0 μg/mL respectively)	[22]
Macrolide	Leiodolide A	Deep-sea sponge, *Leiodermatium* sp.	Anti-*Cryptosporidium* activity against an obligate intracellular parasite *Cryptosporidium parvum* at a half-maximal effective concentration of 103.5 nM	[23]

Macrolide	Tartrolon E (TrtE)	The symbiotic bacterium of marine mollusks (shipworm), *Teredinibacter turnerae*	Broad spectrum anti-apicomplexan activity in vitro against *Plasmodium falciparum* (EC_{50} of 0.1 nM), *Babesia bovis* (EC_{50} of 15.9 nM), *Theileria* (EC_{50} of 0.4 nM), and *Sarcocystis* (EC_{50} of 12.9 nM)	[41]
Macrolide	Tartrolon E (TrtE)	The symbiotic bacterium of marine mollusks (shipworm), *Teredinibacter turnerae* strain T7901	In vitro activity against intracellular stages of *Toxoplasma gondii* with EC_{50} of 3 nM	[41]
Macrolide	Tartrolon E (TrtE)	Symbiotic bacterium of marine mollusks (shipworm), *Teredinibacter turnerae*	In vivo anti-cryptosporidial activity against *Cryptosporidium parvum* infection in neonatal mice with EC_{50} of 3.85 nM	[41]
Bioactive Natural Products from Microorganisms				
Curvicollide family	Curvicollide D	Fungus *Amesia* sp.	In vitro trypanocidal activity against *Trypanosoma brucei* (IC_{50} value of 1.09 μM)	[42]
Cyclic Peptide	Pipecolisporin	Fungal endophyte *Nigrospora oryzae* CF-298113	In vitro activity against *T. cruzi* Tulahuen C4 (IC_{50} of 8.46 μM) and *P. falciparum* (IC_{50} of 3.21 μM)	[43]
Terpenes	Ovalicin	Symbiotic fungus *Pseudallescheria boydii* SNB-CN85 from termites.	In vitro antiplasmodial and anti-trypanosomal activities against *P. falciparum* with an IC_{50} value of 19.8 μM and *T. brucei* with an IC_{50} value of 1.1 μM, respectively	[21]

(Continued)

Table 11.1 (Continued)

Category	Drug/Chemical Name	Source	Activity	References
Secondary metabolites	10-acetyl trichoderonic acid A and 6′-acetoxy-piliformic acid [2],	*Nectria pseudotrichia*, an endophytic fungus isolated from the plant *Caesalpinia echinata*(Brazilwood)	*In vitro* leishmanicidal activities against intracellular amastigote forms of *Leishmania (Viannia) braziliensis*	[44]
Polyether antibiotic	Nigericin	*Streptomyces* sp. ICBG292 *associated* with attine ants	Anti-leishmanial activity against the promastigotes and intracellular amastigotes of *Leishmania donovani* (IC_{50} 0.28 μM and 0.13 μM respectively).	[45]

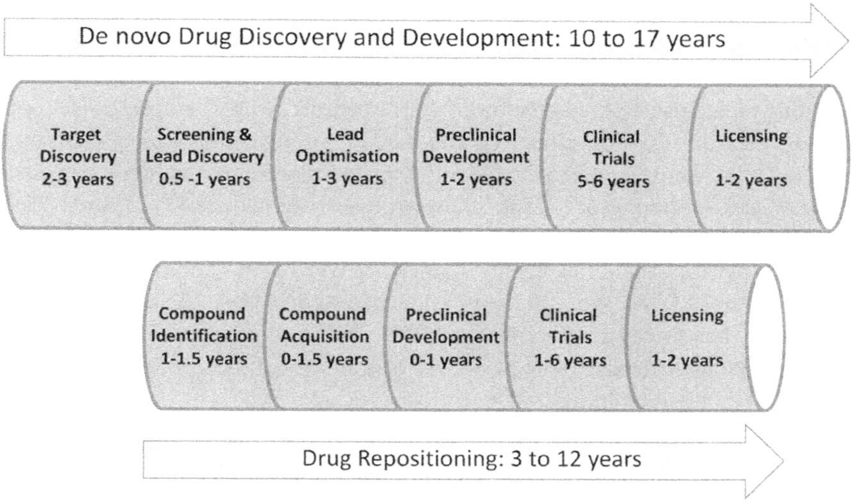

Figure 11.1 Comparative time pathways for de novo drug discovery versus repositioning. (*Source:* [47] / Cambridge University Press / CC BY).

This approach is exploiting the advantage of antifungals produced for the developed world to be used as an antiparasitic drug in the less developed world. Notably, an analog of pentamidine, an antifungal agent, has been successfully optimized for use as a trypanocidal agent, and is currently undergoing clinical trials in Africa. This is an opportunistic approach and it has been successful with clear advantages of cost reduction. This is one of the strategies that was suggested at a meeting of the American Society of Tropical Medicine and Hygiene held in 2004 [48]. Allied to the opportunistic approach is repurposing where some antiparasitic agents presently in use had their origin in veterinary applications. An example is ivermectin which was originally produced by Merck for veterinary purposes and was then spun off for therapeutic use in onchocerciasis. Drug companies are not always ready to invest in an entirely new drug pipeline from discovery to clinical trials, hence repurposing of drugs, remains a prospective route to obtaining new human antiparasitic drugs, especially considering that veterinary drug discovery is more appealing to commercial ventures [4].

Furthermore, several studies have painstakingly adopted non-biased mass screening of large libraries of compounds in a bid to search for well-characterized old or current drugs that could be repurposed for the treatment of parasitic infections, thereby circumventing numerous costly and time-consuming challenges of developing novel antiparasitic drugs. A good example of such remarkable efforts is the screening of over 1000 compounds by Dittmar and co-researchers [49] for their efficacy against *Toxoplasma gondii*. From their findings, about 10% of the compounds screened showed a significant potency as antiparasitic agents, and their already known pharmacokinetic and pharmacodynamic parameters could facilitate their accelerated alternative use. In another related study, a collection of drug-like molecules by Medicine for Malarial Venture (MMV) totalling 400 compounds, meant for the development and delivery of novel and effective antimalarial agents, were screened against the metacestodes of the fox tapeworm *Echinococcus multilocularis* using phosphoglucose isomerase (PGI) assay. Ten new compounds that showed activity against *E. multilocularis* were identified, out of which two antitheilerial agents namely MMV689480 (buparvaquone) and MMV671636 (ELQ-400) gave the most efficient outcome [50]. The drug repurposing strategy has therefore become one of the viable and easy means of obtaining novel antiparasitic drugs. Table 11.2 gives a summary of many other drugs with the potential of being repurposed as antiparasitic agents.

Table 11.2 Typical repurposed or repositioned drugs for antiparasitic use.

Drug	Initial use	Repurposed use	Level of development	References
Anticancer drugs				
Tamoxifen	Prophylactic and therapeutic use forbreast cancer	Antileishmanial	In vivo activity	[51]
Raloxifene	Prophylactic and therapeutic use for breast cancer	Antileishmanial	In vivo activity	[52]
NVP-AEW541 and GSK-J4 HCl	Anti-tumor	Antiparasitic activity against *Toxoplasma gondii*	Both in vitro and in vivo activity	[53]
Antimicrobial drugs				
i) Antifungals				
Itraconazole and Posaconazole	Antifungals	Antiparasitic activity against *L. infantum*, *L. amazonensis*, and *L. donovani*	Both in vitro and in vivo activity	[54]
Terbinafine	Antifungal	Cutaneous leishmaniasis	Phase 1 clinical trials	[55]
ii) Antiparasitics				
Buparvaquone (MMV689480)	Anti-theilerial drug	Treatment of *Echinococcus multilocularis* infection	In vitro activity	[50]

Table 11.2 (Continued)

Drug	Initial use	Repurposed use	Level of development	References
Ferroquine derivatives	Antimalarial agents	Antischistosomal activity	Both in vitro and in vivo activity	[56, 57]
Niclosamide	Anthelmintic drug	Antiparasitic activity against *Toxoplasma gondii*	In vitro and in vivo	[58]
iii) Antibacterials				
(R)-PA-824 & Delamanid	Antituberculosis drug	Antileishmanial activity against visceral leishmaniasis caused by *L. donovani*	Both in vitro and in vivo activity	[59, 60]
Fosmidomycin	Broad spectrum antibacterial drug	Antimalarial agent	Both in vitro and in vivo activity; Phase 2 clinical trial.	[61, 62]
Lipid-lowering drugs				
Rosuvastatin	Reduction of blood cholesterol levels	Chronic toxoplasmosis treatment	In vivo activity	[63]

11.2.3 Advances in Omics-based Techniques and the Discovery of Novel Antiparasitic Drug Targets

The pertinent issue of drug resistance is markedly posing a huge threat to the treatment of major human parasitic infections, hence there is a pressing need for the discovery and development of new antiparasitic agents with novel mechanisms of action (MOA). This can only be facilitated by identifying new drug targets different from the existing ones that have been impaired due to drug resistance [64, 65]. Ordinarily the discovery of a novel drug target may not be essential for drug development, however, it will help in the sorting of potential chemotherapeutic compounds to determine those with novel mechanisms for combating the menace of drug resistance. In other words, having adequate information/knowledge of a compound MOA could facilitate its progression to preclinical evaluation and give reasonable clues to improving its potency and selectivity [64].

In the pre-genomic era, when there was a paucity of information about the whole-genome sequence data of parasites and a shortage of molecular tools to explore them, target identification strategies were limited. As a result, the available methods for determining specific drug targets were based on general phenotypic experiments. As stated by Skinner-Adams and co-researchers [64], several studies on the stage-specific activity of drugs and transport/location of drugs within the cells were usually done to formulate hypotheses on the action and putative drug target, which could be regarded as successful in giving some important MOA information, but they scarcely identify drug target. However, with the advent of technological advancement in the genomic and post-genomic era, antiparasitic drug target identification strategies have been revolutionized. The availability of relevant technology has helped in modifying many of the old pre-genomic techniques for drug target identification to give whole-organism data instead of specific information about a particular pathway or target. The successful whole genome sequencing of many human pathogens, including parasitic agents, has orchestrated the discovery of many new drug targets and helps in validating them [66, 67]. Also, novel techniques are now being used to obtain lots of information about the potency of many currently used and experimental antiparasitic agents [64].

One example is in the case of a malarial parasite where the advancement in mapping the regulatory genes of the parasite has greatly enhanced the possibility of identifying major processes involved in the development of the parasite. In the same vein, the possibility of examining the host genome during infection has equally helped in discovering the consortium of new genes linked with the susceptibility of the host to severe malaria infection. Hence, the advances in genomics have been of help bidirectionally in terms of identifying potential pathways for drugs, either in the parasite or the host [67].The host-directed interventions against severe malaria disease have been regarded as unsuccessful, since no host-directed therapy against the liver stage of malaria has reached the clinical stage [68]. However, the advancement in molecular studies has given insights into host–parasite interactions, thereby necessitating the need to revisit the process to explore its inherent potential [69]. There is increasing knowledge about the network of pathways belonging to the host that are needed by the parasite for its development or need to be deregulated for its survival, which could potentially be targeted. Hence, the ever-increasing range of host factors that contribute to the development of malaria parasite and/or its pathophysiology could be explored to come up with many host-directed therapeutics for the disease. This gives the novel therapeutic approaches more clinical relevance [69].

Furthermore, in this postgenomic period, many potential enzymes and pathways that are unique to different parasitic agents could easily be identified and explored. The advancement in technological know-how could enhance targeted drug discovery through the smart use of software and databases alongside the available biological information [66]. A very good example of the prospect of hitting a novel drug target in parasitic drug development is the use of gene expression techniques for deciphering the effects that interference with lysine acetylation could have in some apicomplexan and kinetoplastid parasites. It has been asserted that the disruption of gene expression by tampering with the lysine acetylation process is detrimental to the survival of some parasites in the aforementioned groups including *Toxoplasma gondii*, *Plasmodium* spp., *Trypanosoma* spp., and *Leishmania* spp [65, 70]. It has also been established that bromodomain proteins are the key gene expression regulators of the lysine acetylation process. Hence, the discovery of a promising new class of drug target has successfully helped in giving birth to the bromodomain inhibitors as novel antiparasitic agents [65, 71–73].

11.3 Challenges of Antiparasitic Drug Discovery and Development

Different challenges are present at virtually all stages of antiparasitic drug discovery and development pipelines, from the phase of discovery and development of new molecular entities (NME) through the phases of preclinical and clinical research and development and up to the phases of Food and Drug Administration (FDA) drug review and post-market drug safety monitoring. However, as stated earlier, the path from the initial discovery of an NME to its market launch is expensive, time-consuming, and challenging. A minimum timeframe of about 10–12 years with resources amounting to over 2 billion USD is usually needed to move a new drug from pre-clinical phases to the market [5, 74]. Parasitic diseases especially do not represent a profitable market and this makes pharmaceutical industries show little interest in new antiparasitic drug discoveries. This lack of perceptible interest shown by pharmaceutical industries as a result of the non-profitability of parasitic drugs is quite a bottleneck. The challenge is even more intense for antibiotics due to their very low annually generated revenue of about 3% of the total cost involved in drug discovery and getting it to the market. Furthermore, the process is entrenched with a high degree of uncertainty associated with its success in reaching the stage of FDA approval [75]. The fact, however, remains that antiparasitic drug discovery is no more expensive or technically more demanding than it is for other indications or infections [3, 48].

Delving into history, before the early twentieth century, most drug-related scientific discoveries were made by academics. However, in the late 1930s, regulatory organisations such as the FDA were established to prevent incidence such as mass poisoning tragedy that came as a result of sulphanilamide use [76, 77]. This was achieved through the introduction and implementation of stringent preclinical and clinical testing of new molecular entities before they are released into the market. In the decades following the 1940s, the discovery of new molecular entities through research efforts by industries increased dramatically [78]. Be that as it may, from 2013, about 55% of new drugs were first discovered through the efforts of academics. Despite these efforts and continual financial investments, new molecular entities are not forthcoming. To this end, the development and implementation of effective business models for the discovery of better and safer drugs are being implemented. The impact of the existence of new public–private partnerships (PPPs), especially in the field of parasitic diseases is also noteworthy. Such PPPs are indeed addressing critical difficulties faced in antiparasitic drug discovery and development. They are established to enhance an overt and collaborative relationship between academia, benevolent organizations, and/or the manufacturing industry in a bid to provide therapeutic drugs to end users. Several PPPs currently assisting in antiparasitic drug discovery include the Drugs for Neglected Diseases initiative (DNDi), Medicines for Malaria Venture (MMV), and Institute for One World Health (IOWH). The presence of PPPs is indispensable in developing new, effective, and inexpensive drugs to treat parasitic diseases worldwide [3, 78].

Another challenge is the limited pharmacokinetics (PK) and pharmacodynamics (PD) studies as well as clinical evidence to validate preclinical research [79]. Pharmacokinetics defines the movement of a drug within and its interaction with the body as it relates to absorption, distribution, metabolism, and excretion. While pharmacodynamics relates to the body's biological response to the administered drug and about how biological processes in the body are impacted by the drug in both dose and time-dependent ways [80]. Understanding the PK and PD properties of any new drug is critical to its development and approval. A reasonable portion of information in a drug label is from PK and PD [81]. Inadequate PK and PD studies slow down the entire process of new drug discovery. Both PD and PK studies are therefore recommended in the early research phases of the new drug development process to inform a proper dose-response effect [82].

In addition, other challenges of human antiparasitic drug discovery include resistance, development of side effects, and reduced efficacy of existing drugs. Drug resistance and reduced efficacy have been a global major challenges in therapy. In antimalarial therapy, for instance, evidence abounds about a reduction in the efficacy of every new antimalarial agent within less than two decades of its introduction [83, 84]. To delay the emergence of resistance in antimalarial treatment, the World Health Organization (WHO) has given the recommendation that all new antimalarial drugs be introduced clinically and administered alongside (i.e., in combination with) another antimalarial agent that has a different MOA [84, 85], which explains why the new class of antimalarials was presented in combination with different companion drug as artemisinin-based combination therapies (ACTs). These combinations are increasingly being reported with delayed clearance of clinical parasitemias. Poor management of treatment therapies exacerbates the occurrence of resistance to treatment therapies. This could be in form of under dosage, or drug use when not required [86]. The increasing emergence of resistance to antimalarial drugs drives home the need for rethinking drug discovery methods to reduce the possibility of parasitic agents developing resistance to them and to ensure the drugs stay in use for a long time.

11.4 Inadequacy of Appropriate In Vivo Models in Novel Drug Evaluation and Certification for Human Use

The use of animal models has played a vital function in the exploration and characterisation of disease pathophysiology and target identification as well as in the in vivo evaluation of novel therapeutic agents. They are quite useful in establishing the antiprotozoal activity of novel drug candidate as well as the pharmacokinetic and pharmacodynamic properties such as absorption, distribution, metabolism, and excretion, as well as their safety profile [79]. However, human clinical test failure of new chemical entities (NCE) that previously showed promising preclinical profiling in animal models has raised questions about the predictability of these models in terms of their efficacy in humans. Although, without a doubt animal, models are required to bridge the possible gap between pre-clinical and clinical research, many challenges are usually connected with the use of animal models as well as their translational values. Challenges such as ethical questions of using animal models, suitability of the animal species when compared to human physiology, and other challenges such as flaws in experimental design, level of validation, and the robustness of models as well as the definition of the endpoint still raise questions about animal models in new drug discovery [87]. A major reservation of using animal models for drug development has to do with whether they can be used to predict the result of a drug for a targeted disease in a human subject when administered.

Animal models such as BALB/c mice, Syrian golden hamsters, dogs, monkeys, and squirrels have been in use [79]. The helminth *Caenorhabditis elegans* has been used in anthelmintic drug research for more than four decades. It has been widely characterized and it is indeed the best characterized organism since 1974 when it was introduced by Brenner as a genetic model system [8]. Although a single species as a model in such diverse phyla of parasites may not be adequate to address all issues associated with the phyla, the free-living lifestyle of *C. elegans* makes it amenable to use as a model, circumventing some limitations associated with the complex lifestyle and fragility of other host-dependent helminths. It is therefore imperative to determine the most suitable animal model that could be used in the drug discovery process for each parasite species or group of species giving utmost consideration to critical aspects such as pharmacokinetic and pharmacodynamic properties as well as drug efficacy [79, 88].

11.5 Conclusion

The current treatment against parasitic infections is suboptimal and riddled with challenges of rising drug resistance, thereby calling for more proactive measures in the development of new drugs, instead of the complacent approach currently being experienced. Since vaccine development is also not progressive, drug therapy remains the way forward. Several research efforts are upcoming, and many prospects are in sight for new antiparasitic drug development. However, there is a need for improved research efforts and expanded partnerships between the public and private entities in such a way as to ensure a more positive impact on new antiparasitic drug development. Although funding for research and development of novel antiparasitic drugs has markedly increased in the last decade, there is still a need for more input by the stakeholders to achieve the sustainable development goal of health for all by the stipulated year. These are poised to facilitate the discovery and development of new antiparasitic agents and promote global health.

References

1 Horn, D. and Duraisingh, M.T. (2014). Antiparasitic chemotherapy: from genomes to mechanisms. *Annual Review of Pharmacology and Toxicology* 54: 71–94.
2 Genta, F.A., Diaz-Albiter, H.M., Salgueiro, P., and Gomes, B. (2016). Control of vector-borne human parasitic diseases. *BioMed Research International* 2016.
3 Pink, R., Hudson, A., Mouriès, M.A., and Bendig, M. (2005). Opportunities and challenges in antiparasitic drug discovery. *Nature Reviews Drug Discovery* 4 (9): 727–740.
4 Goupil, L.S. and McKerrow, J.H. (2014). Introduction: drug discovery and development for neglected diseases. *Chemical Reviews* 114 (22): 11131–11137.
5 Boyd, N.K., Teng, C., and Frei, C.R. (2021). Brief Overview of Approaches and Challenges in New Antibiotic Development: A Focus On Drug Repurposing. *Frontiers in Cellular and Infection Microbiology* 11: 684515.
6 Woods, D.J. and Williams, T.M. (2007). The challenges of developing novel antiparasitic drugs. *Invertebrate Neuroscience* 7 (4): 245–250.
7 Campbell, W.C. (1993). Ivermectin, an antiparasitic agent. *Medicinal Research Reviews* 13 (1): 61–79.
8 Hahnel, S.R., Dilks, C.M., Heisler, I. et al. (2020). *Caenorhabditis elegans* in anthelmintic research–old model, new perspectives. *International Journal for Parasitology: Drugs and Drug Resistance* 14: 237–248.
9 Wink, M. (2012). Medicinal plants: a source of antiparasitic secondary metabolites. *Molecules* 17 (11): 12771–12791.

10 Tagboto, S. and Townson, S. (2001). Antiparasitic properties of medicinal plants and other naturally occurring products. *Advances in Parasitology* 50: 199–295.
11 Newman, D.J. and Cragg, G.M. (2007). Natural products as sources of new drugs over the last 25 years. *Journal of Natural Products* 70 (3): 461–477.
12 Prakash, B., Kujur, A., and Yadav, A. (2018). Drug synthesis from natural products: a historical overview and future perspective. In: *Synthesis of Medicinal Agents from Plants*, (eds A. Tewari and S. Tiwari), 25–46. Elsevier.
13 World Health Organization (2018). *World Malaria Report 2018*. Geneva: World Health Organization.
14 Carolino, K. and Winzeler, E.A. (2020). The antimalarial resistome–finding new drug targets and their modes of action. *Current Opinion in Microbiology* 57: 49–55.
15 Tajuddeen, N. and Van Heerden, F.R. (2019). Antiplasmodial natural products: an update. *Malaria Journal* 18 (1): 1–62.
16 Shen, B. (2015). A new golden age of natural products drug discovery. *Cell* 163 (6): 1297–1300.
17 Sisay, W., Andargie, Y., and Molla, M. (2022). Antimalarial efficacy of hydromethanolic root extract and solvent fractions of *Urtica simensis* Hochst. ex. A. Rich (Urticaceae): An Experimental Study on *Plasmodium berghei*-infected Mice, *Evidence-Based Complementary and Alternative Medicine*. doi: 10.1155/2022/6702733.
18 Draper, S.J., Sack, B.K., King, C.R. et al. (2018). Malaria vaccines: recent advances and new horizons. *Cell Host & Microbe* 24 (1): 43–56.
19 Yahiya, S., Rueda-Zubiaurre, A., Delves, M.J. et al. (2019). The antimalarial screening landscape – looking beyond the asexual blood stage. *Current Opinion in Chemical Biology* 50: 1–9.
20 Ismail, F.M., Nahar, L., Zhang, K.Y., and Sarker, S.D. (2020). Antiparasitic natural products. In: *Annual Reports in Medicinal Chemistry (eds S.D. Sarkar and L. Nahar)*, 55. 115–151. Academic Press.
21 Sorres, J., Hebra, T., Elie, N. et al. (2022). Antiparasitic ovalicin derivatives from *Pseudallescheria boydii*, a mutualistic fungus of French Guiana termites. *Molecules* 27 (4): 1182.
22 Qureshi, K.A., Al Nasr, I., Koko, W.S. et al. (2021). In vitro and in silico approaches for the antileishmanial activity evaluations of actinomycins isolated from novel *Streptomyces smyrnaeus* strain UKAQ_23. *Antibiotics* 10 (8): 887.
23 Bone Relat, R.M., Winder, P.L., Bowden, G.D. et al. (2022). High-throughput screening of a marine compound library identifies anti-cryptosporidium activity of Leiodolide A. *Marine Drugs* 20 (4): 240.
24 Zhang, J.L., Si, H.F., Shang, X.F. et al. (2019). New life for an old drug: in vitro and in vivo effects of the anthelmintic drug niclosamide against *Toxoplasma gondii* RH strain. *International Journal for Parasitology: Drugs and Drug Resistance* 9: 27–34.
25 De Souza, J.H., Michelon, A., Banhuk, F.W. et al. (2020). Leishmanicidal, trypanocidal and antioxidant activity of amyrin-rich extracts from *Eugenia pyriformis* Cambess. *Iranian Journal of Pharmaceutical Research: IJPR* 19 (4): 343.

26 Gonçalves-Santos, E., Vilas-Boas, D.F., Diniz, L.F. et al. (2019). Sesquiterpene lactone potentiates the immunomodulatory, antiparasitic and cardioprotective effects on anti-*Trypanosoma cruzi* specific chemotherapy. *International Immunopharmacology* 77: 105961.

27 Peres, R.B., Fiuza, L.F., da Silva, P.B. et al. (2021). In-vitro phenotypic activity and in silico analysis of natural products from Brazilian biodiversity on *Trypanosoma cruzi*. *Molecules* 26 (18): 5676.

28 Lozano, E.S., Germanó, M.J., Troncoso, M.E. et al. (2022). Therapeutic effect of *Prosopis strombulifera* (LAM) BENTH aqueous extract on a murine model of cutaneous leishmaniasis. *Journal of Traditional and Complementary Medicine* 12 (3): 281–286.

29 Melese Birru, E., Adela Alemu, M., Tsegaw Weredekal, A. et al. (2022). Antimalarial activity of crude extract and solvent fractions of leaves of *Solanum nigrum* L. (Solanaceae) against *Plasmodium berghei* in mice. *Evidence-Based Complementary and Alternative Medicine*. 2022.

30 Ayisi, F., Mensah, C.N., and Borquaye, L.S. (2021). In-vivo antiplasmodial activity and toxicological analyses of the ethanolic leaf and twig extract of *Faurea speciosa* Welw *(Proteaceae)*. *Journal of Parasitology Research*. 2021.

31 Tajuddeen, N., Swart, T., Hoppe, H.C., and van Heerden, F.R. (2021). Antiplasmodial and cytotoxic flavonoids from *Pappea capensis* (Eckl. & Zeyh.) leaves. *Molecules* 26 (13): 3875.

32 Adesina, D.A., Adefolalu, S.F., Jigam, A.A., and Lawal, B. (2020). Antiplasmodial effect and sub-acute toxicity of alkaloid, flavonoid and phenolic extracts of *Sida acuta* leaf on *Plasmodium berghei* infected animals. *Journal of Taibah University for Science* 14 (1): 943–953.

33 Zhang, J., Si, H., Lv, K. et al. (2021). Licarin-B exhibits activity against the *Toxoplasma gondii* RH strain by damaging mitochondria and activating autophagy. *Frontiers in Cell and Developmental Biology, 9:* 1499.

34 Braun, G.H., Ramos, H.P., Candido, A.C. et al. (2021). Evaluation of antileishmanial activity of harzialactone a isolated from the marine-derived fungus *Paecilomyces* sp. *Natural Product Research* 35 (10): 1644–1647.

35 Ogawa, H., Iwasaki, A., Sumimoto, S. et al. (2016). Janadolide, a cyclic polyketide–peptide hybrid possessing a tert-butyl group from an *Okeania* sp. marine cyanobacterium. *Journal of Natural Products* 79 (7): 1862–1866.

36 Iwasaki, K., Iwasaki, A., Sumimoto, S. et al. (2020). Ikoamide, an antimalarial lipopeptide from an *Okeania* sp, marine cyanobacterium. *Journal of Natural Products*. 83 (2): 481–488.

37 Soares, D.C., Szlachta, M.M., Teixeira, V.L. et al. (2016). The brown alga *Stypopodium zonale* (Dictyotaceae): a potential source of anti-leishmania drugs. *Marine Drugs* 14 (9): 163.

38 Murtihapsari, M., Salam, S., Kurnia, D. et al. (2021). A new antiplasmodial sterol from Indonesian marine sponge, *Xestospongia* sp. *Natural Product Research* 35 (6): 937–944.

39 Osei, E., Kwain, S., Mawuli, G.T. et al. (2018). Paenidigyamycin A, potent antiparasitic imidazole alkaloid from the Ghanaian *Paenibacillus* sp. DE2SH. *Marine Drugs* 17 (1): 9.

40 Lu, D., Zhang, N.Z., Yao, Y. et al. (2022). Investigation of antiparasitic activity of two marine natural products, estradiol benzoate, and octyl gallate, on *Toxoplasma gondii* in vitro. *Frontiers in Pharmacology* 13.

41 O'Connor, R.M., Nepveux, V.F.J., Abenoja, J. et al. (2020). A symbiotic bacterium of shipworms produces a compound with broad spectrum anti-apicomplexan activity. *PLoS Pathogens* 16 (5): e1008600.

42 Ortiz-Gonzalez, M., Pérez-Victoria, I., Ramirez-Macias, I. et al. (2022). Curvicollide D isolated from the fungus *Amesia* sp. kills African trypanosomes by inhibiting transcription. *International Journal of Molecular Sciences* 23 (11): 6107.

43 Fernández-Pastor, I., González-Menéndez, V., Annang, F. et al. (2021). Pipecolisporin, a novel cyclic peptide with antimalarial and antitrypanosome activities from a wheat endophytic *Nigrospora oryzae*. *Pharmaceuticals* 14 (3): 268.

44 Cota, B.B., Tunes, L.G., Maia, D.N.B. et al. (2018). Leishmanicidal compounds of *Nectria pseudotrichia*, an endophytic fungus isolated from the plant *Caesalpinia echinata* (Brazilwood). *Memórias Do Instituto Oswaldo Cruz* 113: 102–110.

45 Ortega, H.E., Ferreira, L.L., Melo, W.G. et al. (2019). Antifungal compounds from *Streptomyces* associated with attine ants also inhibit *Leishmania donovani*. *PLoS Neglected Tropical Diseases* 13 (8): e0007643.

46 Paul, S.M., Mytelka, D.S., Dunwiddie, C.T. et al. (2010). How to improve R&D productivity: the pharmaceutical industry's grand challenge. *Nature Reviews Drug Discovery* 9 (3): 203–214.

47 Charlton, R.L., Rossi-Bergmann, B., Denny, P.W., and Steel, P.G. (2018). Repurposing as a strategy for the discovery of new anti-leishmanials: the-state-of-the-art. *Parasitology* 145 (2): 219–236.

48 McKerrow, J.H. (2005). Designing drugs for parasitic diseases of the developing world. *PLoS Medicine* 2 (8): e210.

49 Dittmar, A.J., Drozda, A.A., and Blader, I.J. (2016). Drug repurposing screening identifies novel compounds that effectively inhibit *Toxoplasma gondii* growth. *Msphere* 1 (2): e00042–15.

50 Rufener, R., Dick, L., D'Ascoli, L. et al. (2018). Repurposing of an old drug: in vitro and in vivo efficacies of buparvaquone against *Echinococcus multilocularis*. *International Journal for Parasitology: Drugs and Drug Resistance* 8 (3): 440–450.

51 Trinconi, C.T., Reimao, J.Q., Coelho, A.C., and Uliana, S.R. (2016). Efficacy of tamoxifen and miltefosine combined therapy for cutaneous leishmaniasis in the murine model of infection with *Leishmania amazonensis*. *Journal of Antimicrobial Chemotherapy* 71 (5): 1314–1322.

52 Reimão, J.Q., Miguel, D.C., Taniwaki, N.N. et al. (2014). Antileishmanial activity of the estrogen receptor modulator raloxifene. *PLoS Neglected Tropical Diseases* 8 (5): e2842.

53 Liu, S., Wu, M., Hua, Q. et al. (2020). Two old drugs, NVP-AEW541 and GSK-J4, repurposed against the *Toxoplasma gondii* RH strain. *Parasites & Vectors* 13 (1): 1–3.

54 de Macedo-silva, S.T., Urbina, J.A., De Souza, W., and Rodrigues, J.C. (2013). In vitro activity of the antifungal azoles itraconazole and posaconazole against *Leishmania amazonensis*. *PloS One* 8 (12): e83247.

55 Farajzadeh, S., Esfandiarpour, I., Haghdoost, A.A. et al. (2015). Comparison between combination therapy of oral terbinafine and cryotherapy versus systemic meglumine antimoniate and cryotherapy in cutaneous leishmaniasis: a randomized clinical trial. *Iranian Journal of Parasitology* 10 (1): 1.

56 Keiser, J. and Utzinger, J. (2012). Antimalarials in the treatment of schistosomiasis. *Current Pharmaceutical Design* 18 (24): 3531–3538.

57 Keiser, J., Vargas, M., Rubbiani, R. et al. (2014). In vitro and in vivo antischistosomal activity of ferroquine derivatives. *Parasites & Vectors* 7 (1): 1–7.

58 Zhang, J., Si, H., Li, B. et al. (2019). Myrislignan exhibits activities against *Toxoplasma gondii* RH strain by triggering mitochondrial dysfunction. *Frontiers in Microbiology* 10: 2152.

59 Patterson, S., Wyllie, S., Norval, S. et al. (2016). The anti-tubercular drug delamanid as a potential oral treatment for visceral leishmaniasis. *Elife* 5: e09744.

60 Patterson, S., Wyllie, S., Stojanovski, L. et al. (2013). The R enantiomer of the antitubercular drug PA-824 as a potential oral treatment for visceral leishmaniasis. *Antimicrobial Agents and Chemotherapy* 57 (10): 4699–4706.

61 Wiesner, J., Henschker, D., Hutchinson, D.B. et al. (2002). In vitro and in vivo synergy of fosmidomycin, a novel antimalarial drug, with clindamycin. *Antimicrobial Agents and Chemotherapy* 46 (9): 2889–2894.

62 Mombo-Ngoma, G., Remppis, J., Sievers, M. et al. (2018). Efficacy and safety of fosmidomycin–piperaquine as nonartemisinin-based combination therapy for uncomplicated *Falciparum* malaria: a single-arm, age de-escalation proof-of-concept study in Gabon. *Clinical Infectious Diseases* 66 (12): 1823–1830.

63 Nishi, L., Santana, P.L., Evangelista, F.F. et al. (2020). Rosuvastatin reduced brain parasite burden in a chronic toxoplasmosis in vivo model and influenced the neuropathological pattern of ME-49 strain. *Parasitology* 147 (3): 303–309.

64 Skinner-Adams, T.S., Sumanadasa, S.D., Fisher, G.M. et al. (2016). Defining the targets of antiparasitic compounds. *Drug Discovery Today* 21 (5): 725–739.

65 Jeffers, V., Yang, C., Huang, S., and Sullivan, J.W.J. (2017). Bromodomains in protozoan parasites: evolution, function, and opportunities for drug development. *Microbiology and Molecular Biology Reviews* 81 (1): e00047–16.

66 Rajasekaran, R. and Chen, Y.P. (2015). Potential therapeutic targets and the role of technology in developing novel antileishmanial drugs. *Drug Discovery Today* 20 (8): 958–968.

67 Cowell, A.N. and Winzeler, E.A. (2019). Advances in omics-based methods to identify novel targets for malaria and other parasitic protozoan infections. *Genome Medicine* 11 (1): 1–7.

68 Varo, R., Crowley, V.M., Sitoe, A. et al. (2018). Adjunctive therapy for severe malaria: a review and critical appraisal. *Malaria Journal* 17 (1): 1–8.

69 Glennon, E.K., Dankwa, S., Smith, J.D., and Kaushansky, A. (2018). Opportunities for host-targeted therapies for malaria. *Trends in Parasitology* 34 (10): 843–860.

70 Wang, J., Dixon, S.E., Ting, L.M. et al. (2014). Lysine acetyltransferase GCN5b interacts with AP2 factors and is required for *Toxoplasma gondii* proliferation. *PLoS Pathogens* 10 (1): e1003830.

71 Demont, E.H., Bamborough, P., Chung, C.W. et al. (2014). 1, 3-Dimethyl benzimidazolones are potent, selective inhibitors of the BRPF1 bromodomain. *ACS Medicinal Chemistry Letters* 5 (11): 1190–1195.

72 Martin, L.J., Koegl, M., Bader, G. et al. (2016). Structure-based design of an in vivo active selective BRD9 inhibitor. *Journal of Medicinal Chemistry* 59 (10): 4462–4475.

73 Smith, S.G. and Zhou, M.M. (2016). The bromodomain: a new target in emerging epigenetic medicine. *ACS Chemical Biology* 11 (3): 598–608.

74 DiMasi, J.A., Grabowski, H.G., and Hansen, R.W. (2016). Innovation in the pharmaceutical industry: new estimates of R&D costs. *Journal of Health Economics* 47: 20–33.

75 Plackett, B. (2020). No money for new drugs. *Nature* 586 (7830): S50–2.

76 Skau, K. (2007). Pharmacy is a science-based profession. *American Journal of Pharmaceutical Education* 71 (1).

77 Jarrell, K. (2012). Regulatory history: elixir sulfanilamide. *Journal of GXP Compliance* 16 (3): 12.

78 Preston, S. and Gasser, R.B. (2018). Working towards new drugs against parasitic worms in a public-development partnership. *Trends in Parasitology* 34 (1): 4–6.

79 Alcântara, L.M., Ferreira, T.C., Gadelha, F.R., and Miguel, D.C. (2018). Challenges in drug discovery targeting TriTryp diseases with an emphasis on leishmaniasis. *International Journal for Parasitology: Drugs and Drug Resistance* 8 (3): 430–439.

80 Pacey, S., Workman, P., and Sarker, D. (2008). Pharmacokinetics and pharmacodynamics in drug development. In: *The Encyclopedia of Cancer* (ed. M. Schwab), 2306–2309. Springer.

81 Palmer, M.E., Andrews, L.J., Abbey, T.C. et al. (2022). The importance of pharmacokinetics and pharmacodynamics in antimicrobial drug development and their influence on the success of agents developed to combat resistant gram-negative pathogens: a review. *Frontiers in Pharmacology* 13.

82 Tuntland, T., Ethell, B., Kosaka, T. et al. (2014). Implementation of pharmacokinetic and pharmacodynamic strategies in early research phases of drug discovery and development at Novartis Institute of Biomedical Research. *Frontiers in Pharmacology* 5: 174.

83 McClure, N.S. and Day, T. A theoretical examination of the relative importance of evolution management and drug development for managing resistance. *Proceedings of the Royal Society B: Biological Sciences* 2014 281 (1797): 20141861.

84 Rangel, G.W. and Llinás, M. (2021). Re-envisioning anti-apicomplexan parasite drug discovery approaches. *Frontiers in Cellular and Infection Microbiology* 485.

85 World Health Organization (2006). *Guidelines for the Treatment of Malaria*. Geneva: World Health Organization.

86 De Rycker, M., Baragaña, B., Duce, S.L., and Gilbert, I.H. (2018). Challenges and recent progress in drug discovery for tropical diseases. *Nature* 559 (7715): 498–506.

87 Chatelain, E. and Scandale, I. (2020). Animal models of Chagas disease and their translational value to drug development. *Expert Opinion on Drug Discovery* 15 (12): 1381–1402.

88 Mears, E.R., Modabber, F., Don, R., and Johnson, G.E. (2015). A review: the current in vivo models for the discovery and utility of new anti-leishmanial drugs targeting cutaneous leishmaniasis. *PLoS Neglected Tropical Diseases* 9 (9): e0003889.

12

Phytopharmaceuticals as an Alternative Treatment against Parasites

Rajesh Kumar[1], *Seetha Harilal*[1,*], *Arti Gautam*[2], *Manisha Nigam*[3], *and Abhay Prakash Mishra*[4]

[1] *Faculty of Pharmaceutical Sciences, Kerala University of Health Sciences, Kerala, India*
[2] *Pharmacology Division, National Botanical Research Institute (NBRI-CSIR), Lucknow, Uttar Pradesh, India*
[3] *Department of Biochemistry, School of Life Sciences, H.N.B.G. (Central) University, Srinagar Garhwal, Uttarakhand, India*
[4] *Department of Pharmacology, University of Free State, Bloemfontein, South Africa*
[*] *Corresponding author*

12.1 Introduction

Medicinal herbs have been utilized since ancient times for treating and managing parasitism in various regions of the world [1]. As humans evolved, various parasites have also evolved, which utilize humans for their growth and reproduction. A parasitic organism spends a part or all of its life either on or inside the host for food and reproduction. The disease-causing parasites in humans include helminths and protozoa as well as ectoparasites [2]. The parasitic infections can be diagnosed by the examination of clinical manifestations, and laboratory tests. Laboratory tests include blood tests such as blood smear to observe the parasites causing malaria, filariasis, and babesiosis; and serology tests to observe the antibodies against parasites as a result of the infection. Other laboratory tests include stool examination to detect parasites causing abdominal disease, diarrhea, flatulence, and cramping. Colonoscopy or endoscopy can be utilized if stool examination is unable to detect the parasite. Imaging techniques such as magnetic resonance imaging, x-ray, and computerized axial tomography can be utilized to detect lesions associated with parasitic infections in various organs [3].

Parasitic Infections: Immune Responses and Therapeutics, First Edition.
Edited by Abhay Prakash Mishra and Manisha Nigam.

Various parasitic infections can be zoonotic and transmitted from animals. For example, *Trichinella* and *Cryptosporidium* transmission through food when the pigs and cattle have parasites inside the body; hookworm and roundworm transmission via cats and dogs; and *Baylisascaris* transmission via raccoons. Certain parasites can be transmitted through blood while sharing contaminated syringes or blood transfusions, such as malaria, Chagas disease, toxoplasmosis, African trypanosomiasis, leishmaniasis, and babesiosis. Several blood-borne parasites can be transmitted by insects as vectors and are known as vector-borne infections. Examples include African trypanosomiasis by tsetse flies, Chagas disease by triatomine bugs, malaria by *Anopheles* mosquitoes, leishmaniasis by phlebotomine sandflies, and babesiosis by *Babesia microti*. Some of the parasites, such as helminths and protozoa, can undergo transmission through food. Examples include *Toxoplasma gondii*, *Trichinella*, *Diphyllobothrium*, *Cryptosporidium*, *Anisakis*, *Giardia* intestinalis, *Taenia*, and *Cyclospora cayetanensis*. Many parasites can be transmitted through contaminated water, including amoebiasis, Guinea worm, giardiasis, schistosomiasis, and cryptosporidiosis [4].

Phytotherapy can be considered one of the oldest disciplines of conventional medicine, even though it is widely considered a system of alternative medicine. Medicinal plants have been utilized either whole, as parts, or as extracts of various parts, and as formulations such as galenicals for managing parasitic infections from ancient times and are still used now in many parts of the world as a part of the traditional medicine [5, 6]. From the fifteenth century the process of isolating active phytochemicals from medicinal plants developed, which led to experimental developments in the nineteenth century, leading to the isolation of phytochemicals, and it is continuing today [6]. Treatment of parasitic illnesses such as malaria greatly benefited from such advancements. Then, in the current era, further innovations led to the development of several synthetic drugs and vaccines for the therapy of parasitic infections, but phytotherapy remains in several regions of the world as a part of the traditional medicine serving the underdeveloped, densely populated, and poverty struck areas of the world in the management of several parasitic infections as the plants are widely available and cheap compared to synthetic drugs. Now, the resistance to synthetic therapy, unavailability and affordability of drugs in densely populated underdeveloped regions, and long-term management in chronic conditions all favor the use of phytopharmaceuticals as an alternative or adjunct therapeutic option against parasitic infections [5–7]. Many of the phytopharmaceuticals still remain unexplored. Therefore, further research is of prime importance in the development of phytopharmaceuticals against parasitic infections. This chapter aims to explore the antiparasitic phytopharmaceuticals utilized in various traditional systems of medicine, potential mechanisms, and research advancements specific to various parasitic infections caused by helminths, protozoa, and ectoparasites [8].

12.2 Antiparasitic Phytopharmaceuticals-Mechanisms of Action

Parasites, as eukaryotic organisms, showing similarities in biochemical as well as molecular characteristics with the hosts, pose a difficulty in the research to develop safe and effective antiparasitic therapy. The promising plant extracts and phytochemicals in in vitro studies should exhibit bioavailability and safety in the in vivo and clinical studies to be helpful in treating humans. The selectivity index of the herbal drug can be determined to assess the cytotoxicity against parasites and can be compared with that of human cells [5, 7]. Various targets for antiparasitic drugs include biomembranes, cytoskeletal proteins, RNA, and DNA [5] are shown in Figure 12.1.

12.2.1 Targeting the Biological Membranes

Several plant compounds showing detergent characteristics, such as the triterpenoid as well as steroidal saponins (seen in a vast majority of plants), can alter the permeability as well as the fluidity of the biomembranes and can show antimicrobial as well as cytotoxic activities [5]. The mono, as well as sesquiterpenes, isothiocyanates, and phenylpropanoids of lipophilic nature, mostly deriving from the families Tropaeoleaceae and Brassicaceae, show antimicrobial as well as antiparasitic actions [5]. The phytochemicals of lipophilic nature, including phenylpropanoids and terpenoids, can be seen in many essential oils of plants belonging to the families such as Pinaceae, Asteraceae, Myrtaceae, Lauraceae, Cupressaceae, Lamiaceae, Apiaceae, Burseraceae, Rubiaceae, Rutaceae, and Verbenaceae, can produce a dissolution of the biomembranes altering the fluidity as well as membrane protein function [5, 9, 10].

12.2.2 Inhibitors of Cytoskeletal Proteins

The microtubules as well as actin filaments are vital proteins in the cytoskeleton and are essential for cell structure as well as cell division. The emetine alkaloid from the family Rubiaceae shows DNA intercalation and strongly inhibits the biosynthesis of ribosomal proteins. The alkaloid is utilized in the therapy of amoebial infections [5, 6]. Several plant compounds are known to interact with microtubules, such as sanguinarine from *Bocconia* species and *Macleaya* species, such as in the plant *Sanguinaria canadensis* belonging to the family Papaveraceae, vinblastine from *Catharanthus roseus* belonging to the family Apocynaceae, rotenone from *Lonchocarpus*, and *Derris* belonging to the family Fabaceae, podophyllotoxin from *Podophyllum* species belonging to the family Berberidaceae, and *Linum* species belonging to the family Linaceae, colchicine from the *Colchicum* species such

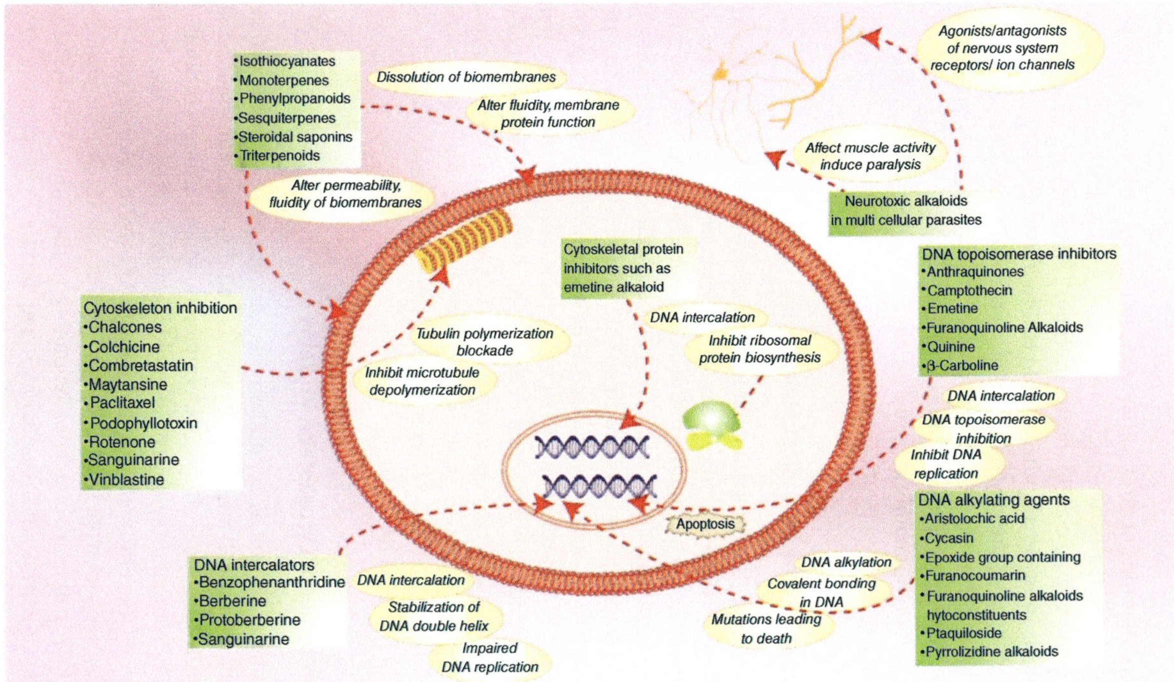

Figure 12.1 Antiparasitic phytopharmaceuticals with various mechanisms of action.

as in *Gloriosa superba* belonging to the family Colchicaceae, maytansine from *Maytenus* species belonging to the family Celastraceae, chalcones, and combretastatin from *Combretum caffrum* belonging to the family Combretaceae and can produce blockade of tubulin polymerization to microtubules or produce inhibition of microtubule depolymerization [5]. An example is paclitaxel, and some of these plant compounds are currently used in cancer chemotherapy [5, 11, 12], even though some may have antiparasitic potential. Several parasites living in the intracellular environment or in the blood show unique adaptation to the microenvironment, and such receptors, enzymes, etc., can be targeted for antiparasitic therapy [5].

12.2.3 Targeting the RNA and DNA

Several plant drugs producing DNA damage show cytotoxic as well as antiparasitic activities. A mechanism of DNA damage produced by plant compounds includes DNA intercalation. The compounds that can intercalate with the DNA can produce stabilization of the double helical structure of DNA and result in an impairment of the DNA replication. The plant compounds can be planar, aromatic, as well as hydrophobic, and intercalate between the nucleotide pairs, particularly the GC pairs. The compounds can produce deletions as well as frameshift mutations resulting in cell death [11]. The plant compounds include alkaloids such as benzophenanthridine, protoberberine, sanguinarine and berberine [13, 14]. Such alkaloids are commonly produced by the plants belonging to the families Menispermaceae, Papaveraceae, Ranunculaceae, and Berberidaceae. These plants are believed to have antiviral, antimicrobial, and antiparasitic properties. Various DNA intercalating plant compounds can produce inhibition of enzymes such as DNA topoisomerase I and II affecting the DNA replication and leading to apoptosis [5, 11]. This can also happen in the protozoa, and is therefore considered antiparasitic activity [15]. The DNA topoisomerase inhibitors include camptothecin, an indole alkaloid belonging to the plants *Camptotheca acuminata, Mostuea brunonis, Notapodytes* sp., *and Ophiorrhiza* sp. [5, 13]. Several alkaloids such as furanoquinoline alkaloids belonging to the family Rutaceae, anthraquinones of the family Rhamnaceae and Polygonaceae, β-carboline of the family Nitrariaceae such as in *Peganum harmala*, quinoline including quinine, furanocoumarins of the family Fabaceae and Apiaceae, and emetine of the family Rubiaceae such as in *Cephaelis acuminata* are believed to have DNA intercalating properties [5, 13].

Several drugs can produce DNA alkylation by the transfer of alkyl groups as well as the formation of covalent bonds in the DNA. If the enzymes facilitating DNA repair cannot reverse the DNA alkylation, mutations such as deletions, point, and frameshift mutations can occur during DNA replication, killing the parasite. The plant compounds producing DNA alkylation include phytochemicals possessing an epoxide group and compounds such as furanoquinoline alkaloids from Rutaceae species, pyrrolizidine alkaloids from Crotalarieae, Boraginaceae, and Asteraceae species, aristolochic acid from *Aristolochia* species, furanocoumarins from Fabaceae, as well as Apiaceae species, ptaquiloside from *Pteridium aquilinum* species, and cycasin from Cycadaceae species [5, 13].

12.2.4 Targeting the Nervous System

Several of the alkaloids show neurotoxic activities [5, 13, 16] and can act as antagonists or agonists of nervous system receptors as well as ion channels and can be utilized as potential anthelmintic drugs [13, 16]. Various multicellular parasites possess a nervous system with vital neurotransmitters as well as receptors, including acetylcholine and its receptors for regulating muscle activity. The potassium and sodium channels are essential for neurotransmission and the ions play a vital role in the process [16]. All of the above targets can be utilized to induce paralysis of the parasitic worms. The arrest of the muscles can result in the death of the parasites as they cannot remain attached to the gastrointestinal tract walls and get rapidly eliminated with the feces with laxatives. The neurotoxic drug can easily affect the tapeworms as well as Cestodes in the intestine compared to those in the internal organs. The dosage should be ideal for killing the parasite and having no effect on the treated patient. Such compounds should not be taken up by the intestinal cells as that can lead to undesirable effects [5].

12.3 Phytopharmaceutical Therapy

Table 12.1 classifies and describes various parasites, infections caused, mode of transmission, clinical manifestation, current as well as phytopharmaceutical therapy.

12.3.1 Helminths

Several categories of worms such as Cestoda, Nematoda, and Trematoda can infect humans and produce infections such as schistosomiasis, onchocerciasis, lymphatic filariasis, and loiasis that affect millions of people worldwide. About two billion people worldwide are affected by a minimum of one helminth at any point in their life [5, 17, 18]. The infection cases are seen more in the tropical and subtropical parts of the world, mainly in the developing countries with hygiene issues and poor access to newer effective therapies. Several in vitro as well as in vivo assay techniques have been studied for screening drugs against helminths as well as filarial infections that evaluate the viability as well as motility of the organism [5, 18]. Several medicinal plants, extracts, and phytochemicals have also been studied, and many are utilized in the community as part of the traditional medicine against helminths in the affected regions of the world. Since a plethora of parasites infects humans, sufficient experimental protocols and assays are unavailable regarding each parasite or its important stages of life [5].

12.3.1.1 Cestodes

12.3.1.1.1 Taeniasis

Taeniasis is an infection in the intestine caused due to adult tapeworms *Taenia solium and Taenia asiatica*, which are known as the pork tapeworm, and *Taenia saginata*, which is known as the beef tapeworm. Taeniasis as well as cysticercosis can spread through undercooked or raw infected pork, and the severe disease can lead to neurocysticercosis. The drugs for treatment include niclosamide and praziquantel. The *Hedychium coronarium* root oil and *Hedychium spicatum* from the family Zingiberaceae showed good anti-helminthic activities, even better than synthetic drugs such as piperazine phosphate [19, 20]. The extracts of *Glinus lotoides* utilized in the folklore therapy of taeniasis in Ethiopia were reported to be safe in an animal study and therefore need to be explored in clinical studies [21]. A study reported the usefulness of pumpkin seeds as well as areca nut extracts against taeniasis and produced synergism resulting in the eradication of tapeworms [22, 23]. The study mentioning the effectiveness of pumpkin seeds reported the cure of infection in an infected patient [23]. Another study reports on the essential oils of certain plants possessing strong anthelmintic activity and these include *Artabotrys odoratissimus*, *Artemisia pallens*, *Capillipedium foetidum*, *Cymbopogon martini* grass, and *Eupatorium triplinerve* [19].

12.3.1.1.2 Echinococcosis

Echinococcosis is caused by larval parasites belonging to the genus *Echinococcus*. Echinococcosis can be classified into alveolar echinococcosis as well as cystic echinococcosis [24]. Alveolar echinococcosis is produced by larval *Echinococcus multilocularis*, a tapeworm of size 1–4 mm, and is seen in dogs and coyotes, as well as foxes, which can be considered definitive hosts. Small rodents can be considered intermediate hosts. Human infections are rare and can be more dangerous than cystic echinococcosis, and can lead to parasitic tumors in the lungs, liver, brain, and various organs. The disease, if not treated, can be lethal, with a mortality rate of 50–75% [24, 25]. Cystic echinococcosis (hydatid disease) is produced by larval *Echinococcus granulosus*, a tapeworm of size 2 to 7 mm, and is seen in dogs, which can be considered a definitive host. The goats, sheep, pigs, and cattle can be considered intermediate hosts. Even though asymptomatic, the infection can produce cysts that can enlarge in the lungs, liver, and various organs [24].

Table 12.1 Current phytopharmaceutical therapies for various parasitic infections.

Causative Organism (vector)	Diseases/ Infection	Mode of Transmission	Epidemiology and Distribution	Clinical Manifestation	Current Therapy	Phytopharmaceutical Therapy	References
1. Helminths							
(A) Platyhelminths (Flatworms)							
(i) Cestodes (tapeworms)							
Gastrointestinal infections							
Dipylidium caninum (intermediate host: dog lice and fleas infecting humans; main host: dogs and cats)[a]	Dog tapeworm infection	Swallowing of flea that is infected with a tapeworm larva	Worldwide with the exception of Antarctica	Anal pruritus, anorexia, diarrhea, dullness, weight loss	Praziquantel, niclosamide	*Filix mass*, *Mallotus philippinensis*, *Cassia angustifolia*, *Trachyspermum ammi* L., *Foeniculum vulgare* and Barbrung combination has been reported to expel *Dipylidium caninum*	[19]
Hymenolepis diminuta (intermediate host: fleas, beetles; main host: rodents)[a]	Rat tapeworm infection/ hymenolepiasis	Fecal-oral transmission	Worldwide but less frequent	Abdominal pain, anorexia, diarrhea, headaches, weakness	Praziquantel, niclosamide, nitazoxanide	*Cucurbita mexicana* seed extract (in vitro); *Adhatoda vasica* (leaf), *Butea* *minor* (seeds), *Solanum myriacanthum* (fruits), *Strobilanthes discolor* (leaf), *Trifolium repens* (shoots) and *Zanthoxylum rhetsa* (leaf) have moderate to high degree activity (in vitro)	[19]
Hymenolepis nana (fleas, beetles intermediate hosts: insects; main host: rodents)[a]	Dwarf tapeworm infection/ hymenolepiasis	Fecal-oral transmission	Worldwide, higher incidence in institutionalized groups and children of temperate areas	Abdominal pain, anorexia, diarrhea, headaches, weakness	Praziquantel, niclosamide, nitazoxanide	*Diospyros mollis* extract, *Leucana leucocephala* infusion established anthelmintic activity (in vivo)	[19]

Taenia saginata/ beef tapeworm *heterophyes* (intermediate host: cattle; main host: humans)[a]	Taeniasis	Fecal-oral transmission	*Taenia saginata* infections can be found in areas of Eastern Africa, Asia, Latin America, Eastern Europe, Russia. Even though the occurrence of taeniasis is rare in the United States, it is prevalent in regions of poor sanitation with wider population of cattle and humans	*Taenia saginata* infections cause diarrhea, hunger pangs, flatulence, mild epigastric discomfort, nausea, and weight loss. In taeniasis, active excretion of proglottids in feces, hence found visible in the anal region and rarely wedged in the regions of appendix, ducts of, pancreas and bile	Praziquantel, niclosamide, albendazole	*Hedychium coronarium* root oil and *Hedychium spicatum*, extracts of *Glinus lotoides* utilized in the folklore therapy of taeniasis in Ethiopia/ (safe in preclinical studies), *Cucurbita maxima* +*Areca catechu* extracts produced synergism and eradication of tapeworms, *Cucurbita maxima* seeds (reported cure in Tibetan woman); a combination of *Filix mass*, *Mallotus philippinensis*, *Cassia angustifolia*, *Trachyspermum ammi* L., *Foeniculum vulgare* and Barbrung combination has been reported to expel *Taenia* species	[19, 20]
Taenia solium/ pork tapeworm (intermediate host: pigs; main host: humans, less common are cats, dogs, sheep, primates)[a]	Taeniasis, Cysticercosis	Fecal-oral route	*Taenia solium* infections occur at a higher rate in the regions of Asia (India), Eastern Europe, Latin America, and sub-Saharan Africa. Taeniasis caused by *Taenia solium* can be seen in Latin American immigrants residing regions of the United States and underdeveloped communities lacking proper sanitation as well as consumption of undercooked or raw pork	*Taenia solium* infections have lesser symptoms compared to that of *T. saginata* but can result in the development of a severe disease, cysticercosis	Treatment for taeniasis is the same as mentioned above. Treatment for cysticercosis is albendazole which might be effective/ superior to praziquantel	Same as above	[20]

(Continued)

Table 12.1 (Continued)

Causative Organism (vector)	Diseases/ Infection	Mode of Transmission	Epidemiology and Distribution	Clinical Manifestation	Current Therapy	Phytopharmaceutical Therapy	References
Extraintestinal/tissue infections							
Echinococcus granulosus (intermediate host: cattle, goats, pigs, sheep; main host: dogs)[a,b]	Human echinococcosis/ Hydatid disease	Hand to mouth transfer or via ingestion of contaminated food, water, soil	Worldwide occurrence but precise mapping not possible due to lack of accurate genotyping and case reports. Human incidence can go beyond 50 per 100,000 person/ year in endemic regions, prevalence 5–10% in regions of Argentina, Central Asia, China, East Africa, Peru	Cyst rupture can produce reaction in the host including anaphylactic shock, eosinophilia, fever, urticaria. Clinical manifestations most commonly observed are that of organs such as liver, lungs where development of cysts occur, and less frequently can also affect other organs such as bone, heart, kidney, spleen and central nervous system	In preoperative cases praziquantel can be useful. Chemotherapy with albendazole (first choice), and mebendazole (second choice) is preferred treatment and more effective in combination with surgery. Third option, PAIR (percutaneous aspiration, injection of protoscolicidal chemical solutions and reaspiration) are also effective. Best option is surgery for liver cysts and for those located secondarily in other organs	*Salvia officinalis*, seeds of *Nigella sativa, Thymus vulgaris*, and *Zingiber officinale* antiprotoscolices activities in in vitro and in vivo models. Various phytochemicals such as ampelopsin, berberine, carvacrol, gallic acid, genistein, menthol, thymol, thymoquinone possess protoscolicidal efficacy; ampelopsin extracted from *Ampelopsis grossedentata* is effective against *Echinococcus granulosus* and has potential to treat echinococcosis	[7, 31, 32]
Echinococcus multilocularis (intermediate host: humans; main host: red foxes (*Vulpes vulpes*). Other competent hosts are canids, namely raccoon dogs (*Nyctereutes procyonoides*) wolves, and domestic dogs)[a]	Alveolar echinococcosis	Ingestion of contaminated food and water	Distributed in Central Asia, central and northern regions of Europe, northern Japan, northern Russia, northwestern Alaska, and northwestern Canada, and north-central United States with prevalence in Asia, northern Europe, and North America; mortality rate is 50–70%	Can remain asymptomatic for 5–15 years; gradually develops into a primary tumor-like lesion in the liver. Other clinical features include abdominal pain, malaise, weight loss, liver failure. Larval metastases can occur adjacent to liver or via lymphatic and blood into organs such as brain or lungs. If left unattended can progress and be fatal	Albendazole as prophylactic anti-infective following radical surgery	Ampelopsin extracted from *Ampelopsis grossedentata* is effective against *Echinococcus multilocularis* and has potential to treat echinococcosis	[32]

Taenia solium /pork tapeworm[b]	Neurocysticercosis	Entry of ingested *Taenia solium* eggs into the central nervous system	In endemic regions or countries, neurocysticercos is estimated to be the reason for 30% of epilepsy cases	Can be clinically asymptomatic or show varied symptoms (dependent on immune response of the host or the pathological changes) and these include blindness, chronic headaches, hydrocephalus, epileptic seizures, elevated intracranial hypertension associated symptoms	Chemotherapy using praziquantel and albendazole. Praziquantel is preferred treatment for women and children. Albendazole consumption showed superior or equivalent reduction of live cysticerci compared to praziquante. Reduction in generalized seizures frequency when treated with albendazole and dexamethasone. Consumption of albendazole and praziquantel in multiple brain cysts patients demonstrated elevated cysticidal effects	*Melia azedarach* L. extract possess considerable anthelmintic activity. Essential oils of certain plants possess strong anthelmintic activity and these include *Artabotrys odoratissimus*, *Artemisia pallens*, *Capillipedium foetidum*, *Cymbopogon martini* grass, *Eupatorium triplinerve*	[19]
(ii) Trematodes (flukes)							
Gastrointestinal infections							
Fasciolopsis buski (Intermediate host: snails genera *Biomphalaria*; main hosts: humans, pigs)[a]	Faciolopsiasis	Ingestion of contaminated raw water plants	Southeastern and southern regions of Asia	Abdominal pain, anemia, allergic reactions, diarrhea, fever, intestinal obstruction, nausea, vomiting	Praziquantel	*Cucurbita mexicana* seed, and *Flemingia vestita* root tuber peel extract, and the plant, *Lysimachia ramosa* Wall possess anthelmintic efficacy (in vitro); Crude extracts of *Cannabis sativa* possess trematocidal activity; *Alpinia nigra* crude extracts possess flukicidal activity (in vitro)	[19]
Heterophyes heterophyes (first intermediate host: snails genera *Cerithidia and Pironella*; second intermediate host: brackish water fish; main host: humans,cats, dogs, birds)[a]	Heterophyiasis	Ingestion of contaminated raw fish	Middle East, Egypt, southeast Europe, Sudan. In endemic regions, higher intensity of occurrence in people living near river banks	Colicky pain in abdomen, and diarrhea are the common symptoms, Development of fatal valvular or myocardial damage can occur if the eggs migrate to the heart	Praziquantel	Combination of *Areca catechu* and *Cucurbita maxima* seeds extract possess excellent activity (in vivo)	[135]

(Continued)

Table 12.1 (Continued)

Causative Organism (vector)	Diseases/ Infection	Mode of Transmission	Epidemiology and Distribution	Clinical Manifestation	Current Therapy	Phytopharmaceutical Therapy	References
Metagonimus yokogawai/ Japanese fluke (first intermediate host: snails genera *Semisulcospira*; second intermediate host: brackish water fish; main host: humans,cats, dogs, birds)[a]	Metagonimiasis	Ingestion of contaminated raw fish	Israel, Manchuria, states in Balkans, Siberia, Spain	Colicky abdominal pain, diarrhea	Praziquantel	Not available	–
Systemic/tissue infections							
Schistosoma mansoni (Intermediate host: snails genera *Biomphalaria*; main host: wild primates), *S. japonicum* (Intermediate host: snails genera *Oncomelania*; main host: dogs, cats, cattle, goats, horses, pigs, rodents), *S. mekongi* (Intermediate host: *Neotricula aperta*; main host: dogs), *S. intercalatum*, and *S. haematobium* (Intermediate host: snails genera *Bulinus*)[a,b]	Schistosomiasis (bilharziasis)	When a schistosomiasis patient contaminates freshwater source, parasite egg gets released from the patient excreta and hatch in the water body	Prevalent in subtropical and tropical regions of the world. 90% of schistosomiasis cases seeking treatment reside in Africa. Intestinal schistosomiasis caused by *Schistosoma mansoni* is distributed throughout Africa, Caribbean, Middle East, South American regions such as Brazil, Suriname, Venezuela; *S. japonicum* in certain regions of Southeast Asia, China, and Indonesia; *S. mekongi* in some areas of Laos and Cambodia; *S. intercalatum* in West and Central Africa. Urogenital schistosomiasis caused by *S. haematobium* is distributed throughout Africa, Corsica and certain regions of Middle East	Intestinal schistosomiasis: abdominal pain, bloody stool, diarrhea. In progressive cases, enlargement of liver with frequent association of ascites and portal hypertension which may also cause conditions such as splenomegaly. Urogenital schistosomiasis: haematuria is common. In progressive cases, kidney damage, ureter and bladder fibrosis are occasionally diagnosed. In later stages, bladder cancer can be a possible complication. In men, prostrate, seminal vesicles and other organ pathological condition may be present. In women, occurrence of vulvar nodules, vaginal bleeding, genital lesions, and painful intercourse	Praziquantel	Antimalarial drugs (utilized artemisinin or its derivatives, quinoline alkaloids, oxamniquine). Curcumin considered to have parasiticidal properties, *Momordica balsamina* excellent schistosomicidal actions (in vitro), *Sanguinaria canadensis, Jatropha elliptica, Plectranthus neochilus, Schefflera vinosa, Baccharis trimera, Asparagus stipularis, Allium sativum, Clerodendrum umbellatum, Hemerocallis fulva, Cleome droserifolia, Curcuma longa, Persea americana*, and *Artemisia annua* showed inhibition of the growth as well as locomotion of the adult worms, essential oils (*Plectranthus neochilus* and *Calyptridium umbellatum* extracts) promotes inhibition of females laying eggs; *Euphorbia milii* latex possess schistosomastatic effects (in vitro)	[7, 43]

Extraintestinal infections							
Clonorchis sinensis/ Chinese or oriental liver flukes (first intermediate host: Freshwater snail *Parafossarulus manchouricus* and many varieties; second intermediate host: variety of freshwater fish; main host: humans and others include badgers, buffaloes, cats, camels, dogs, pigs, rats, weasels)[a]	Clonorchiasis	Ingestion of contaminated uncooked/ raw fishes	Third most prevalent in the world. Endemic to east Asia, eastern Russia, Asian immigrants residing regions of United States	In mild worm burden, mild abdominal symptoms are manifested. In high worm burdens, abdominal pain, nausea, and diarrhea. For infections that lasts for a longer duration, more severe symptoms such as malnutrition and hepatomegaly may occur. Development of liver abscesses, biliary complications and pancreatitis are rare	Praziquantel or albendazole	Alantolactone, dihydroalantolactone, and isoalantolactone (*Inula helenium* root) and alantolactone, isoalantolactone and β-cyclocostunolide (*Saussurea lappa* root) are proven to have clonorchicidal effect	[136]
Opisthorchis viverrini/ Southeast Asian liver flukes (First intermediate host: freshwater snail; second intermediate host: freshwater fish; main host: humans' cats, dogs, pigs, rats, weasels)[a]	Opisthorchiasis	Ingestion of infected raw or undercooked fish and other aquatic products	Cambodia, southern and central Vietnam, northeast Thailand	In mild cases, abdominal pain, constipation or diarrhea, indigestion may be present. In severe cases, additionally nausea can occur	Praziquantel, alternative: Albendazole	*Thunbergia laurifolia* aqueous extract has been reported to have antiinflammatory and antioxidant activity against hamster opisthorchiasis (in vivo)	[137]
Paragonimus westermani/ Lung flukes (First intermediate host: freshwater snail; second intermediate host: freshwater fish; main host: humans)[a]	Paragonimiasis	Ingestion of infected raw or undercooked crawfish or crab	Southeast and southwestern Asia	Initially, abdominal pain, constipation or diarrhea	Praziquantel, alternative: Triclabendazole	Not available	–

(Continued)

Table 12.1 (Continued)

Causative Organism (vector)	Diseases/ Infection	Mode of Transmission	Epidemiology and Distribution	Clinical Manifestation	Current Therapy	Phytopharmaceutical Therapy	References
Fasciola hepatica (first intermediate host: Freshwater snail; second intermediate host: freshwater fish; main host: humans)[a]	Fasciolosis	Ingestion of infected raw or undercooked aquatic plants	Southeast and southern regions of Asia	Initially, occurrence of abdominal pain, and diarrhea. With heavy burden of infection, nausea, vomiting, abdominal pain, fever, and intestinal obstruction can occur. Initially, abdominal pain, constipation or diarrhea	Praziquantel	*Stryphnodendron adstringens*, *Harpagohytum procumbens*, *Psidium guajava* L., *Eugenia uniflora* were highly effective in controlling *F.hepatica* egg hatching	[138]
(B) Nematodes (Roundworms)							
Gastrointestinal infections							
Ascaris lumbricoides (main host: humans, swine)[a]	Ascariasis	Ingestion of infective eggs	Most common intestinal worm infections occurring in regions of poor sanitation practices and human feces being used as fertilizer. Occur worldwide with majority cases from subtropical and tropical areas of the Americas, Asia, and sub-saharan Africa	In high worm burden, intestinal obstruction and abdominal pain may occur	Albendazole, mebendazole, ivermectin	Santonin (*Artemisia santonica*) found to be useful, ascoridole (anti-ascariasis), deaspidin, aspidin, and filixic acid (treatment of veterinary praxis) shows vermicidal activity, arecoline and pelletierine alkaloid can be utilized as anthelmintics and can target the acetylcholine receptors, santonin from various plant sources can also be utilized as anthelmintics; *Cucurbita mexicana* seed, *Carica papaya* seed extract has a good anthelmintic property (in vitro); essential oils of certain plants possess strong anthelmintic activity and these include *Artabotrys odoratissimus*, *Artemisia pallens*, *Capillipedium foetidum*, *Cymbopogon martini* grass, *Eupatorium triplinerve*	[19, 20]

Capillaria philippinensis (main host: humans)[a]	Human intestinal capillariasis	Ingestion of undercooked fish	Endemic to Philippines, epidemics occurred in the regions of northern Luzon. About 10% of cases are fatal	Gastrointestinal/ abdominal disease manifest initially, if left untreated due to autoinfection	Mebendazole, albendazole	*Allium sativum* aqueous extract increases mortality rate in *Capillaria* species (in vitro)	[139]
Enterobius vermicularis/ Pinworm (main host: humans)[a]	Enterobiasis	Fecal-oral route	Worldwide occurrence and most frequent in crowded or schooling children	Frequently asymptomatic, perianal pruritus during night, Occasion vulvovaginitis or peritoneal granulomas	Mebendazole, pyrantel pamoate, albendazole	A clinical evaluation of Elisevi kudineer (*Merremia emarginata, Moringa oleifera, Terminalia chebula, Terminalia bellirica, Phyllanthus emblica, Embelia ribes*) according to Siddha aspect was performed on children yielding encouraging results	[140]
Trichuris trichiura/ Whipworms (vector: non-biting cyclorrhaphan flies; main host: humans)[a]	Trichuriasis	Fecal-oral route	Frequent in tropical climate and unhygienic conditions and estimated to be the third most common among roundworms infecting about 800 million of people worldwide	Light infections are asymptomatic, heavy infections may develop rectal prolapse, or in children may develop impairment in growth and cognitive development as well as severe anemia	Albendazole, mebendazole, ivermectin	A trial on Philippine children using a mixture of oil of *Chenopodium album* and tetrachloroethylene which was found effective in 30 among 100 children and further cured 82 on three doses and on fifth dose all were cured.	[141]
Systemic/tissue infections							
Loa loa (vector: *Chrysops silacea, C. dimidiate*; intermediate host: fly; main host: humans)[a]	Loaisis	Transmitted via repeated bites of deerflies	Found in central and west Africa	Majority people are asymptomatic. Most commonly known manifestations are cala bar swellings, non-painful itchy swellings commonly found near the joints, eyeworms and sometimes worm crawling under the skin	Diethylcarbamazine, albendazole	Different extracts of *Piper umbellatum* L. and *Tragia benthami* demonstrated selective filaricidal activity/ highly active against *L.loa*	[142]
Dracunculus medinensis (intermediate host: Cyclops; main host: humans)[a]	Dracunculiasis	Transmitted via ingestion of unfiltered water containing copepods	Found in remote areas of Africa and in 2021 the cases have come down from millions to 15 patients	Diarrhea, dizziness, itchy rash, fever, nausea, vomiting	No drug treatment or vaccine	Euphol, tirucallol (leaf of *Elaeophorbia drupifer*) effective against guinea worm	[59, 143]

(Continued)

Table 12.1 (Continued)

Causative Organism (vector)	Diseases/ Infection	Mode of Transmission	Epidemiology and Distribution	Clinical Manifestation	Current Therapy	Phytopharmaceutical Therapy	References
Onchocerca volvulus (main host: humans)[a]	Onchocerciasis (river blindness)	Transmitted via repeated bites of blackflies (genus *Simulium*)	Found in regions of tropical climate and affect about 31 countries of sub-Saharan Africa, Yemen and few regions of South America. At least 20.9 million people worldwide get infected among which about 14.6 million will have skin disease and 1.15 million vision loss	Itching, disfiguring of skin, visual impairment	Ivermectin for killing microfilariae; doxycycline for killing macrofilariae	*Aloe barteri* referred for the treatment of *O. volvulus*-induced skin conditions, carapolide A and oliverine (fruit and seed of *Carapa procera*) (in vitro), plant extract of *Cassia aubrevellei* active against *O. volvulus* microfilariae (in vitro), Stem bark of *Erythophleum ivorense* used, fruit of *Pachyelasma tessmanii*, stem bark of *Pachypodanthium staudtii*, fruit and leaf of *Raphia farinifera* (in vitro)	[59]
Strongyloides stercoralis (main host: humans)[a]	Strongyloidiasis	Direct penetration via direct contact of infected larvae with the human skin	Global prevalence is unknown but estimation of 30–100 million infected human cases worldwide. Most common in the subtropical, tropical and warm temperate areas across globe	Majority people are asymptomatic. Those of who are symptomatic havecommon complaints of abdominal (bloating, heartburn, stomachache), allergic reactions (an itchy rash at the site of worm entry and recurrence in the buttocks and thigh regions), and respiratory (throat irritation and dry cough) issues	Ivermectin as first line therapy and albendazole as an alternative therapy in both acute as well as chronic strongyloidiasis whereas for hyperinfection syndrome ivermectin is the drug of choice.	Extracts of *Artocarpus altilis*, *Cuscuta americana*, *Mimosa pudica* leaves, *Salvia serotina*, *Stachytarpheta jamaicensis*, had a constant larval inactivation effect (in vitro); ethanolic extract of *Spondias mombin* L. showed potent effect against strongyloidiasis giving an impression of further exploring its phytoconstituents which could be used for treatment against *Strongyloides stercoralis* (in vitro); Another study reports *Lawsonia inermis* as an alternative to flubendazole and requires further in vivo testing	[144]

Trichinella spiralis (hosts: humans, pigs, cats, rats and mice)[a,b]	Trichinosis (trichinellosis)	Ingestion of contaminated raw or undercooked animal meat	Estimation of 10,000 cases occur per year. *Trichinella spiralis* has the most global distribution among *Trichinella* species	Abdominal pain, diarrhea, nausea, vomiting occurs in human intestinal invasion. Facial or eyelid oedema, conjunctivitis, itching, muscle pain may occur on migration of larvae into the muscle tissues	Albendazole, mebendazole	*Commiphora molmol*, *Lipidium sativum* monotreatment (in vivo) showed significant mortality rate of the parasite and recommends the former as complementary to albendazole in the eradication of trichinellosis larvae stage; Due to effectiveness of *Artemisia annua* in experimentally induced trichinellosis, there is an open door with the potential to treat intestinal trichinellosis (in vivo); Another study states that *Allium sativum* was effective against muscle larvae and adult worms of *T. spiralis* and possess a promising anthelmintic property (in vitro)	[145–147]

(Continued)

Table 12.1 (Continued)

Causative Organism (vector)	Diseases/ Infection	Mode of Transmission	Epidemiology and Distribution	Clinical Manifestation	Current Therapy	Phytopharmaceutical Therapy	References
Wuchereria bancrofti (vector: genera *Aedes*, *Anopheles*, *Culex* and *Mansonia*; intermediate host: mosquitoes; main host: humans)[a,b]	Lymphatic filariasis (LF)/ elephantiasis	Human to human transmission via mosquito bites	Most common in sub-tropical and tropical regions of the world	Edema, severe lymphedema, hydrocele	Diethylcarbamazine citrate	Tetraphyllin b, Epitetraphyllin b, gumminferol and proanthocyanidins (*Adenia gummifera*), 19-epischolaricine, 19- vellesamine (latex, bark of *Alstonia scholaris*), alusenone, hirsutanonol, hirustenone, and platyphyllenone (leaves of *Alnus nepalensis*), repenone, repenol (immature shoots of *Boerhavia repens*), Cardenolide (whole plant of *Calotropis procera*), Calotropin, calctin, alpha and β-calotropeols, calotropises, quiterpenol, calotropin, calotropogenin (leaf and latex *Calotropis gigantea*), 7,4′dihydroxyflavon, apigenin, luteonin, tetracosanoic acid (inner bark of *Dichrostachys cinerea*), demethylwedlo-lactone, stigmasterol, wedelolactone (whole plant of *Eclipta alba*) *strebloside* and asperoside (from stem bark of *Streblus asper*) (in vitro and in vivo)	[59, 65, 69]

(C) Acanthocephalins (Thorny-headed worms)							
Gastrointestinal infections							
Macrocanthorynchus hirudinaceus (main host: swine)[a]	Acanthocephaliasis	Accidental ingestion of the intermediate host such as dung beetles leads to transmission in humans	Widely distributed. Occurs in regions where there are domestic or wild swine	Early infection may be asymptomatic. Symptoms commonly occurring are abdominal pain and other digestive problems	Ivermectin, lefluronomide	A study mentions *Allium sativum* extract in reducing the intensity of Acanthocephala infection	[148]
Moniliformis moniliformis (main host: rat)[a]	Acanthocephaliasis	Ingestion of arthropods containing infective cystacanths.	Occurrence is not known.	Early infection may be asymptomatic. Symptoms commonly occurring are abdominal pain and other digestive problems	Pyrantel pamoate	Not available	[149]
2. Protozoa							
Gastrointestinal infections							
Entamoeba histolytica (Main host: humans and other primates)[a]	Amoebiasis	Fecal-oral contamination or accidental ingestion of *E.histolytica* contaminated water or food	Occurs worldwide, with a prevalence in subtropical and tropical countries, mostly in developing countries where fresh water source is contaminated with human feces. Risk groups in industrialized countries include homosexual men, immunocompromised people, recent immigrants, travelers, institutionalized populations	Luminal amoebiasis is asymptomatic. Invasive intestinal amoebiasis shows severe dysentery and associated complications, and when chronic can lead to further complications such as amoeboma, peritonitis, perforations. Extraintestinal amoebiasis commonly manifests amoebic liver abscesses, additionally other manifestations such as brain abscess, perianal skin with necrotic lesions, and pleuropulmonary abscess	Metronidazole, nitazoxanide, tinidazole, dehydroemetine, chloroquine	*Artemisia ludoviciana* extracts and leaf infusions, *Zanthoxylum liebmannianum* preparations in the management of the disease. Asarinin in *Z. liebmannianum* is more effective against *Entamoeba histolytica* compared to other constituents	[74, 76]

(Continued)

Table 12.1 (Continued)

Causative Organism (vector)	Diseases/ Infection	Mode of Transmission	Epidemiology and Distribution	Clinical Manifestation	Current Therapy	Phytopharmaceutical Therapy	References
Cryptosporidium species (host: mammals including humans, birds, reptiles)[a]	Cryptosporidiosis	Fecal-oral contamination or ingestion of contaminated water or food	Found worldwide including United States.	Commonly observed symptom include dehydration, fever, nausea, stomach cramps, vomiting, watery diarrhea, weight loss	Nitazoxanide	In a study, *Allium sativum*, *Zingiber officinale*, and Mirazid (a *Commiphora molmol* derived oleo gum resin) were evaluated against *Cryptosporidium* (in vivo). *Allium sativum* showed eradication of oocysts of the infected mice from the intestine and stool; *Zingiber officinale* supplementation suggested that the immunomodulatory, anti-inflammatory as well as antioxidant property might have contributed to the correction of elevated inflammatory risk factors. In short, all three extracts showed symptomatic improvement which were significant. In a highthroughput phenotypic screening, baicalein and cedrelone were found to be more effective and able to kill the parasite *Cryptosporidium parvum*. Another study suggests the effectiveness of a mixture of *Ficus carica* and *Olea europaea* leaves extract against *Cryptosporidium*	[150–152]

Giardia species (host varies with different *Giardia* species, e.g., *G. lamblia*: humans, dog, wild species; *G. muris*: rodents)[a]	Giardiasis	Ingestion of contaminated water or food	Worldwide occurrence with prevalence in children and warm climatic regions	Dehydration, diarrhea, foul-smelling, greasy stools that tend to float, flatulence, stomach cramps, nausea, and vomiting are short term symptoms. With progressiveness, increasing fatigue and weight loss can be observed	Metronidazole, nitazoxanide, and tinidazole, are effective medications against *Giardia* infections. Furazolidone, quinacrine, and paromomycin are other medications that can also be used	*Artemisia ludoviciana* extracts and leaf infusions, *Zanthoxylum liebmannianum* preparations in the management of the disease. Asarinin in *Z. liebmannianum* more effective against *Giardia duodenalis* compared to other constituents	[73]
Cyclospora cayetanensis (main host: humans)[a]	Cyclosporosis	Fecal-oral contamination	Commonly found in the subtropical and tropical regions	In disease-endemic regions, infections can be asymptomatic. Symptomatic manifestations can be varying severity in watery diarrhea, which can further develop abdominal symptoms such as cramps, and sometimes low-grade fever and headache	CDC recommends treatment with a combination of two antibiotics namely sulfamethoxazole and trimethoprim	Not available	–

(Continued)

Table 12.1 (Continued)

Causative Organism (vector)	Diseases/ Infection	Mode of Transmission	Epidemiology and Distribution	Clinical Manifestation	Current Therapy	Phytopharmaceutical Therapy	References
Urogenital Tract Infections							
Trichomonas vaginalis (main host: humans)[a]	*Trichomonas vaginalis* infection/ trichomoniasis	About 70% people are infected via sexual contact and minority can be via nonsexual transmission of trichomonas through sharing of formites	Distributed worldwide and has higher prevalence in people with venereal diseases or multiple sexual partners.	In men, frequently asymptomatic but occasional symptoms might occur such as epididymitis, prostatitis, and urethritis. Unlike men, in women infection causes frequent symptoms to occur, prominently vaginitis with a pus like discharge along with abdominal pain, cervical and vulvar lesions, dyspareunia and dysuria	Recommended regimen of metronidazole and an alternative regimen of tinidazole	α-tomatine (*Solanum lycopersicum* leaves), betulinic acid and ursolic acid (*Platanus acerifolia* barks, and *Malus domestica* peels), candimine (*Hippeastrum morelianum*), curcumin (*Curcuma longa*), dalcochinin-8′-O-β-D-glucoside (Thai plants), geraniol (*Amomum tsao-ko* Crevost and Lemari'e essential oil), eucalyptus (*Eucalyptus camaldulensis*), lycorine (plants of Amaryllidaceae family), 2-O-lauroyllycorine (*Hippeastrum santacatarina* bulbs), procyanidin, prodelphindin (*Manilkara rufula* leaves), quercetin, quercetrin (*Dorstenia* species), resveratrol, and various extracts of *Arbutus unedo*, *Carica papaya* and *Cocos nucifera*, and *Cussonia* species	[5, 78, 79]

Systemic Infections							
Plasmodium falciparum (intermediate host: humans; main host: female *Anopheles* mosquitoes)[a,b]	Malaria	Transmitted by the bite of mosquito, blood transfusion	*P.falciparum* is more frequent in terms of infection compared to other *Plasmodium* species. Restricted to subtropical, tropical regions, the Korean peninsula, and altitudes below 1,500 m	Abdominal pain, chills, diarrhea, fever, headache. muscle or joint pain, nausea and vomiting.	Chloroquine phosphate, Hydroxychloroquine, Primaquine phosphate, Tafenoquine In 2021, WHO has recommended RTS,S/AS01 being the first malarial vaccine to be used in sub-Saharan Africa regions having moderate to high transmission	Artemisinin (*Artemisia annua*), cinchonine, cinchonidine, quinine, quinidine (*Cinchona officinalis* bark), *Feretia apodanthera*, *Securidaca longepedunculata*, *Guiera senegalensis*, *Combretum micranthum*, *Morinda citrifolia*, and *Pycnanthus angolensis* utilized in traditional medicine for the treatment of malaria. Various extracts of the plants such as *Combretum collinum*, *Anthocleista nobilis*, *Securinega virosa*, *Lophira lanceolata*, *Terminalia avicenoides*, *Tapinanthus dodoneifolius*, *Celtis integrifolia*, *Ficus capraefolia*, *Oppilia celtidifolia*, and *Cordia myxa*	[5]

(Continued)

Table 12.1 (Continued)

Causative Organism (vector)	Diseases/ Infection	Mode of Transmission	Epidemiology and Distribution	Clinical Manifestation	Current Therapy	Phytopharmaceutical Therapy	References
Trypanosoma brucei (Vector: *Glossina* genus (tsetse flies); main reservoir: humans)[a,b]	Sleeping sickness (African trypanosomiasis)	Transmission via tsetse fly	Eastern and southeastern African rural regions. About less than 100 cases were reported in 2020 to WHO out of which 89 cases infected by *T. b. rhodesiense* are from Malawi	East African trypanosomiasis-acute, progresses within 1–3 weeks to the next stage, chancre development post tsetse fly bite. West African trypanosomiasis has a slower progression of about three years compared to that of East African trypanosomiasis. Chancre is rarely seen in West African trypanosomiasis. Common symptoms include extreme fatigue, fever, joint and muscle ache, irritability, severe headaches, lymph nodes swelling, and sometimes skin rashes in few. In the second stage where there is central nervous system invasion, neurologic problems including personality changes, and progressive confusion may develop. If left untreated can become fatal.	First stage *T. b. gambiense* infection: pentamidine, suramin; *T. b. rhodesiense*: suramin. Second stage *T. b. gambiense*: combination of nifurtimox-eflornithine; *T. b. rhodesiense*: melarsoprol accompanied by prednisolone is the only drug that can be used for treatment and to lessen the risk of encephalopathy	*Garcinia huillensis, Warburgia salutaris, Acalypha hispida, Vernonia auriculifera, Securinega virosa, Diospyros verrucosa, Moringa stenopetala, Albizia gummifera, Lantana camara, Azadirachta indica, Caralluma penicillata, Caralluma russelliana, Caralluma tuberculate, V. subuligera, Argimone mexicana, Tinospora bakis, Aristolochia bracteolate*, and *Ehretia amoena* have trypanocidal actions. *Kerivoula africana* and leaves of *Annona senegalensis* (in vivo) were found effective in *Trypanosoma brucei rhodesiense* and *Trypanosoma brucei brucei* respectively.	[5]

Tissue Infections							
Trypanosoma cruzi (intermediate host: humans and triatomine insects; main host: female *Anopheles* mosquitoes)[a,b]	Chagas' disease (American trypanosomiasis)	Transmission via feces/urine contaminated with triatomine bugs. It can also transmit via transplacentally, transplantation, blood transfusions or laboratory errors	About 6–7 million are infected worldwide. Rural areas of America, few regions of west Africa and central Africa rural regions. About less than 600 cases were reported in 2020 to WHO	Acute phase: Asymptomatic or mild symptoms such as anorexia body aches, diarrhea, fatigue, fever, headache, vomiting. Additionally, chagoma hepatomegaly, lymphadenopathy splenomegaly can appear, but can be severe in case of immunocompromized patients such as those infected with HIV or undergoing chemotherapy. Chronic phase: Can be asymptomatic, about 30% develop megacolon, megaesophagus, arrhythmia, cardiomegaly, cardiac failure or arrest	Benznidazole and nifurtimox	Extracts of *Crinum erubescens*, *Rhodophiala andicola*, *Ageratum conyzoides*, *Handroanthus impetiginosa*, identified as highly active and specific against *T. cruzi* (in vitro)	[114, 115]

(Continued)

Table 12.1 (Continued)

Causative Organism (vector)	Diseases/ Infection	Mode of Transmission	Epidemiology and Distribution	Clinical Manifestation	Current Therapy	Phytopharmaceutical Therapy	References
Acanthamoeba species (Host: humans)	*Acanthamoeba* keratitis[a]	*Acanthamoeba* keratitis can be transmitted to eye via skin wounds, contact lens usage, or inhalation into the lungs	85% cases occur in contact lens wearers and one to two cases per million contact lens users have chances of occurrence of the disease in the United States	Amoebic keratitis: Blurred vision, eye pain and redness, light sensitivity, excessive tearing	Amoebic keratitis: chlorhexidine and polyhexamethylene biguanide (PHMB) (0.02%) individually or in combination; chlorhexidine+ natamycin is also effective; topical adjuncts such as ketoconazole, itraconazole are also used	*Curcuma longa*, *Eryngium planum*, *Peucedanum palimbioides*, *Croton ericoides*, *Teucrium polium*, *Peucedanum caucasicum*, *Croton isabelli*, *Tanacetum vulgare*, *Pterocaulon polystachyum*, *Solidago virgaurea*, *Rubus chamaemorus*, *Teucrium chamaedrys*, *Peucedanum chryseum*, *Pancratinum maritimum*, *Croton pallidulus*, *Pouzolzia indica*, *Peucedanum longibracteolatum*, *Solidago graminifolia*, *Arachis hypogaea*, *Pueraia lobata*, and *Eryngium maritimumi*; Yaundentine hydrochloride (isolated from Betulinic acid), botulin (isolated from *Yemeni lawsoniainermis* L), and lupeol, kolavenic acid, quercetin (isolated from the crude extract of *Caesalpinia pulcherrima* and *Polyalthia longifolia* var *pendula*) possess anti-acanthamoebic activities	[6, 18]
	Granulomatous amoebic encephalitis (GAE)[c]	GAE occurs in immunocompromized people; might enter the body through skin/lungs and via bloodstream reach the brain	In the United States, Hispanic Americans are most infected	GAE: Headaches, lethargy (tiredness), fever (low-grade), nausea. light sensitivity, neck stiffness or pain in neck and head, vomiting	GAE: Treatment with flucytosine, fluconazole pentamidine, and sulfadiazine, in addition to this clarithromycin or azithromycin followed by surgical resection to clear CNS lesion		

Leishmania species (Vectors: Sandflies; Host: mammals including humans)[a]	Leishmaniasis	Transmitted by the bite of phlebotomine sand flies	Over 1.3 million cases of leishmaniasis occurs annually with a 20,000 to 30,000 deaths, commonly affecting poor countries, 70% eastern Mediterranean, 23% America, 4% Africa, 2% Europe, 1% South East Asia. About 700,000 to 1.2 million or more new cases are reported per year for cutaneous leishmaniasis, but is estimated to have decreased to 100,000 or less in visceral leishmaniasis which ranged upto 400,000 previously. Occurs about 39% in South East Asia, 24% Africa, 19% eastern Mediterranean, 15% America, 2% western Pacific, 1% Europe	Cutaneous leishmaniasis can be asymptomatic, for symptomatic cases one or more painless or painful sores develop which can change into papules or nodules and may turn out as ulcers which may be covered by scab or crust. Some develop swollen glands near the sores. Visceral leishmaniasis can be asymptomatic, for symptomatic cases being more severe, there is development of anemia, fever, hepatomegaly, loss of weight, leukopenia, splenomegaly, thrombocytopenia Mucosal leishmaniasis is less common and be a consequence of cutaneous leishmaniasis, developing mucous membrane destruction	Cutaneous leishmaniasis Systemic therapy: amphotericin B deoxycholate, pentamidine isethionate, miltefosine, ketoconazole, itraconazole, and fluconazole. Local therapy: cryotherapy, thermotherapy, topical application of paromomycin. Visceral leishmaniasis Liposomal amphotericin B, miltefosine, paromomycin, pentamidine isethionate Mucosal leishmaniasis systemic therapy: amphotericin B deoxycholate	*Alstonias cholaris*, *Acorus calamus*, *Berberis* and several natural products such as curcumin, coumarins, flavonoids, saponins, lignans, iridoids, polyketides, cyclic peptides, monoterpenes, chalcones, toxoids, and quinoline alkaloids aristate, were found to possess excellent anti-leishmanial action; mahanine (carbazole alkaloid) from *Murraya koenigii* produced apoptosis via the arrest of the cell cycle, fragmentation of DNA, as well as the externalization of phosphatidyl serine in *L. donovani* cultures	[34, 119, 120, 122, 123]
Toxoplasma gondii (Main host: domestic cats and their relatives)[a]	Toxoplasmosis	Fecal-oral contamination, transplacentally, blood transfusions and organ transplantations, undercooked food or food contaminated by contaminated knives, utensils or cutting boards	More than 40 million in the United States are infected. Regions of lower altitudes, humidity and heat are better environment for oocysts survival hence can be found infected the most. Preference for undercooked meat has brought up a high prevalence of infection in France whereas stray cats frequency is a reason for high prevalence in Central America	Body aches, fatigue, fever, headache, lymphadenopathy	Pyrimethamine and sulfadiazine, plus folinic acid in combination	Turraenthin (*Turraeanthus africanus* stem bark extracts), xanthoxyletin (*Balsamocitrus camerunensis*), *Vernonia colorata*, *Lippia multiflora*, and *Combretum micranthum* shows marked inhibitory properties against the growth of *Toxoplasma*	[122]

(Continued)

Table 12.1 (Continued)

Causative Organism (vector)	Diseases/ Infection	Mode of Transmission	Epidemiology and Distribution	Clinical Manifestation	Current Therapy	Phytopharmaceutical Therapy	References
3. Ectoparasites							
Fleas (Host: Humans, animals such as dogs, cats, rats)[a]	Flea-borne typhus, plague, or cat scratch disease.	Fecal-oral contamination	Flea-borne typhus can occur in the subtropical and tropical climates around the world	Body aches and muscle pain, cough, fever and chills, loss of appetite nausea, rash, stomach pain, vomiting	Doxycycline	Prospective double-blind randomized placebo-controlled study of plant based biotics (containing hydroalcoholic extracts of *Thymus vulgaris*, *Salvia rosmarinus*, *Melissa officinalis*, *Trigonella foenum*, *Artemisia absinthium*, and *Cymbopogon citratus*) supplemented for five consecutive months reduced flea burden	[153]
Sarcoptes scabiei (Mites)[a]	Scabies	Skin-to-skin contact with an infected person	Distribution is worldwide about 200 million, affecting almost all races and socioeconomic classes in all climatic conditions	Itching and a skin rash	0.5% malathion or 5% permethrin in aqueous base, 10–25% benzyl benzoate emulsion or 5–10% sulphur ointment	*Aegle marmelos*, *Aloe barbadensis* Mill, *Azadirachta indica*, *Capsicum annuum*, *Cinnamomum camphor*, *Curcuma longa* *Eupatorium perfoliatum*, *Ficus carica* L, *Fumaria indica* L, *Heliotropium indicum* *Lavandula* oil, *Rosmarinus officinalis*, *Solanum hannemanii*, *Solanum nigrum* L, *Syzygium aromaticum* oil *Tinospora cordifolia*	[154]

Demodex folliculorum (follicle mite)[d]	Demodicosis/ Demodex folliculitis	Human to human spreading	Rarely occurring	Burning sensation, itchy, pustules, redness, roughness on skin, scales, sensitive skin. appearance of white sheen on skin or eyelashes	Benzyl benzoate, crotamiton, ivermectin, metronidazole, permethrin, salicylic acid, selenium sulfide, and sulfur products	Weekly tea tree oil (50%) and daily tea tree shampoo can effectively eradicate ocular *Demodex*; High efficacy and lower side effects of the essential oils obtained from Camphor oil, clove oil, Galangal oil, neem oil, peppermint oil, sage oil, and tea tree oil along with its active ingredient, terpinen-4-ol	[155, 156]
Pediculus capitis (Head lice)[a]	Pediculosis	Direct contact with an infested individual or sharing of combs, clothes or brushes of infected person (head-to-head contact)	Distribution is worldwide, and usually observed in regions of poverty, war, or homelessness	Acne-like blemishes, dryness, eye redness, flushing, itching, skin redness, rough-feeling skin	Pyrethrins, permethrin or malathion	A randomized controlled, blinded, monocentric study of a mineral oil shampoo when compared with the control product (pyrethrum extract (0.3%) + piperonyl butoxide (0.7%)), kept 78% patients' lice free on single application which lasted throughout the observation period	[157]
Phthirus pubis (pubic lice)[a]	Pthiriasis	Spread via sexual contact	Distribution is worldwide occurring in all levels of the society, races and ethnic groups	Itchy, watery eyes	1% permethrin or a mousse containing pyrethrins and piperonyl butoxide, 0.5% malathion, ivermectin (oral and topical)	Studies not available	–

[a]https://www.cdc.gov.
[b]https://www.who.int.
[c]https://www.wikidoc.org/index.php/Granulomatous_amoebic_encephalitis_epidemiology_and_demographics.
[d]https://my.clevelandclinic.org/health/diseases/22775-demodex-face-mites.

The current therapeutic approaches in cystic echinococcosis depend on the characteristics of the cyst and include surgery, cyst puncture, as well as chemotherapy. Percutaneous aspiration, chemical injection (chemical protoscolicidal solutions), as well as reaspiration are newer strategies employed instead of surgery. Chemotherapy utilizes benzimidazoles, mebendazole and albendazole. Albendazole in a dosage of 10–15 mg per kg body weight and mebendazole in a dosage of 40–50 mg per kg for several months is considered effective for the infection. Regarding alveolar echinococcosis, chemotherapy along with surgery, or chemotherapy alone, is utilized, and benzimidazoles are given for a minimum of two years with the patients under observation for about ten years to prevent disease progression and recurrence [25].

The medicinal plants which can be utilized for the management of the disease include *Salvia officinalis*, and seeds of *Nigella sativa, Thymus vulgaris*, and *Zingiber officinale;* they were studied for antiprotoscolices [7, 26–28] and found promising. Studies report that the plant extracts can be effective against the cysts with the help of an immunomodulatory action influencing the activity of the macrophages, production of the interleukins, and Kupffer cell proliferation, leading to a decrease in the viability of protoscolices [5, 7, 29, 30]. According to a systematic review by Ali et al., various phytochemicals such as ampelopsin, berberine, carvacrol, gallic acid, genistein, menthol, thymol, and thymoquinone possess protoscolicidal efficacy that can be used to treat echinococcosis [31]. Another study suggested the potential of ampelopsin, used clinically against cancer and other diseases, which can be extracted from *Ampelopsis grossedentata* has potential to treat echinococcosis due to its potent action against both *Echinococcus granulosus* and *E. multilocularis* without causing hepatotoxicity [32].

12.3.1.2 Trematodes

12.3.1.2.1 Schistosomiasis

Schistosomiasis (bilharzia), infection is caused by the parasitic worms *Schistosoma mansoni, Schistosoma japonicum*, or *Schistosoma haematobium* [33], and is treated utilizing praziquantel and its combination with albendazole as well as ivermectin. Antimalarial drugs, including artemisinin or its derivatives, quinoline alkaloids, as well as oxamniquine, have also been utilized [17]. Curcumin can be considered to have parasiticidal properties and can be utilized against *Schistosoma* [34]. A plant, *Momordica balsamina*, showed excellent schistosomicidal actions [35, 36]. Several indigenous plants of Africa, such as *Sanguinaria canadensis, Jatropha elliptica, Plectranthus neochilus, Schefflera vinosa, Baccharis trimera, Asparagus stipularis, Allium sativum, Clerodendrum umbellatum, Hemerocallis fulva, Cleome droserifolia, Curcuma longa, Persea americana*, and *Artemisia annua* have been studied for their activities against schistosomiasis and showed inhibition of the growth as well as locomotion of the adult worms. They also disrupted the worm pairing and inhibited their ability to lay eggs [7, 37]. The essential oils derived from *Plectranthus neochilus* and *Calyptridium umbellatum* extracts were believed to promote females to be separated from the males leading to inhibition of egg laying [38]. The plants such as *Jatropha curcas* from the family Euphorbiaceae having phorbolesters, *Rumex dentatus*, and *Rheum palmatum* belonging to the family Polygonaceae having Anthraquinones, as well as saponins, show activities against molluscs including schistosomiasis vectors such as *Biomphalaria, Oncomelania*, and *Bulinus* [39].

Several African plants were explored as molluscicides as they can control the vector snails causing the disease. The plants include *Swartzia madagascariensis, Phytolacca dodecandra, Combretum imberbe, Jathropha curcas, Gardenia thunbergia, Warburgia salutaris, Sapindus saponaria, Balanites maughamii, Apodytes dimidiate, Solanum nodiflorum, Berkheya speciosa, Euclea natalensis, Combretum molle, Warburgia ugandensis*, and *Balanites aegyptiaca* [7, 40]. Other molluscicidal indigenous African plants include *Lannea edulis, Dicoma anomala, Ricinus communis, Pterocarpus angolensis, Terminalia brachystemma, Securidaca longipedunculata, Abrus precatorius, Ximenia caffra, Maytenus senegalensis, Elephantorrhiza goetzei, Vernonia amygdalina, Ozoroa insignis, Swartzia madagascariensis, Dracaena fragrans, Ambrosia maritime, Jatropha curcas, Croton machrostachys, Anacardium occidentale, Terminalia mollis, Tetrapleura tetraptera, Acioa emenii, Bombax costatum, Rothmania whitefieldii, Cryptogonone argentea, Morinda lucida, Dialium guineense, Xiris anceps, Ozoroa insignis, Rauvolfia caffra*, and *Combretum* species [7, 41, 42]. According to a study conducted by Augusto and team, *Euphorbia milii* latex can demonstrate strong schistosomastatic effects thereby reducing the chances of transmission of schistosomiasis [43].

12.3.1.3 Nematodes

12.3.1.3.1 Ascariasis

Ascariasis is an infection produced by the roundworm *Ascaris lumbricoides* affecting the small intestine. About one billion people all over the world are affected, leading to annual mortality of 60,000 people [20, 44, 45]. The mode of transmission is via contaminated water, food, and soil containing *Ascaris lumbricoides* eggs. The *Artemisia santonica* consists of santonin, a sesquiterpene lactone found to be useful against *Ascaris lumbricoides* even though there is a chance of toxicity. The *Chenopodium ambrosioides* consist of ascoridole, which is found to show anti-ascariasis activities [46]. In vitro studies were conducted exploring the anthelmintic activity of *Lippiano diflora* [47] Various plants with anti-ascariasis constituents include *Zanthoxylum zanthoxyloides, Clausena anisata*, as well as *Punica granatum*. The extracts of these plants showed marked anthelmintic action against *Ascaris suum* in a study [48]. Plants such as *Matricaria chamomilla, Ricinus communis, Syzygium aromaticum, Juglans nigra, Artemisia vulgaris, and Dryopteris filixmas*, have also been utilized for managing intestinal worms in various parts of the world [49]. Essential oils of certain plants such as *Artabotrys odoratissimus*, *Artemisia pallens*, *Capillipedium foetidum*, *Cymbopogon martini* grass, and *Eupatorium triplinerve* possess strong anthelmintic activity against *A. lumbricoides* [19]. The phytochemical arecoline from *Areca catechu* belonging to the family Arecaceae and the plant *Punica granatum* belonging to the family Lythraceae having pelletierine alkaloid can be utilized as anthelmintics and can act on the acetylcholine receptors [13, 50]. Filixic acid, a phytochemical from *Dryopteris* ferns is utilized as an anthelmintic compound in the treatment of veterinary praxis, and the plant *Dryopteris filix-mas* is a fern belonging to the family Dryopteridaceae having phloroglucinols including deaspidin and aspidin, as well as filixic acid showing vermicidal activity and can be utilized against cestodes infecting the intestine resulting in muscle paralysis in the worms [51]. Several other medicinal plants such as *Artemisia abrotanum* belonging to the family Asteraceae, *Butea frondosa* belonging to the family Fabaceae, *Thymus vulgaris* belonging to the family Lamiaceae, *Zanthoxylum liebmannianum* belonging to the family Rutaceae, *Teloxys graveolens* belonging to the family Amaranthaceae, *Embelia schimperi* belonging to the family Myrsinaceae, *Albizzia anthelmintica*, *Millettia thonningii*, and *Artemisia maritima* having santonin as a constituent, can also be utilized as anthelmintics [18].

12.3.1.3.2 Onchocerciasis

Onchocerciasis, subcutaneous filariasis caused by *Onchocerca volvulus* and transmitted by *Simuliidae*, the blackfly [7], can be treated with albendazole and ivermectin, but no therapies have been found effective against dracunculiasis [17]. A small number of plants, including African medicinal plants, have been studied for activities against dracunculiasis as well as onchocerciasis [18, 52], resulting in the screening of plant extracts, but the active phytochemicals have not yet been elucidated. Several indigenous plants of Africa such as *Discoglypremna caloneura*, *Ocimum gratissimum*, *Carapa procera*, *Margaritaria discoidea*, *Alstonia boonei*, *Morinda lucida*, *Anogeissus leiocarpus*, *Polyalthia suaveolens*, *Nauclea latifolia*, *Khaya senegalensis*, *Euphorbia hirta*, *Piliostigma thonningii*, *Homalium africanum*, *Annona senegalensis*, *Craterispermum laurinum*, and *Parquetina nigrescens* have been utilized and showed action against microfilariae as well as the adult worms, and therefore require further exploration to develop potential therapies against onchocerciasis. Few phytochemicals on which in vitro/in vivo studies were conducted providing evidence of effectiveness against onchocerciasis include anthraquinone, carapolide A, oliverine, and chrysophanic acid [7, 53–59].

12.3.1.3.3 Lymphatic Filariasis

Lymphatic filariasis is caused by the nematodes *Wuchereria bancrofti*, *Brugia timori*, and *Brugia malayi*, transmitted with the help of the vector *Anopheles*, *Mansonia*, *Aedes*, and *Culex* mosquitoes [60]. In lymphatic filariasis, ivermectin, as well as diethylcarbamazine along with albendazole, can be considered the drug of choice [17]. The drugs can be regarded as helpful in treating microfilariae but are not efficient regarding adult worms and can also produce adverse effects such as itching, cough, dizziness, fever, hypertension, rapid breathing, and tachycardia [61]. Newer therapies are urgently needed for treating filariasis resistant to standard drugs.

Several traditional medicinal plants from Chinese traditional medicine as well as from Ayurveda and other indigenous systems of medicine have been studied, and potential natural products are also being studied [18, 62]. The plant *Streblus asper* belonging to the family Moraceae, having cardiac glycosides such as strebloside and asperoside, shows activity against the filarial parasite during in vitro as well as in vivo assays and is being clinically studied [63–65]. *Streblus asper* showed efficacy in treating filarial lymphedema, and its decoction was found effective in comparison with other plants, including *Argyreia nervosa, Crataeva nurvala*, and *Butea monosperma* [18]. A study exploring *Piper betel* reported its anti-leishmanial as well as anti-filarial actions. The anti-filarial studies were done in mice models infected with *Brugia malayi*, and good effects were shown by the methanolic as well as n-hexane leaf extracts which also exhibited immunomodulatory activities. The plant roots also showed anti-filarial actions [66]. A study performed on *Setaria cervi* reported that *Butea monosperma* showed marked anti-filarial activity [20]. The study exploring *Vitex negundo* root extracts as well as *Aegle marmelos* leaf extracts done with the microfilariae *Brugia malayi* reported the loss of its motility [67]. The *Senecio nudicaulis*, as well as *Mallotus phillippensis* hydro-alcoholic leaf extracts, showed efficacy in blocking the motility of *Setaria cervi* nerve and muscle preparations; and the aqueous extract of the plants showed a blockade of stimulatory acetylcholine response in the motility of the worms [68]. Another study done with *Setaria digitata* reported that *Oldenlandia herbacea* showed good macrofilaricidal actions [61]. In a study by Choudhari [59] and by Maurya et al., [69] it was mentioned that various other phytochemical constituents contribute to the treatment of elephantiasis including 19-epischolaricine, 19-vellesamine (latex, bark of *Alstonia scholaris*), alusenone, hirsutanonol, hirustenone, and platyphyllenone (leaves of *Alnus nepalensis*), repenone, repenol (immature shoots of *Boerhavia repens*), 7,4′dihydroxyflavon, apigenin, luteonin, tetracosanoic acid (inner bark of *Dichrostachys cinerea*), demethylwedlolactone, stigmasterol, and wedelolactone (whole plant of *Eclipta alba*). (2s)-1, 2-di-o-linolenoyl-3-o-a-galactopyranosyl-(1/6)-o-b-galactopyranosyl glycerol [1], (2s)-1-ooleoyl-2-o-palmitoyl-3-o-a-galactopyranosyl-(1/6)-o-b-galactopyranosyl glycerol obtained from leaves of *Bauhinia racemose* belonging to the family Leguminosae are effective against lymphatic filarial parasites [59, 69].

The extracts of various plants such as *Adenia gummifera, Adansonia digitata, Citrus medica, Mimosa pudica, Newbouldia laevis, Sapium grahamii, Raphia farinifera, Leonotis* sp., *Physedra longipes, Kigellia africana, Afstonia boonei, Calotropis procera* (Cardenolide), *Sargentodoxa cuneate, Cleodendrum capitum, Cassia alata, Cayaponia martiana, Aloe barteri, Euphorbia balsamifera, Lantana camara, Boerhavia diffusa, Ricinus communis, Pachypodanthium staudtii, Hilleria latifolia, Alstonia congensis, Mussaenda elegans, Richiea caparoides, Cleistopholis glauca, Odyendea gabunensis, Carapa procera*, and *Delonix elata*, have been utilized widely in sub-Saharan Africa for the management of filariasis but lacks scientific evidence; therefore an exploration of these plants could lead to potential therapeutic options in the management of filariasis [7, 59, 69].

12.3.2 Protozoa

12.3.2.1 Giardiasis and Amoebiasis

Giardiasis is caused by *Giardia duodenalis*, a protozoan infecting the intestine, particularly the small intestine as well as the duodenum. Giardiasis is currently treated with the help of drugs such as tinidazole, metronidazole, nitazoxanide, furazolidone, quinacrine, and paromomycin [70]. Amoebiasis is caused by the protozoa *Entamoeba histolytica* infecting the large intestine. Amoebiasis is treated with the help of drugs such as tinidazole, metronidazole, paromomycin, diloxanide furoate, and iodoquinol [71].

Medicinal plants can also be utilized to tackle these diseases. A study reported the utilization of North American *Artemisia ludoviciana* extracts in the management of giardiasis and amoebiasis. The leaf infusions were believed to produce good results [72]. Another study reported the utilization of a Central American plant *Zanthoxylum liebmannianum* preparations, to manage amoebiasis as well as giardiasis [73]. The leaf extracts show good inhibitory activities and consist of flavonoids such as beta-sitosterol, hyperin, and asarinin, the latter showing most effectiveness against *Giardia duodenalis* and *Entamoeba histolytica*. Various plants such as *Piper betle, Zingiber zerumbet, Eclipta prostrata, Piper chaba*, and *Boesenbergia pandurate* extracts can also be utilized in the management of giardiasis and amoebiasis [6, 74–76].

12.3.2.2 *Trichomonas vaginalis*

Trichomoniasis is a disease produced by *Trichomonas vaginalis* and can be considered as a sexually transmitted infection caused by protozoa that predominantly affects women, and men to a lesser extent, producing approximately 170 million infections all over the world [5]. The therapy of trichomoniasis utilizes 5-nitroimidazoles, including tinidazole, as well as metronidazole, even though some strains of *Trichomonas* have started showing resistance against standard therapies. The potential phytochemicals, including alkaloids such as berberine, saponins, dibenzofurans, diterpenes, anthraquinones, and polyacetylenes, need further studies [77]. Some of the phytochemicals which have undergone in vitro/in vivo studies and found to have anti-trichomonas activity which are promising include α-tomatine (*Solanum lycopersicum* leaves), betulinic acid and ursolic acid (*Platanus acerifolia* barks, and *Malus domestica* peels), candimine (*Hippeastrum morelianum*), curcumin (*Curcuma longa*), dalcochinin-8′-O-β-D-glucoside (Thai plants), geraniol (*Amomum tsao-ko* Crevost and Lemari'e essential oil), eucalyptus (*Eucalyptus camaldulensis*), lycorine (plants of Amaryllidaceae family), 2-O-lauroyllycorine (*Hippeastrum santacatarina* bulbs), procyanidin, prodelphindin (*Manilkara rufula* leaves), quercetin, quercetrin (*Dorstenia* species), resveratrol, and various extracts of *Arbutus unedo, Carica papaya, Cocos nucifera,* and *Cussonia species* [78, 79]. Clinical trial on *Commiphora molmol* showed 84.6% anti-trichomonas potential [79]. Trichomoniasis also shows an association with HIV as well as cervical malignancies [77].

12.3.2.3 Malaria

Malaria is an infection caused by four species of parasite: *Plasmodium falciparum, Plasmodium ovale, Plasmodium vivax, and Plasmodium malariae,* belonging to the *Plasmodium* genus and the phylum Apicomplexa [80]. The mode of transmission is via the vector *Anopheles* mosquito. The disease affects about 250 million people resulting in the mortality of over a million people annually, and a majority of them are children [5]. *Plasmodium* species, particularly *Plasmodium falciparum,* evolved over the years, becoming resistant to various synthetic drugs, and the drug resistance can be attributed to several mechanisms, one of which is the overexpression of ABC transporters resulting in pumping out the drug from the parasitic cell and is an ATP-dependent process [5]. Artemisinin is a sesquiterpene isolated from *Artemisia annua* belonging to the family Asteraceae and is considered an important event in developing drugs for treating malaria. Artemisinin shows excellent activity even against multidrug-resistant *Plasmodium falciparum* and is seen in the Mediterranean as well as the Asian region; the plant is also utilized in Chinese traditional medicine [5, 81]. Artemisinin has also been utilized for the development of semisynthetic derivatives such as artesunate, which is the water-soluble form and is still utilized today [82]. The first effective drug in the treatment of malaria was extracted from *Cinchona officinalis* as well as other *Cinchona* species belonging to the family Rubiaceae growing mostly in south and central America. The quinoline alkaloids, including cinchonine, quinine, cinchonidine, and quinidine, are isolated from the bark of *Cinchona* and are utilized in treating malaria. Quinine was effective against *Plasmodium* during its stages in the blood [5, 81] and was utilized as the lead structure to synthesize various antimalarial drugs, including mefloquine, proguanil, chloroquine, atovaquone, and pyrimethamine. Today quinine is utilized either alone or in a combined form with antibiotics such as clindamycin, tetracycline, or doxycycline for the therapy of infections caused by *Plasmodium falciparum.* Regarding antimalarial drug development, the compound having an IC_{50} lower than 11 μM can be considered suitable for further research in the case of *Plasmodium falciparum* [5].

About 1200 medicinal plants have been utilized for the treatment of malaria and fever in various parts of the globe [5]. Several plants are known to possess phytochemicals with marked antimalarial action, but there is a lack of clinical trials [81]. Various plants such as *Feretia apodanthera, Securidaca longepedunculata, Guiera senegalensis, Combretum micranthum, Morinda citrifolia*, and *Pycnanthus angolensis* were utilized in traditional medicine for the treatment of malaria [7, 83]. Antimalarial action is widely seen in phenolics such as epigallocatechin gallate, xanthones, ellagic acid, curcumin, flavonoids, and coumarins; terpenoids such as triterpenes, sesquiterpenes, iridoids, and diterpenes; alkaloids such as isoquinoline, indole, and indolizidine; quinones, quassinoids, naphthopyrones, polyacetylenes, and cucurbitacins from the family Cucurbitaceae [5, 81, 84]. Various extracts of the plants such as *Combretum collinum, Anthocleista nobilis, Securinega virosa, Lophira lanceolata, Terminalia avicenoides, Tapinanthus dodoneifolius, Celtis integrifolia, Ficus capraefolia, Oppilia celtidifolia*, and *Cordia myxa*, have been reported to exhibit excellent antimalarial actions against *Plasmodium falciparum* resistant to chloroquine [85]. A formulation consisting of *Allium sativum* extract combined with other plant ingredients such as *Lepidium sativum, Sida schimperi, Withania somnifera, Zingiber officinale, Solanum dasyphyllum, Datura stramonium, Ocimum basilicum, Ruta chalepensis, Girardinia diversifolia, Artemisia afra, Schinus molle*, and *Otostegia integrifolia*, could be used with beneficial effects [20, 86]. The combinations of phytochemicals in the plant extracts are believed to show synergistic action against the malarial parasite [5]. An example is a polyphenol, epigallocatechin-3-gallate, which in combination with a saponin, digitonin can act synergistically and produce a reduction in the survival as well as motility of *Plasmodium berghei* [87]. Digitonin can help increase the uptake of epigallocatechin 3-gallate. Therefore, developing synergistic combinations from plant compounds can be better than pointing to a single molecule during the drug development against the malarial parasite [87]. A study documented the anti-plasmodial actions of plant extracts from *Aegle marmelos, Leucas aspera, Phyllanthus amarus, Lantana camara, Piper nigrum*, and *Momordica charantia* [88]. Another study reported significant action of plants products from *Khaya senegalensis, Cassia siamea, Aloe schweinfurthii*, and *Piliostigma thonningii* against *Plasmodium falciparum* gametocytes [89].

The plants such as *Oxygonum sinuatum, Hypoxis colchicifolia, Mimusops obtusifolia*, and *Mimusops caffra* were utilized by South African communities for the management of malaria [7]. The plants such as *Ageratum conyzoides, Momordica balsamina, Momordica charantia*, and *Diospyros monbuttensis* have been utilized in Nigeria for their antimalarial activities [90, 91], and a study showed that the plant extracts possess excellent anti-plasmodial activities [92]. A study documented about 200 indigenous plants utilized traditionally in Cameroon for the management of malaria [93] but only 26 of these plants were further explored for their efficacy as well as cytotoxicity. A study showed that the Cameroonian plants such as *Vernonia amygdalina*, *Cuviera longiflora*, *Coula edulis*, *Eucalyptus globules*, and *Dacryodes edulis* exhibit marked anti-plasmodial action against *Plasmodium falciparum* [7, 94, 95]. The plants such as *Lippia javanica*, *Microglossa pyrifolia*, *Strychnos henningnsii*, *Kigelia africana*, *Carissa edulis*, *Strychnos usambarensis*, *Maytenus heterophylla*, and *Periploca linearifolia* were utilized by the Kenyan communities for the management of malaria [96]. The hydrophilic as well as methanolic extracts of *Microglossa pyrifolia*, *Periploca linearifolia*, and *Lippia javanica* show significant antiparasitic actions against *Plasmodium falciparum* strains sensitive as well as resistant to chloroquine. *Strychnos henningnsii* methanolic extracts showed maximum activity against *Plasmodium falciparum* sensitive to chloroquine [96]. The Kenyan communities also utilize plants such as *Cyperus articulatus* rhizomes, *Boscia salicifolia*, *Clutia robusta*, *Rhus natalensis*, *Artemisia afra*, *Catharanthus roseus*, *Ximenia americana*, *and Cyathula schimperiana* for the management of malaria [7, 97, 98].

12.3.2.4 Intestinal Helminths

Helminths infecting the intestine can be treated utilizing vermifuges and vermicides. Ivermectin is being utilized as an anthelmintic drug. Mebendazole is another synthetic drug that can be utilized. Its mechanism of action is the inhibition of microtubules in the worms as well as the activity in the intestinal cells, leading to its degeneration. The plant *Chenopodium ambrosioides* belonging to the family Amaranthaceae having ascaridole as a constituent can be utilized against hookworms, but there are toxic effects, including mutagenicity [5, 13, 99].

12.3.2.5 Trypanosomiasis

Trypanosomes are parasites of less threat to humans and commonly known to infect other animal species via bloodstream [5]. Based on host range, distribution and pathogenicity, trypanosomes can be divided into different species namely *Trypanosoma brucei*, *T. equiperdum*, and *T. evansi*. *T. brucei* are widely distributed in the Tropical African regions and are transmited via tsetse-cyclical transmission utilising both wild as well as domestic animals. Based on the variation in gene expression *T. brucei* is further subdivided into *T. b. brucei*, *T. b. gambiense*, and *T. b. rhodesiense*. In addition to domestic and wild mammals, the latter two subspecies are known for human infectivity responsible for causing Human African trypanosomiasis (HAT) or sleeping sickness, whereas *T. b. brucei* causes Nagana in cattle, and hence is used in research for screening assays. *T. equiperdum* is distributed worldwide and is transmitted mainly in horses via copulation. *T. evansi* is also distributed in South America, Asia and northern regions of Africa and is mainly transmitted in bovids, camels, dogs and horses via bloodsucking flies [100]. Currently treatment of HAT is done using synthetic compounds such as pentamidine, suramin, pafuramidine, melarsoprol and eflornithine, which are few in number with more side effects, against which some strains of trypanosomes have developed resistance, which directs to the urgency of new drug development. However, screening of many medicinal plants as well as their metabolites has not yet passed clinical trials or turned from bench-side to bed-side experience. Some of the natural products such as cardiac glycosides, alkaloids, polyacetylenes, phenolics, and saponins found in Araliaceae, Asteraceae and Apiaceae have a submicromolar range of activity and selectivity and so far have been explored only by Wink [5]. Determination of the mechanism via which the phytoconstituents as well as the currently available anti-trypanosomal drugs could produce therapeutic effects is still challenging and needs to be explored [101].

Various plants with significant trypanocidal actions include *Garcinia huillensis, Warburgia salutaris, Acalypha hispida, Vernonia auriculifera, Securinega virosa, Diospyros verrucosa, Moringa stenopetala, Albizia gummifera, Lantana camara, Azadirachta indica, Caralluma penicillata, Caralluma russelliana, Caralluma tuberculate, V. subuligera, Argimone mexicana, Tinospora bakis, Aristolochia bracteolate*, and *Ehretia amoena* [5, 102–108]. The plant *Kerivoula africana* from Kenya was reported to be active against *Trypanosoma brucei rhodesiense* in animal experiments [5]. In a study, the *Annona senegalensis* leaves were able to act against the cerebrospinal fluid as well as the blood stages of *Trypanosoma brucei brucei* in animal experiments [109]. Several African plants from Nigeria, such as *Piliostigma reticulatum, Anchomanes difformis, Lannea kerstingii, Securidaca longepedunculata, Terminalia avicenoides, Parkia clappertioniana, Khaya senegalensis, Striga* sp.*, and Cassytha* sp. shows anti-trypanocidal action against *Trypanosoma congolense* and *Trypanosoma brucei brucei* [110, 111].

12.3.2.6 Chagas Disease

It is estimated that about 28 million people are at risk of Chagas disease with 15 million people infected with an annual mortality of 12,500 [5, 112]. Chagas disease is caused by the organism *Trypanosoma cruzi*, and the mode of transmission is via triatomine bugs, water or food contaminated with the organism, organ transplantation as well as blood transfusion. The therapy utilizes nifurtimox and benznidazole, which show serious side effects. The ergosterol biosynthesis antagonists, as well as inhibitors of protease x trans-sialidase, can also be utilized [112]. Many phytochemicals that have been proved effective against *Trypanosoma brucei* can also be tested and utilized for the treatment of *Trypanosoma cruzi* [15, 112, 113]. An in vitro study performed on extracts of *Ageratum conyzoides, Handroanthus impetiginosa*, and *Ruta graveolens* against *T. cruzi* concluded that intracellular amastigotes multiplication was significantly reduced and these extracts could be potential candidates against trypanosomiasis [114]. From a study including the screening of Amaryllidaceae plants, 37 extracts out of 79 were found to be active against *T. cruzi* and on further progression of the study, extracts of *Crinum erubescens* and *Rhodophiala andicola* showed excellent anti-amastigote activity. However, this study has opened doors for identifying and characterizing newer alkaloids with specific antiparasitic property [115].

12.3.2.7 *Acanthamoeba* Infections

Acanthamoeba causes three infections: granulomatous amoebic encephalitis, *Acanthamoeba* keratitis, and disseminated infection affecting various organs [116]. Several medicinal plant extracts were studied for the management of *Acanthamoeba* infections. A Mexican plant, *Buddleia cordata* fruit, bark, and root extracts consisting of linarine have been utilized for treating cutaneous *Acanthamoeba* infections, and a study showed good results while combining the plant extract with metronidazole in the treatment of dermatitis produced by *Acanthamoeba castellanii* [117]. Various plants studied and utilized include *Curcuma longa*, *Eryngium planum*, *Peucedanum palimbioides*, *Croton ericoides*, *Teucrium polium*, *Peucedanum caucasicum*, *Croton isabelli*, *Tanacetum vulgare*, *Pterocaulon polystachyum*, *Solidago virgaurea*, *Rubus chamaemorus*, *Teucrium chamaedrys*, *Peucedanum chryseum*, *Pancratinum maritimum*, *Croton pallidulus*, *Pouzolzia indica*, *Peucedanum longibracteolatum*, *Solidago graminifolia*, *Arachis hypogaea*, *Pueraia lobata*, and *Eryngium maritimum* [6]. According to a study, betulinic acid, botulin (isolated from *Yemeni lawsoniainermis* L), and lupeol, kolavenic acid, quercetin (isolated from the crude extract of *Caesalpinia pulcherrima* and *Polyalthia longifolia* var *pendula*) possess anti-acanthamoebic activities [118].

12.3.2.8 Leishmaniasis

Leishmaniasis can be classified into visceral, cutaneous, diffuse, and mucocutaneous types. Visceral leishmaniasis can be considered a lethal disease, producing about 60,000 deaths annually [5, 119]. Leishmaniasis occurs when macrophages in the host get invaded by protozoa belonging to the genus *Leishmania*. Leishmaniasis can occur along with HIV leading to a poor prognosis. The synthetic drugs utilized for the treatment include meglumine, stibogluconate, pentamidine, and amphotericin B, which can result in various side effects [112, 120]. Various newer drugs include paromomycin, miltefosine, and 8-aminoquinoline sitamaquinen [119]. Regarding the research in finding newer drugs, analyzing the amastigote form in the macrophages can be complex compared to analyzing the promastigotes residing inside the insects. Recent advances utilize transgenic amastigote that can show an expression of the reporter genes, including GFP [119, 121].

A study explored the in vitro effect of the leaf extracts of ten plants against *L. donovani* and reported a reduction in the viability of the parasites. Three of the plants, including *Alstonias cholaris, Acorus calamus*, and *Berberis aristate*, were found to possess excellent anti-leishmanial action [122]. Several natural products under investigation include curcumin, coumarins, flavonoids, saponins, lignans, iridoids, polyketides, cyclic peptides, monoterpenes, chalcones, toxoids, and quinoline alkaloids for potential anti-leishmanial actions [34, 119, 120]. The mahanine (carbazole alkaloid) from *Murraya koenigii* produced apoptosis via the arrest of the cell cycle, fragmentation of DNA, as well as the externalization of phosphatidyl serine in *L. donovani* cultures. The macrophages which were infected showed anti-amastigote actions with the help of the generation of reactive oxygen species while undergoing treatment with mahanine [123]. Phytochemical berberine is also considered a promising drug against leishmaniasis [112, 120]. *Cedrus deodara* leaf extracts were also studied for their action against leishmaniasis and among them the benzene extracts showed excellent anti-leishmanial as well as immunomodulatory properties [124]. The plants *Kalanchoe pinnata* belonging to the family Crassulaceae, and *Baccharis retusa*, belonging to the family Asteraceae, show the presence of flavonoids as well as flavanones that can be utilized against leishmaniasis [119]. Several indigenous plants were studied for their utilization in leishmaniasis. The plants possessing leishmanicidal activities include *Warbugia ugandensis, Moringa stenopetala, Allium sativum, Kalanchoe pinnata, Aloe secundiflora, Plumbago capensis, Callistemon citrinus, Baccharis retusa, Coriandrum sativum, Acacia nilotica, Lippia sidoides*, and *Ambrosia miratima* [5, 125–128].

12.3.2.9 Toxoplasmosis

Infection by the intracellular parasite *Toxoplasma gondii*, produce toxoplasmosis in about 30% of the population worldwide and is a leading cause of mortality [7, 129]. The disease is treated with pyrimethamine, sulfadiazine, and folinic acid [130]. Medicinal plants are also used indigenously for the management of the disease. Stem bark extracts from the plant *Turraeanthus africanus* showed good antiparasitic activity and inhibited the growth of the parasite by 55%, which can be attributed to the compound turraenthin [131]. The plant *Balsamocitrus camerunensis* was utilized indigenously by the people of the Central African Republic and Cameroon to manage the disease. A study reported that the xanthoxyletin isolated from the plant could produce significant inhibition of parasitic growth by 82.12 percent [132]. The other anti-plasmodial plants include *Vernonia colorata, Lippia multiflora*, and *Combretum micranthum*, shows marked inhibitory properties against the growth of *Toxoplasma* [133, 134].

12.4 Conclusion and Future Directions

Many parasites infecting humans now show resistance to chemotherapy, pointing to the importance of developing alternatives that can be considered the need of the hour. In many cases, vaccines have not yielded fruitful results, reiterating the importance of screening to find small molecules active against parasites. A vast number of medicinal plants, as well as phytochemicals, have been studied in the case of trypanosomiasis and malaria but studies regarding other parasitic infections are very few. Most studies exploring the antiparasitic activities of medicinal plant extracts as well as phytochemicals are limited to in vitro experiments pointing to the importance of an urgent need for in vivo studies to translate the results into clinical trials and further drug development. The use of computational methods to screen phytochemicals as potential antiparasitic compounds can expedite the antiparasitic drug discovery process. Exploration of combinatorial effects of antiparasitic phytopharmaceuticals can probably utilize the synergistic and multi-targeted mechanisms leading to potential novel antiparasitic formulations. Even though there are many approved synthetic drugs, the wide availability and cheapness of plants means phytotherapy is utilized in several regions of the world as a part of traditional medicine, helping the underdeveloped, densely populated, and poverty struck regions of the world in the management of several parasitic infections. The resistance to synthetic therapy, affordability and unavailability of drugs in densely populated underdeveloped areas, and long-term management in chronic conditions all favor the utilization of phytopharmaceuticals as an adjunct or alternative therapeutic option against parasitic infections. An in-depth exploration of medicinal plants for antiparasitic drug development can probably lead to the development of potent antiparasitic drugs as many of the traditional medicinal plants are known to have antiparasitic actions; therefore, such efforts can hopefully repeat the triumph of artemisinin.

References

1 Athanasiadou, S., Githiori, J., and Kyriazakis, I. (2007). Medicinal plants for helminth parasite control: facts and fiction. *Animal* 1 (9): 1392–1400.
2 CDC - Parasites - About Parasites (2022). Available from: https://www.cdc.gov/parasites/about.html (accessed 12 August acca 2022).
3 CDC - Parasites - Resources - Diagnosis (2022). Available from: https://www.cdc.gov/parasites/references_resources/diagnosis.html (Accessed 12 August 2022).

4 CDC - Parasites - Parasitic Transmission (2021). Available from: https://www.cdc.gov/parasites/transmission/index.html. [Accessed 12 August 2022].

5 Wink, M. (2012). Medicinal plants: a source of anti-parasitic secondary metabolites. *Molecules* 17 (11): 12771–12791.

6 Derda, M. and Hadaś, E. (2015). The use of phytotherapy in diseases caused by parasitic protozoa. *Acta Parasitologica* 60 (1): 1–8.

7 Mwangi, V.I., Mumo, R.M., Nyachieo, A., and Onkoba, N. (2017). Herbal medicine in the treatment of poverty associated parasitic diseases: a case of sub-Saharan Africa. *Journal of Herbal Medicine* 10: 1–7.

8 Paloque, L., Triastuti, A., Bourdy, G., and Haddad, M. (2018). Natural products as antiparasitic agents. *Natural Antimicrobial Agents* 19: 215.

9 Wink, M. (2008). Evolutionary advantage and molecular modes of action of multi-component mixtures used in phytomedicine. *Current Drug Metabolism* 9 (10): 996–1009.

10 Anthony, J.P., Fyfe, L., and Smith, H. (2005). Plant active components – a resource for antiparasitic agents? *Trends in Parasitology* 21 (10): 462–468.

11 Wink, M. (2007). Molecular modes of action of cytotoxic alkaloids: from DNA intercalation, spindle poisoning, topoisomerase inhibition to apoptosis and multiple drug resistance. In: *The Alkaloids: Chemistry and Biology*. (ed. G.A. Cordell), 1–47. Academic Press.

12 Stanton, R.A., Gernert, K.M., Nettles, J.H., and Aneja, R. (2011). Drugs that target dynamic microtubules: a new molecular perspective. *Medicinal Research Reviews* 31 (3): 443–481.

13 Wink, M. and Schimmer, O. (2010). Molecular modes of action of defensive secondary metabolites. In: *Annual Plant Reviews Volume 39: Functions and Biotechnology of Plant Secondary Metabolites* (Ed. J. Roberts), 21–161. Wiley.

14 Schmeller, T., Latz-Brüning, B., and Wink, M. (1997). Biochemical activities of berberine, palmatine and sanguinarine mediating chemical defence against microorganisms and herbivores. *Phytochemistry* 44 (2): 257–266.

15 Rosenkranz, V. and Wink, M. (2008). Alkaloids induce programmed cell death in bloodstream forms of trypanosomes (Trypanosoma b. brucei). *Molecules* 13 (10): 2462–2473.

16 Wink, M. (2000). Interference of alkaloids with neuroreceptors and ion channels. In: *Studies in Natural Products Chemistry*. (ed. Atta-ur-rahman), 3–122. Elsevier. (Bioactive Natural Products (Part B); vol. 21).

17 Mullner, A., Helfer, A., Kotlyar, D. et al. (2011). Chemistry and pharmacology of neglected helminthic diseases. *Current Medicinal Chemistry* 18 (5): 767–789.

18 Murthy, P.K. and Joseph, S.K. (2011). Plant products in the treatment and control of filariasis and other helminth infections and assay systems for antifilarial/anthelmintic activity. *Planta Medica* 77 (6): 647–661.

19 Tandon, V., Yadav, A.K., Roy, B., and Das, B. (2011). *Phytochemicals as Cure of Worm Infections in Traditional Medicine Systems*. 351–378.

20 Singh, A., Mishra, A., Chaudhary, R., and Kumar, V. (2020). Role of herbal plants in prevention and treatment of parasitic diseases. *Journal of Scientific Research* 64 (01): 50–58.

21 Demma, J., Gebre-Mariam, T., Asres, K. et al. (2007). Toxicological study on *Glinus lotoides*: a traditionally used taenicidal herb in Ethiopia. *Journal of Ethnopharmacology* 111 (3): 451–457.

22 Li, T., Ito, A., Chen, X. et al. (2012 November 1). Usefulness of pumpkin seeds combined with areca nut extract in community-based treatment of human taeniasis in northwest Sichuan Province, China. *Acta Tropica* 124 (2): 152–157.

23 Ito, A., Li, T., Chen, X. et al. (2013 June). Mini review on chemotherapy of taeniasis and cysticercosis due to *Taenia solium* in Asia, and a case report with 20 tapeworms in China. *Tropical Biomedicine* 30 (2): 164–173.

24 Prevention CC for DC and. CDC - Echinococcosis. (2021). [Accessed 14 August 2022]. Available from: https://www.cdc.gov/parasites/echinococcosis/index.html.

25 Prevention CC for DC and. CDC - Echinococcosis - Resources for Health Professionals (2020). [Accessed 14 August 2022]. Available from: https://www.cdc.gov/parasites/echinococcosis/health_professionals/index.html.

26 Baqer, N.N., Khuder, M.H., and Amer, N. (2014). Antiprotoscolices effects of ethanolic extract of *Zingiber officinale* against *Echinococcus granulosus* in vitro and in vivo. *International Journal of Advanced Research* 2 (10): 11.

27 Yones, D.A., Taher, G.A., and Ibraheim, Z.Z. (2011). In vitro effects of some herbs used in Egyptian traditional medicine on viability of protoscolices of hydatid cysts. *The Korean Journal of Parasitology* 49 (3): 255–263.

28 Al-Mayah, K.S., Al-Bashir, N.M., and Al-Azzawi, B.M. (2012). In vivo efficacy of *Nigella sativa* aqueous seed extract against metacestode of echinococcus granulosus. *Medical Journal of Babylon* 9 (1).

29 Haq, A., Lobo, P.I., Al-Tufail, M. et al. (1999). Immunomodulatory effect of *Nigella sativa* proteins fractionated by ion exchange chromatography. *International Journal of Immunopharmacology* 21 (4): 283–295.

30 Salem, M.L. (2005). Immunomodulatory and therapeutic properties of the *Nigella sativa* L. seed. *International Immunopharmacology* 5 (13): 1749–1770.

31 Ali, R., Khan, S., Khan, M. et al. (2020 October 13). A systematic review of medicinal plants used against *Echinococcus granulosus*. *PLOS ONE* 15 (10): e0240456.

32 Xin, Q., Yuan, M., Li, H. et al. (2019). In vitro efficacy of ampelopsin against *Echinococcus granulosus* and *Echinococcus multilocularis*. *Journal of Veterinary Medicine* 81: 1853–1858.

33 CDC – Schistosomiasis (2021). [Accessed 14 August 2022]. Available from: https://www.cdc.gov/parasites/schistosomiasis/index.html.

34 Haddad, M., Sauvain, M., and Deharo, E. (2011). Curcuma as a parasiticidal agent: a review. *Planta Medica* 77 (6): 672–678.

35 Bandeira, S.O., Gaspar, F., and Pagula, F.P. (2001). African ethnobotany and healthcare: emphasis on Mozambique. *Pharmaceutical Biology* 39 (sup1): 70–73.

36 Ramalhete, C., Magalhães, L.G., Rodrigues, V. et al. (2012). In vitro schistosomicidal activity of Balsaminol F and Karavilagenin C. *Planta Medica* 78 (18): 1912–1917.

37 Neves, B.J., Andrade, C.H., and Cravo, P.V.L. (2015). Natural products as leads in schistosome drug discovery. *Molecules* 20 (2): 1872–1903.

38 Ndjonka, D., Rapado, L.N., Silber, A.M. et al. (2013). Natural products as a source for treating neglected parasitic diseases. *International Journal of Molecular Sciences* 14 (2): 3395–3439.

39 Liu, S.Y., Sporer, F., Wink, M. et al. (1997). Anthraquinones in *Rheum palmatum* and *Rumex dentatus* (Polygonaceae), and phorbol esters in *Jatropha curcas* (Euphorbiaceae) with molluscicidal activity against the schistosome vector snails *Oncomelania*, *Biomphalaria*, and *Bulinus*. *Tropical Medicine and International Health* 2 (2): 179–188.

40 Ojewole, J.A.O. (2004). Indigenous plants and schistosomiasis control in south africa: molluscicidal activity of some zulu medicinal plants. *The Boletin Latinoamericano Y Del Caribe de Plantas Medicinales 3*.

41 Hostettmann, K. (1984). On the use of plants and plant-derived compounds for the control of schistosomiasis. *Naturwissenschaften* 71 (5): 247–251.

42 Ndamba, J., Nyazema, N., Makaza, N. et al. (1994). Traditional herbal remedies used for the treatment of urinary schistosomiasis in Zimbabwe. *Journal of Ethnopharmacology* 42 (2): 125–132.

43 Augusto, R.D.C., Friani, G., Vasconcellos, M.C.D. et al. (2015). *Schistosoma mansoni*: phytochemical effect on aquatic life cycle. *Open Journal of Veterinary Medicine* 5 (6): 127–132.

44 Fahim, S.M., Das, S., Gazi, M. et al. (2018). Association of intestinal pathogens with faecal markers of environmental enteric dysfunction among slum-dwelling children in the first two years of life in Bangladesh. *Tropical Medicine and International Health* 23 (11): 1242–1250.

45 Darlington, C.D. and Anitha, G.F.S. (2018). Ascaridial volvulus: an uncommon cause of ileal perforation. *Iranian Journal of Medical Sciences* 43 (4): 432–435.

46 Romero-Benavides, J.C., Ruano, A.L., Silva-Rivas, R. et al. (2017). Medicinal plants used as anthelmintics: ethnomedical, pharmacological, and phytochemical studies. *European Journal of Medicinal Chemistry* 129: 209–217.

47 Rahmatullah, M., Jahan, R., Azam, F.S. et al. (2011). Folk medicinal uses of Verbenaceae family plants in Bangladesh. *African Journal of Traditional, Complementary, and Alternative Medicines* 8 (5 Suppl): 53–65.

48 Williams, A.R., Soelberg, J., and Jäger, A.K. (2016). Anthelmintic properties of traditional African and Caribbean medicinal plants: identification of extracts with potent activity against *Ascaris suum* in vitro. *Parasite* 23: 24.

49 Bahmani, M., Saki, K., Rafieian-Kopaei, M. et al. (2014). The most common herbal medicines affecting Sarcomastigophora branches: a review study. *Asian Pacific Journal of Tropical Medicine* 7: S14–21.

50 Wink, M. and Van Wyk, B.E. (2008). *Mind-altering and Poisonous Plants of the World*. Timber Press.

51 van Wyk, B.E. and Wink, M. (2018). *Medicinal Plants of the World*. 521 p. CABI.

52 Ibrahim, M.A., Nwude, N., Ogunsusi, R.A., and Aliu, Y.O. (1984). Screening of West African plants for anthelmintic activity. *ILCA Bull.* 17.

53 Samje, M., Metuge, J., Mbah, J. et al. (2014). In vitro anti-*Onchocerca ochengi* activities of extracts and chromatographic fractions of *Craterispermum laurinum* and *Morinda lucida*. *BMC Complementary and Alternative Medicine* 14 (1): 325.

54 Titanji, V.P., Evehe, M.S., Ayafor, J.F., and Kimbu, S.F. (1990). Novel *Onchocerca* volvulus filaricides from *Carapa procera*, *Polyalthia suaveolens* and *Pachypodanthium staudtii*. *Acta Leidensia* 59 (1–2): 377–382.

55 Cho-Ngwa, F., Abongwa, M., Ngemenya, M.N., and Nyongbela, K.D. (2010). Selective activity of extracts of *Margaritaria discoidea* and *Homalium africanum* on *Onchocerca ochengi*. *BMC Complementary and Alternative Medicine* 10 (1): 62.

56 Ndjonka, D., Agyare, C., Lüersen, K. et al. (2011). In vitro activity of Cameroonian and Ghanaian medicinal plants on parasitic (*Onchocerca ochengi*) and free-living (*Caenorhabditis elegans*) nematodes. *Journal of Helminthology* 85 (3): 304–312.

57 Fakae, B.B., Campbell, A.M., Barrett, J. et al. (2000). Inhibition of glutathione S-transferases (GSTs) from parasitic nematodes by extracts from traditional Nigerian medicinal plants. *Phytotherapy Research* 14 (8): 630–634.

58 Nyasse, B., Ngantchou, I., Nono, J.J., and Schneider, B. (2006). Antifilarial activity in vitro of polycarpol and 3-O-acetyl aleuritolic acid from Cameroonian medicinal plants against *Onchocerca gutturosa*. *Natural Product Reports* 20 (4): 391–397.

59 Choudhari, J.K. (2018). An perspective: potential medicinal plant resources for filariasis disease. *CSVTU International Journal of Biotechnology, Bioinformatics and Biomedical* 3 (3): 41–49.

60 Lourens, G.B. and Ferrell, D.K. (2019 June 1). Lymphatic Filariasis. *Nurse Clinical* 54 (2): 181–192.

61 Behera, D.R. and Bhatnagar, S. (2018). Filariasis: role of medicinal plant in lymphatic filariasis. *International Journal of Herbal Medicine* 7.

62 Comley, J.C. (1990). New macrofilaricidal leads from plants? *Trop Med Parasitol Off Organ Dtsch Tropenmedizinische Ges Dtsch Ges Tech Zusammenarbeit GTZ* 41 (1): 1–9.

63 Rastogi, S., Kulshreshtha, D.K., and Rawat, A.K.S. (2006). Streblus asper Lour. (Shakhotaka): a review of its chemical, pharmacological and ethnomedicinal properties. *Evidence-based Complementary and Alternative Medicine* 3 (2): 217–222.

64 Chatterjee, R.K., Fatma, N., Murthy, P.K. et al. (1992). Macrofilaricidal activity of the stembark of *Streblus asper* and its major active constituents. *Drug Development Research* 26 (1): 67–78.

65 Chamariya, R., Raheja, R., Suvarna, V., and Bhandare, R. (2022). A critical review on phytopharmacology, spectral and computational analysis of phytoconstituents from *Streblus asper* Lour. *Phytomedicine Plus* 2 (1): 100177.

66 Singh, M., Shakya, S., Soni, V.K. et al. (2009). The n-hexane and chloroform fractions of *Piper betle* L. trigger different arms of immune responses in BALB/c mice and exhibit antifilarial activity against human lymphatic filarid Brugia malayi. *International Immunopharmacology* 9 (6): 716–728.

67 Sahare, K.N., Anandhraman, V., Meshram, V.G. et al. (2008). Anti-microfilarial activity of methanolic extract of *Vitex negundo* and *Aegle marmelos* and their phytochemical analysis. *Indian Journal of Experimental Biology* 46: 128–131.

68 Singh, R., Singhal, K.C., and Khan, N.U. (1997). Antifilarial activity of *Mallotus philippensis* Lam. on *Setaria cervie* (Nematoda: Filarioidea) in vitro. *The Indian Journal of Physiology and Pharmacology* 41 (4): 397–403.

69 Maurya, S.K., Singh, A.K., and Seth, A. (2015). Potential Medicinal Plants for Lymphatic Filariasis: a review. *Journal of Critical Reviews* 2 (1).

70 (2021). Diagnosis and treatment information for medical professionals | giardia | parasites | *CDC*. [Accessed 14 August 2022]. Available from: https://www.cdc.gov/parasites/giardia/medical-professionals.html.

71 Gastrointestinal amebiasis guide: causes, symptoms and treatment options. Drugs.com. [Accessed 14 August 2022]. Available from: https://www.drugs.com/health-guide/gastrointestinal-amebiasis.html.

72 Said Fernández, S., Ramos Guerra, M.C., Mata Cárdenas, B.D. et al. (2005). In vitro antiprotozoal activity of the leaves of *Artemisia ludoviciana*. *Fitoterapia* 76 (5): 466–468.

73 Arrieta, J., Reyes, B., Calzada, F. et al. (2001). Amoebicidal and giardicidal compounds from the leaves of *Zanthoxylum liebmannianun*. *Fitoterapia* 72 (3): 295–297.

74 Sawangjaroen, N., Subhadhirasakul, S., Phongpaichit, S. et al. (2005). The in vitro anti-giardial activity of extracts from plants that are used for self-medication by AIDS patients in southern Thailand. *Parasitology Research* 95 (1): 17–21.

75 Sawangjaroen, N., Phongpaichit, S., Subhadhirasakul, S. et al. (2006). The anti-amoebic activity of some medicinal plants used by AIDS patients in southern Thailand. *Parasitology Research* 98 (6): 588–592.

76 Mr, L. (2008). Ipecacuanha: the South American vomiting root. *The Journal of the Royal College of Physicians of Edinburgh* 38 (4).

77 Gehrig, S. and Efferth, T. (2009). Development of drug resistance in *Trichomonas vaginalis* and its overcoming with natural products. *The Open Bioactive Compounds Journal* 2.

78 Hashemi, N., Ommi, D., Kheyri, P. et al. (2021). A review study on the anti-*Trichomonas* activities of medicinal plants. *International Journal for Parasitology: Drugs and Drug Resistance* 15: 92–104.

79 Bala, V. and Chhonker, Y.S. (2018). Recent developments in anti-*Trichomonas* research: an update review. *European Journal of Medicinal Chemistry* 143: 232–243.

80 (2020). Prevention CC for DC and CDC - malaria - about malaria - biology. Available from: https://www.cdc.gov/malaria/about/biology/index.html. [Accessed 12 August 2022].

81 Willcox, M. (2011). Improved traditional phytomedicines in current use for the clinical treatment of malaria. *Planta Medica* 77 (6): 662–671.

82 Thomas, K. and Ying, W. (2008). Artemisinin — an innovative cornerstone for anti-malaria therapy. In: *Natural Compounds as Drugs: Volume II*, (ed. F. Petersen and R. Amstutz), 383–422. Basel: Birkhäuser.

83 Ancolio, C., Azas, N., Mahiou, V. et al. (2002). Antimalarial activity of extracts and alkaloids isolated from six plants used in traditional medicine in Mali and Sao Tome. *Phytotherapy Research* 16 (7): 646–649.

84 Bero, J. and Quetin-Leclercq, J. (2011). Natural products published in 2009 from plants traditionally used to treat malaria. *Planta Medica* 77 (6): 631–640.

85 Sanon, S., Gansane, A., Ouattara, L.P. et al. (2013). In vitro antiplasmodial and cytotoxic properties of some medicinal plants from western Burkina Faso. *African Journal of Laboratory Medicine* 2 (1): 81.

86 Alebie, G., Urga, B., and Worku, A. (2017). Systematic review on traditional medicinal plants used for the treatment of malaria in Ethiopia: trends and perspectives. *Malaria Journal* 16 (1): 307.

87 Hellmann, J.K., Münter, S., Wink, M., and Frischknecht, F. (2010). Synergistic and additive effects of epigallocatechin gallate and digitonin on *Plasmodium* sporozoite survival and motility. *PLOS ONE* 5 (1): e8682.

88 Kamaraj, C., Kaushik, N.K., Rahuman, A.A. et al. (2012). Antimalarial activities of medicinal plants traditionally used in the villages of Dharmapuri regions of South India. *Journal of Ethnopharmacology* 141 (3): 796–802.

89 Amoah, L.E., Kakaney, C., Kwansa-Bentum, B., and Kusi, K.A. (2015). Activity of herbal medicines on *Plasmodium falciparum* gametocytes: implications for malaria transmission in Ghana. *PLOS ONE* 10 (11): e0142587.

90 Ezzat, S.M., Salem, M.A., and Zayed, A. (2021). Plants against malarial and typhoid fever. In: *Phytochemistry, the Military and Health*. (ed. A.G. Mtewa and C. Egbuna), 285–312. Elsevier.

91 Azas, N., Laurencin, N., Delmas, F. et al. (2002). Synergistic in vitro antimalarial activity of plant extracts used as traditional herbal remedies in Mali. *Parasitology Research* 88 (2): 165–171.

92 Olasehinde, G.I., Ojurongbe, O., Adeyeba, A.O. et al. (2014). In vitro studies on the sensitivity pattern of *Plasmodium falciparum* to anti-malarial drugs and local herbal extracts. *Malaria Journal* 13 (1): 63.

93 Titanji, V.P., Zofou, D., and Ngemenya, M.N. (2008 April 10). The antimalarial potential of medicinal plants used for the treatment of malaria in Cameroonian folk medicine. *African Journal of Traditional, Complementary, and Alternative Medicines* 5 (3): 302–321.

94 Zofou, D., Kengne, A.B.O., Tene, M. et al. (2011). In vitro antiplasmodial activity and cytotoxicity of crude extracts and compounds from the stem bark of *Kigelia africana* (Lam.) Benth (Bignoniaceae). *Parasitology Research* 108 (6): 1383–1390.

95 Cimanga, K., De Bruyne, T., Pieters, L. et al. (1997). In vitro and in vivo antiplasmodial activity of cryptolepine and related alkaloids from *Cryptolepis sanguinolenta*. *Journal of Natural Products* 60 (7): 688–691.

96 Ayuko, T.A., Njau, R.N., Cornelius, W. et al. (2009). In vitro antiplasmodial activity and toxicity assessment of plant extracts used in traditional malaria therapy in the Lake Victoria Region. *Memorias Do Instituto Oswaldo Cruz* 104: 689–694.

97 Kokwaro, J.O. (1993). Medicinal plants of East Africa.: published and printed by Kenya literature bureau. *Nairobi Kenya* 62 (2): 176–177.

98 Gathirwa, J.W., Rukunga, G.M., Njagi, E.N.M. et al. (2008). The in vitro anti-plasmodial and in vivo anti-malarial efficacy of combinations of some medicinal plants used traditionally for treatment of malaria by the Meru community in Kenya. *Journal of Ethnopharmacology* 115 (2): 223–231.

99 Kliks, M.M. (1985). Studies on the traditional herbal anthelmintic *Chenopodium ambrosioides* L.: ethnopharmacological evaluation and clinical field trials. *Social Science & Medicine* 21 (8): 879–886.

100 Gibson, W. (2003). Species concepts for trypanosomes: from morphological to molecular definitions? *Kinetoplastid Biology and Disease* 2 (1): 10.

101 Thomas, J.A., Baker, N., Hutchinson, S. et al. (2018). Insights into antitrypanosomal drug mode-of-action from cytology-based profiling. *PLOS Neglected Tropical Diseases* 12 (11): e0006980.

102 Ngure, R.M., Ongeri, B., Karori, S.M. et al. (2013). Anti-trypanosomal effects of *Azadiracta indica* (neem) extract on *Trypanosoma brucei rhodesiense*-infected mice. *Eastern Journal of Medicine* 14 (1): 2–9.

103 Freiburghaus, F., Kaminsky, R., Nkunya, M.H.H., and Brun, R. (1996). Evaluation of African medicinal plants for their in vitro trypanocidal activity. *Journal of Ethnopharmacology* 55 (1): 1–11.

104 Freiburghaus, F., Ogwal, E.N., Nkunya, M.H.H. et al. (1996). In vitro antitrypanosomal activity of African plants used in traditional medicine in Uganda to treat sleeping sickness. *Tropical Medicine and International Health* 1 (6): 765–771.

105 Freiburghaus, F., Jonker, S.A., Nkunya, M.H.H. et al. (1997). In vitro trypanocidal activity of some rare Tanzanian medicinal plants. *Acta Tropica* 66 (2): 79–83.

106 Mekonnen, Y., Yardley, V., Rock, P., and Croft, S. (1999). In vitro antitrypanosomal activity of *Moringa stenopetala* leaves and roots. *Phytotherapy Research* 13 (6): 538–539.

107 Nibret, E., Ashour, M.L., Rubanza, C.D., and Wink, M. (2010). Screening of some Tanzanian medicinal plants for their trypanocidal and cytotoxic activities. *Phytotherapy Research* 24 (6): 945–947.

108 Abdel-Sattar, E., Shehab, N.G., Ichino, C. et al. (2009). Antitrypanosomal activity of some pregnane glycosides isolated from *Caralluma* species. *Phytomedicine* 16 (6): 659–664.

109 Ogbadoyi, E.O., Abdulganiy, A.O., Adama, T.Z., and Okogun, J.I. (2007). In vivo trypanocidal activity of *Annona senegalensis* Pers. leaf extract against *Trypanosoma brucei brucei*. *Journal of Ethnopharmacology* 112 (1): 85–89.

110 Atawodi, S.E. (2003). Occurrence of preformed volatile nitrosamines in preparations of some Nigerian medicinal plants: a preliminary report. *Food and Chemical Toxicology* 41 (4): 551–554.

111 Atawodi, S.E., Ameh, D.A., Ibrahim, S. et al. (2002). Indigenous knowledge system for treatment of trypanosomiasis in Kaduna state of Nigeria. *Journal of Ethnopharmacology* 79 (2): 279–282.

112 Nussbaum, K., Honek, J., Cadmus, C.V.C.C.M., and Efferth, T. (2010). Trypanosomatid parasites causing neglected diseases. *Current Medicinal Chemistry* 17 (15): 1594–1617.

113 Merschjohann, K., Sporer, F., Steverding, D., and Wink, M. (2001). In vitro effect of alkaloids on bloodstream forms of *Trypanosoma brucei* and *T. congolense*. *Planta Medica* 67 (7): 623–627.

114 Teixeira, T.L., Teixeira, S.C., da Silva, C.V., and de Souza, M.A. (2014). Potential therapeutic use of herbal extracts in trypanosomiasis. *Pathogens and Global Health* 108 (1): 30–36.

115 Martínez-Peinado, N., Cortes-Serra, N., Tallini, L.R. et al. (2021). Amaryllidaceae plants: a potential natural resource for the treatment of Chagas disease. *Parasites & Vectors* 14 (1): 337.

116 (2021). CDC - acanthamoeba infection [Accessed 14 August 2022]. Available from: https://www.cdc.gov/parasites/acanthamoeba/index.html.

117 Rodríguez-Zaragoza, S., Ordaz, C., Avila, G. et al. (1999). In vitro evaluation of the amebicidal activity of *Buddleia cordata* (Loganiaceae, H.B.K.) on several strains of *Acanthamoeba*. *Journal of Ethnopharmacology* 66 (3): 327–334.

118 Siddiqui, R., Akbar, N., Khatoon, B. et al. (2022). Novel Plant-Based Metabolites As Disinfectants against *Acanthamoeba castellanii*. *Antibiotics* 11 (2): 248.

119 Tempone, A.G., de Oliveira, C.M., and Berlinck, R.G.S. (2011). Current approaches to discover marine antileishmanial natural products. *Planta Medica* 77 (6): 572–585.

120 Polonio, T. and Efferth, T. (2008). Leishmaniasis: drug resistance and natural products (review). *International Journal of Molecular Medicine* 22 (3): 277–286.

121 Dube, A., Gupta, R., and Singh, N. (2009). Reporter genes facilitating discovery of drugs targeting protozoan parasites. *Trends in Parasitology* 25 (9): 432–439.

122 Sidana, A. and Farooq, U. (2015). Evaluation of antileishmanial activity of plants used in Indian traditional medicine. *The Bangladesh Journal of Pharmacology* 10 (2): 423–426.

123 Roy, S., Dutta, D., Satyavarapu, E.M. et al. (2017). Mahanine exerts in vitro and in vivo antileishmanial activity by modulation of redox homeostasis. *Scientific Reports*. 7 (1): 4141.

124 Narayan, S., Thakur, C.P., Bahadur, S. et al. (2017). *Cedrus deodara*: in vitro antileishmanial efficacy & immumomodulatory activity. *Indian Journal of Medical Research* 146 (6): 780–787.

125 Wabwoba, B.W., Anjili, C.O., Ngeiywa, M.M. et al. (2010). Experimental chemotherapy with *Allium sativum* (Liliaceae) methanolic extract in rodents infected with *Leishmania major* and *Leishmania donovani*. *Journal of Vector Borne Diseases 47:* 160–167.

126 Mariita, R.M., Okemo, P.O., Orodho, J.A. et al. (2010). Efficacy of 13 medicinal plants used by indigenous communities around Lake Victoria, Kenya, against tuberculosis, diarrhoea causing bacteria and *Candida albicans*. *International Journal of Pharmacy and Technology* 2 (3): 771–791.

127 Kinuthia, G.K., Kabiru, E.W., Gikonyo, N.K. et al. (2012). In vitro activity of aqueous and methanolic crude extracts of *Callistemon citrinus* against *Leishmania major* [PhD Thesis]. *Doctoral Dissertation*. Kenyatta University.

128 Ngure, P.K., Tonui, W.K., Ingonga, J. et al. (2009). In vitro antileishmanial activity of extracts of *Warburgia ugandensis* (Canellaceae), a Kenyan medicinal plant. *Journal of Medicinal Plant Research* 3 (2): 061–6.

129 Lüder, C.G.K., Bohne, W., and Soldati, D. (2001). Toxoplasmosis: a persisting challenge. *Trends in Parasitology* 17 (10): 460–463.

130 CDC – Toxoplasmosis (2021). Available from: https://www.cdc.gov/parasites/toxoplasmosis/index.html. [Accessed 14 August 2022].

131 Vardamides, J.C., El Alaoui, H., Massoma, D.L. et al. (2008). Anticoccidial constituents from the stem bark of *Turraeanthus africanus*. *Chemistry of Natural Compounds* 44 (6): 696–700.

132 Happi, E.N., Mbosso, E.J.T., Nguemfo, E.L. et al. (2014). Anti-*Toxoplasma gondii* activity of constituents from *Balsamocitrus camerunensis* L (Rutaceae). *African Journal of Biotechnology* 13 (52).

133 Oketch-Rabah, H.A., Christensen, S.B., Frydenvang, K. et al. (1998). Antiprotozoal properties of 16,17-Dihydroxybrachycalyxolide from *Vernonia brachycalyx*. *Planta Medica* 64 (6): 559–562.

134 Benoit-Vical, F., Santillana-Hayat, M., Kone-Bamba, D. et al. (2000). Anti-*Toxoplasma* activity of vegetal extracts used in West African traditional medicine. *Parasite* 7 (1): 3–7.

135 Mahmoud, L.H., Basiouny, S.O., and Dawoud, H.A. (2002). Treatment of experimental heterophyiasis with two plant extracts, areca nut and pumpkin seed. *Journal of the Egyptian Society of Parasitology* 32 (2): 501–506.

136 Rhee, J.K., Baek, B.K., and Ahn, B.Z. (1985). Structural investigation on the effects of the herbs on *Clonorchis sinensis* in rabbits. *American Journal of Chinese Medicine* 13 (01n04): 119–125.

137 Wonkchalee, N., Boonmars, T., Laummaunwai, P. et al. (2013). A combination of praziquantel and the traditional medicinal plant *Thunbergia laurifolia* on *Opisthorchis viverrini* infection and cholangiocarcinoma in a hamster model. *Parasitology Research* 112 (12): 4211–4219.

138 Marques, L.T., Guedes, R.A., Rodrigues, W.D. et al. (2020). Chemical composition of various plant extracts and their in vitro efficacy in control of *Fasciola hepatica* eggs. *Ciencia Rural* 50.

139 Khalil, A.I., Radwan, N.A., Osman, G.Y. et al. (2016). In vitro effect of LC90 of albendazole and *Allium sativum* water extract on the fine structure of *Capillaria* sp. (Capillaridae Nematoda). *Journal of Bioscience and Applied Research* 2 (2): 125–135.

140 Elamathi, L. (2016). *A Clinical Evaluation of Elisevi Kudineer a Siddha Drug in the treatment of Keerinoi (Thread Worm - Enterobius vermicularis) in Children*. Masters thesis, National Institute of Siddha, Chennai, India.

141 Del Mundo, F. and Talabis, S.M. (1963). A tetrachlorethylene-chenopodium preparation in trichuriasis. *Journal of the American College of Cardiology* 39 (9): 710–715.

142 Cho-Ngwa, F., Monya, E., Azantsa, B.K. et al. (2016 August 30). Filaricidal activities on Onchocerca ochengi and Loa loa, toxicity and phytochemical screening of extracts of *Tragia benthami* and *Piper umbellatum*. *BMC Complementary and Alternative Medicine* 16 (1): 326.

143 Ayitey-Smith, E., Ashitey, G.A., Richardson, R., and Adjei, S. (1989). *Comparative clinical study of the effectiveness of Elaeophorbia drupifera* and Metronidazole in Guinea Worm Infection. *Ghana Medical Journal* 23: 139–141.

144 Robinson, R.D., Williams, L.A., Lindo, J.F. et al. (1990). Inactivation of *Strongyloides stercoralis* filariform larvae in vitro by six Jamaican plant extracts and three commercial anthelmintics. *West Indian Medical Journal* 39 (4): 213–217.

145 Abuelenain, G.L., Fahmy, Z.H., Elshennawy, A.M. et al. (2022). Phenotypic Changes of *Trichinella spiralis* Treated By *Commiphora molmol*, *Lepidium sativum*, and Albendazole: in Vitro Study. *Helminthologia* 59 (1): 37–45.

146 El-kady, A.M., Abdel-Rahman, I.A.M., Sayed, E. et al. (2022). A potential herbal therapeutic for trichinellosis. *Frontiers in Veterinary Science 9*.

147 Abou Hussien, N., Faheem, M., Sweed, E., and Ibrahim, A. (2022). Ultrastructural tegumental changes of *Trichinella spiralis* adult and larval stages after in vitro exposure to *Allium sativum*. *Experimental Parasitology* 239: 108314.

148 Rosny, H.S., Hossain, M.M.M., Hasan-Uj-Jaman, M. et al. (2016). Dietary supplementation of garlic (*Allium sativum*) to prevent *Acanthocephala* infection in aquaculture. *International Journal of Fisheries and Aquatic Studies* 4 (3): 188–192.

149 Mathison, B.A., Mehta, N., and Couturier, M.R. (2021). Human acanthocephaliasis: a thorn in the side of parasite diagnostics. *Journal of Clinical Microbiology* 59 (11): e02691–20.

150 Abouel-Nour, M.F., El-Shewehy, D.M.M., Hamada, S.F., and Morsy, T.A. (2016). The efficacy of three medicinal plants; garlic, ginger and mirazid and a chemical drug metronidazole against *Cryptosporidium parvum* : II- histological changes. *Journal of the Egyptian Society of Parasitology* 240 (3298): 1–16.

151 Jin, Z., Ma, J., Zhu, G., and Zhang, H. (2019). Discovery of novel anti-cryptosporidial activities from natural products by in vitro high-throughput phenotypic screening. *Frontiers in Microbiology 10*.

152 Abd El-Hamed, W.F., Yousef, N.S., Mazrou, Y.S.A. et al. (2021). Anticryptosporidium efficacy of *Olea europaea* and *Ficus carica* leaves extract in immunocompromised mice associated with biochemical characters and antioxidative system. *Cells* 10 (9): 2419.

153 Moog, F., Plichart, G.V., Blua, J.L., and Cadiergues, M.C. (2020). Evaluation of a plant-based food supplement to control flea populations in dogs: a prospective double-blind randomized study. *International Journal for Parasitology: Drugs and Drug Resistance* 12: 35–38.

154 Akram, M., Riaz, M., Noreen, S. et al. (2020). Therapeutic potential of medicinal plants for the management of scabies. *Dermatologic Therapy* 33 (1): e13186.

155 Gao, Y.Y. (2005). In vitro and in vivo killing of ocular *Demodex* by tea tree oil. *British Journal of Ophthalmology* 89 (11): 1468–1473.

156 Huang, C.X., Li, H.G., Luo, H.Q. et al. (2021). Essential oils for the treatment of *Demodex*. *E3S Web of Conferences* 271: 04034.

157 Wolf, L., Eertmans, F., Wolf, D. et al. (2016). Efficacy and safety of a mineral oil-based head lice shampoo: a randomized, controlled, investigator-blinded, comparative study. *PLOS ONE* 11 (6): e0156853.

13

Nanoparticles for Antiparasitic Drug Delivery

Abdulkadir Mohammed Danyaro[1], *Habibu Tijjani*[1,*], *Swinder Jeet Singh Kalra*[2], *and Ahmed Olatunde*[3]

[1] Department of Biochemistry, Bauchi State University, Gadau, Bauchi, Nigeria
[2] Department of Chemistry, D.A.V College, Kanpur, India
[3] Department of Medical Biochemistry, Abubakar Tafawa Balewa University, Bauchi, Bauchi, Nigeria
* Corresponding author

13.1 Introduction

Parasitic diseases continue to pose enormous challenges on the health of humans and animals, especially in tropical and sub-tropical areas in the world. These challenges are borne on the problems associated with the prevention and control of parasites [1]. These causative agents (parasites) are protozoans that is, single celled animals such as *Plasmodium* and amoebae; helminths or parasitic worms such as ascaris and hookworms; and arthropods like lice and flies [2]. They cause infections by either directly associating with the host, like in the case of *Plasmodium falciparum* that causes malaria or through the secretion of toxins [3]. Some are found only in animals while others are zoonotic, that is, they can be spread between animals and humans. The drugs used for their treatment are decades old, unrealistic and have become harmful to humans and animals, because of developed resistance, efficacy loss, long treatment courses, and risks of severe adverse effects, to mention just a few [1]. Others with their target parasites not yet resistant may likely become so owing to high incidences of infections and uncontrolled usage [4, 5]. To overcome these setbacks, scientists embarked on studies to develop new antiparasitic agents with high bioavailability, efficacy and fewer side effects [6]. Meanwhile, producing new drugs is an expensive and time-consuming task. Luckily, these problems associated with existing drugs can be overcome by using nanocarriers (nanoparticle systems used for drugs delivery) [6–8] brought

Parasitic Infections: Immune Responses and Therapeutics, First Edition.
Edited by Abhay Prakash Mishra and Manisha Nigam.

about by advancements made in nanotechnology and medicine. These nanoparticles are small particles that are either particulate dispersions or solid colloidal structures obtained or extracted from living organisms (natural), produced by chemical processes (synthetic), or by chemical modification of natural polymers (semi-synthetic), in which a drug can be entrapped, encapsulated, adsorbed, or attached chemically to the nanomaterial. They present great biological applications like the improved and target-selective delivery of drugs, vaccines and antioxidants because of their biodegradability, biocompatibility, and application versatility [7]. In addition, these particles are of different sizes, shapes, surface charges and surface hydrophobicities, which affects the absorption and cellular uptake of nanoparticle-conjugated drugs better than that of free drugs and give them added advantages like bioavailability, improved drug encapsulation and pharmacokinetics among others, over other forms of drug delivery systems (DDS) [9, 10]. Hence, this chapter was developed to provide insights on better ways to deliver antiparasitic drugs using nanoparticles, and as a contribution towards improving the treatment and eradication of parasitic diseases.

13.2 Parasitic Diseases (Parasitosis)

Parasitosis are chronic sicknesses brought about and spread by organisms that live in other organisms (hosts), that is endoparasites or on the host, ectoparasites, and use the host for food, shelter, development and reproduction at the expense of the host [2]. These infectious diseases could manifest either as infections that may present with fever (including malaria, African trypanosomiasis, and invasive amoebiasis); illnesses presenting lesions of the skin and/or soft tissues (such as filariasis, and cutaneous leishmaniasis); diseases involving the liver and/or gastrointestinal tract (GIT) (opisthorchiasis, and clonorchiasis); or gut worms (ascaris, hookworm, and trichuris) [11]. Their causative agents are parasites [2]. Parasitic diseases are more harmful than bacterial infections because they are rarely diagnosed early, thereby bringing about enormous health burden and economic loss [12]. This is because they pass through different life cycles and often have more than one host, starting from their vectors like arthropods before they are finally transmitted to higher hosts such as humans and animals through blood sucking, ingestion of contaminated food, and exposure of wounds (Table 13.1) [13, 14]. Their switching into different stages of life owing to favorable conditions mostly occurs in tropical/sub-tropical regions that are prevalent with malnutrition, contaminated food, poor hygiene and health status among other factors, all of which contribute to the complexities in their growth, distribution and transmission and make them difficult to prevent, control or be treated, thereby increasing incidences of infections and deaths [12, 14–17].

Table 13.1 Some parasitic diseases, causative organisms and hosts.

S/N	Parasitic diseases	Family	Causative organisms	Transmitting agent (vector)
1)	Malaria	Protozoan	*Plasmodium* spp.	Mosquito
2)	Leishmaniasis	Trypanosomatidae	*Leishmania* spp.	Sand fly
3)	Chagas disease	Trypanosomatidae	*Trypanosome cruzi*	Triatomid bugs
4)	African trypanosomiasis	Trypanosomatidae	*Trypanosome brucei*	Tsetse fly
5)	Babesiosis	Babestidae	*Babesta* spp.	Tick
6)	Toxoplasmosis	Felidae	*Toxoplasma gondii*	Parasite oocyst
7)	Trichomoniasis	Trichomonididae	*Trichomonas vaginalis*	Genital contacts
8)	Giardiasis (beaver fever)	Hexamitidae	*Giardia* spp.	Fecal-oral
9)	Cryptosporidiosis	Cryptosporididae	*Cryptosporidium* spp.	Oral ingestion
10)	Amoebiasis	Entamoebidae	*Entamoeba histolytica*	Fecal-oral
11)	Schistosomiasis	Schistosomatidae	*Schistosoma* spp.	Snail
12)	Onchocerciasis (river blindness)	Onchocercidae	*Onchocerca vovulus*	Roundworm
13)	Lymphatic filariasis (elephantiasis)	Onchocercidae	*Brugia malayi*, *Wuchereria bancrofti*, and *Brugia timori*	Mosquito (*Culex pipiens*)
14)	Nematode infection:	–	–	Oral ingestion
	a) Ascariasis b) Trichuriasis c) Hookworm disease d) Enterobiasis (human pin-worm) e) Strongyloidiasis f) Trichinosis g) Dirofilariasis	Ascarididae Trichuridae Ancylostomatidae Ancylostomatidae Oxyuridae Strongylidae Trichinellidae Onchocercidae	*Ascaris* spp. *Trichuris trichiura* *Ancylostoma duodenale*, *Necator americanus* *Enterobius vermicularis* *Strongyloides stercoralis* *Trichinella spiralis* *Diroflaira* spp.	Fecal-oral Oral ingestion Penetration of skin Penetration of skin Fecal-oral Penetration of human skin Oral ingestion Mosquito

Source: Pisarski, 2019; Tijjani et al., 2022c

Malaria is a common parasitic disease pinpointed by the World Health Organization (WHO) to be in the forefront among the estimated three billion infections and it causes approximately one million deaths globally [18, 19]. But collectively, zoonotic parasitic diseases that constitute the neglected tropical diseases (NTDs) constitute most of the global burden of disease (GBOD), accounting for about 60% of nearly (400) identified infectious diseases. However, the top five [5] diseases include cryptosporidiosis, intestinal nematode infection, leishmeniasis, schistosomaniasis, and lymphatic filariasis [19]. Years of heavily funded researches have not yet produced vaccines for parasitic diseases and their control is based on effective elimination of vectors and animal reservoirs as well as avoidance strategies like sanitary disposal of feces, handwashing, proper cooking of food and provision of purified water [20, 21]. Today, efficient control of parasitic diseases can be achieved by their prevention, monitoring, elimination and eradication approaches through mass drug administration to an entire population in an area, intensified disease management, limiting or eradicating vectors, application of veterinary science to improve the well-being of potential animal hosts, provision of clean water, and promoting sanitation and hygiene [22]. Antiparasitic drugs that are formulated as tablets, powders and injections [12] are used for the prevention of parasitic diseases and for their treatment. But their efficacy is hampered by their limited absorption due to their reduced solubility and stability within the gastrointestinal tract, susceptibility to enzymatic degradation or inactivation, and poor penetration of membrane barriers [23, 24]. Usage of large and/or multiple doses to obtain satisfactory effects also adds to the problems by producing resistance to antiparasitic drugs and toxicity [4]. Nevertheless, these drugs remain the mainstream in treatment of parasitosis because these problems can be overcome to improve their therapeutic effects by using nanocarriers (nanoparticles delivery systems) brought about by the advancement of nanotechnology in the field of medicine (nanomedicine). In addition, plant secondary metabolites (phytocompounds) such as terpenoids, alkaloids, flavonoids and phenolic compounds among others are also used for treating parasitic diseases because of their relative safety compared to conventional synthetic antiparasitic drugs as well as their profound therapeutic benefits and affordability to low-income earners [14, 25–27]. A good example is artemisinin (a sesquiterpene lactone) that has been reported to possess profound antimalarial activity [28]. The next sections of this chapter focus on nanoparticles and their applications in the delivery of antiparasitic drugs.

13.3 Nanoparticle Drug Delivery Systems

Nanoparticles drug delivery systems (NDDS) are systems of lipids, polymeric, metallic or miscellaneous compositions, having dimension of 100nm or less, used to specifically target pathogens, the immune system and/or other bodily tissues or cells for effective drugs delivery [8, 29, 30]. NDDS possess unique properties like biocompatibility and biodegradability that make them suitable for drug delivery [31]. They can also increase the size and solubility of the drug molecule, thereby reducing its renal clearance and immunogenicity by masking it [32]. They can be synthesized from natural or synthetic monomer subunits and drugs can be either conjugated onto their surfaces or physically encapsulated within them depending to the method of their preparations [33, 34], which could be lipid-based delivery systems (LBDS), polymeric delivery systems or metallic/inorganic delivery systems. Some of these NDDS provides initial burst drug release while others afford sustained targeted drug release suitable for chronic infections [35]. This is to enhance the bioavailability and efficacy of drugs for treatment of specific or multiple diseases by increasing their specificity, selectivity as well as to limit their side effects by reducing their dose and regularity of intake [36–38]. Researches proved that nanoparticles facilitate specificity by binding directly to over-expressed receptors on targeted cells [29] or by diffusion to target tissue through passive permeability, persistence of which enhances bioavailability [39]. They also exhibit multiple synergistic functions, as demonstrated when nifurtimox in combination with eflornithine was used to treat Human African trypanosomiases, which positively influenced the host immune response to parasites and promoted subsequent reduction of the parasite load [40, 41]. The ability of nanoparticles to encapsulate conventional drugs and directly convey it to targeted tissues, to circulate freely as macromolecular conventional drugs, and their capacity to sustainably and controllably discharge drugs over time, offer greater benefits in the treatment of chronic diseases like parasitic infections [42–44]. The NDDS applicable to nanomedicine include lipid nanoemulsion (NE), liposomes, polymeric micelles, solid lipid nanoparticles (SLNs), dendrimers, silica nanoparticles, ceramic nanoparticles, magnetic nanoparticles, gold nanoparticles, quantum dots, chitosan, among others (Table 13.2).

Table 13.2 Types of nanoparticles.

Nanoparticles		
Lipid-based Systems	**Polymeric Systems**	**Metallic/Inorganic Systems**
1) Liposomes	1) Nanopsheres	1) Nanospheres
2) Micelles	2) Nanocapsules	2) Nanocapsules
3) Niosomes	3) Nanofibres	3) Nanowires
4) Transferosomes	4) Nanotubes	4) Nanotubes
Lipid particulate systems	5) Nanodiscs	5) Nanodiscs
1) Nanodisc	6) Micelles	
2) Solid lipid nanoparticles (SLNs)		
3) Nanostructured lipid carriers (NLCs)		
4) Lipid conjugates		
Emulsions		
1) Nanoemulsion		

Source: Shrestha et al., 2014; Volpedo et al., 2019.

13.4 Lipid-based Drug Delivery Systems

Lipid-based drug delivery systems (LBDDS) are lipid formulations developed in various forms to provide specific and time-controlled delivery of drugs [45]. Studies have shown that, they have the ability to amplify the bioavailability and solubility of drugs. Their biocompatibility and versatility have made them gain increasing importance in repositioning therapeutic agents for curing chronic illnesses since they can traverse the blood–brain barrier [46–48]. LBDDS comprises emulsions including nanoemulsions, vesicular systems like liposomes, as well as lipid particulate systems such as solid lipid nanoparticles (SLNs).

13.4.1 Liposomes

Liposomes were the first studied drug nanodelivery systems [49]. They are artificial spherical shaped vesicles of about 80 nm composed of steroids (mostly cholesterol) and non-toxic phospholipids that enhances drugs' solubility, safety, biotransformation, and how the body reacts to it [49–51]. Drugs are usually incorporated into engineered liposomes by an encapsulation process and their discharge is determined by the carriers' building blocks, surrounding environment, osmotic gradient and pH [49]. Their hydrophilic and hydrophobic properties made them more stable than other polymeric nanoparticles. Also, a modified form of liposomes referred to as transferosomes have been designed with the ability to elicit deformation of the membrane bilayer and wriggle through nooks and crannies of the body such as skin. These made them among the most promising systems for drugs delivery [49, 52, 53]. Generally liposomes have abilities to improve biocompatibility and bioavailability, lower immunogenicity, prevent degradation of the active components, prolong half-life, control the release and raise the safety of drugs [49, 54].

13.4.2 Lipid Nanoemulsion

Lipid nanoemulsions (NE), also known as submicron emulsions, ultrafine emulsions, or miniemulsions are nano-sized colloidal particulate systems considered thermodynamically and kinetically stable dispersions (for instance water and oil) stabilized by suitable surfactants and co-surfactants, to form a single phase [55, 56]. Presently, nanoemulsions are of three types; water in oil (w/o) type, oil in water (o/w) type, as well as bicontinuous (microdomains of water and oil interdispersed within the system) type [56]. NE as drug delivery systems increase the rate of absorption of the drugs, reduce variability in absorption of drugs, protect drugs from oxidation and hydrolysis, deliver lipophilic drugs after solubilization, enhance bioavailability, incorporate both lipophilic and hydrophilic drugs, and enhance efficacy by reducing total dose and side effects. It is non-toxic and non-irritant to skin and mucous membranes, and controls the release of the drugs [56–58].

13.4.3 Solid Lipid Nanoparticles

Solid lipid nanoparticles (SLN) are condensed LBDDS at normal body temperature and lower, that can enhance the target specificity of drugs and decrease their toxicity [30]. They are formulated using water, surfactants, and lipids that require heating to melt. SLNs were first developed in 1996 for hydrophobic drugs delivery. The SLNs, when coated with poly(ethyleneglycol) (PEG), raised the stability and solubility of such hydrophobic drugs. They also increase a drug's ability to cross of membrane bilayers and enhance their clearance by the reticuloendothelial system [59, 60]. They are usually grouped into lipid-drug conjugates (LDCs), SLNs, and nanostructured lipid carriers (NLCs), each of which can adequately protect drugs embedded in them. Both NLCs and LDCs are modified forms developed to overcome the limitations of SLNs because they have liquid lipid inside them and do not crystallize during storage. NLCs were produced to increase payload and prevent drug expulsion, while LDCs were produced for use as delivery systems for drugs that are lipophobic [49].

13.5 Polymeric Drug Delivery Systems

Polymeric drug delivery systems (PDDS) are condensed nanocarriers formulated using molecules capable of building a polymer (very large molecule) [30] that can improve delivery, efficiency and safety of drugs by dictating the timing, amount, and point of their discharge [61, 62]. The methods of their preparations could favor the formation of round polymers like nano-particles and nanosphere/micelles, or those that have an oily or watery core, surrounded by thin polymer membrane known as nanocapsules [63, 64]. Among the common PDDS used are natural polymeric drug delivery systems (nPDDS) like chitosan derivatives, arginine derivatives, cyclodextrin derivatives, poly (glycolic acid), polymeric micelles, and synthetic polymeric drug delivery systems (sPDDS) like poly(N-isopropyl acrylamide).

13.5.1 Polymeric Micelles

Polymeric micelles are self-assembled spherical mono-layered amphiphilic co-polymers, composed of hydrophilic shell that encapsulates a hydrophobic core formed in an aqueous environment that targets poorly soluble drugs for possible incorporation [65]. They are formulated mostly with poly(esters) and poly (L-amino acids) and the commonly used among them include PEG and poly (glutamic acid) (PGA) that can enhance solubility and stability, prolong half-life, reduce immune response, and ensure specific targeting of drugs. These benefits arise from the properties of the proteins and small molecules used for their formations [66].

13.5.2 Dendrimer Nanofibers

These are polymers that contain a core single atom or molecule, dendrons (monomers that attach to the core), and active groups on their surfaces that are crucial to their biocompatibility and physicochemical properties [54, 67, 68]. Dendrimers encapsulate drugs within their internal structure especially when the drugs are labile, toxic or poorly soluble [69]. They chemically attach drugs through electrostatic interactions and covalent conjugation or by physical adsorption [9, 70], so as to lower their discharge rate, prevent their cumulation at target site, and lower toxicity at other cell or tissues [69, 71].

13.5.3 Chitosan Derivatives

Chitosan nanoparticles are range of nano-sized polymers obtained from chitin by chemical or biochemical deacetylation [72]. They are bunch of nanoparticles, 10–80 nm in dimension that enhance drug delivery [72]. Unfortunately, they are not sufficiently stable in the body and this limits their utilization, unless they are modified into novel forms like chitosan-phthalate polyphosphoric acid cross-linked nanoparticles [73].

13.6 Metallic/Inorganic Drug Delivery Systems

Metallic/inorganic drug delivery systems are nanoparticles composed of different metals, metal oxides and inorganic elements like gold, silver, silica, iron and others designed as colloidal suspensions to adsorb or covalently bind drugs for target specific delivery and controlled release [9]. Most of them are undergoing clinical trials and are not fully approved as DDS [9, 30].

13.6.1 Silica Nanoparticles

Silica nanoparticles are amorphous nano-sized silica substances of various forms employed for the controlled release of drugs, which is made possible by their highly porous framework and biocompatibility [49, 74]. They are the most widely used inorganic nanoparticles for biological purposes owing to their adjustable electrical, optical, and mechanical capabilities [74]. Among their various types include xerogels, mesoporous silica (2–50nm), nanohelicles, nanozigzags, nanotubes, and nanoribbons. But the xerogels and mesoporous silica are the most used forms because they possess high porosity and enormous surface areas that easily adsorbs drugs [49, 74]. However, investigations indicate they show toxicity and adverse effects like lipid peroxidation and cell membrane damage through generation of reactive oxygen species (ROS) in cells, decrease of glutathione levels as well as induction of antioxidant enzymes [75–78] and cell apoptosis [78].

13.6.2 Magnetic Nanoparticles

These are composed of materials like iron and chemical components that can be magnetically manipulated to achieve target specific drug delivery [49, 79]. The biocompatibility of MNPs and their magnetic properties made them ideal for selective targeting of desired cells or tissues under the guidance of an external magnetic field. They are projected as promising DDS essential for the treatment of diseases. They may be in a form of a polymer, gold, and silica packed around a magnetic inner space [79]. Iron oxide (Fe_3O_4) nanoparticles are the only approved MNPs for clinical use. This may be because they naturally exist in human heart, liver and spleen as magnetite and maghemite, indicating their biocompatibility and non-toxicity at physiological concentrations [80, 81].

13.6.3 Quantum Dots Nanoparticles

Quantum dot nanoparticles (QDN) are semiconductor nanocrystals, about 2–10nm in diameter that control the discharge of drugs in response to light, heat, magnetic or electromagnetic stimulus [9]. Cai et al., [82] demonstrated that zinc oxide-QDNs decorated with PEG enhanced doxorubicin's anti-cancer activity by making it stable at physiological conditions and ensuring target specific release only at acidic intercellular conditions of tumor cells. Generally, QDNs improve the pharmacokinetics of drugs as well as their biochemical, physicochemical and therapeutic properties [83]. They are formed through a top-down approach where a bulk material is ground into very small nanomaterials like in electron beam lithography as well as through a bottom-up approach which is the self-assembly of thin layers of the basic units into semiconductor nanocrystals [84].

13.7 Nanoparticle Drug Delivery Systems for Parasitic Diseases

Advancements in nanomedicine offer promising developments in the use of nanoparticles like liposomes, SLNs, polymeric micelles and nanoparticles, nano-suspensions, and gold nanoparticles (Table 13.3), to improve the bioavailability, sustained and/or controlled release, and membrane penetrability of antiparasitic drugs [12, 85]. Investigations advocate control of the size of whatever nanoparticle is chosen, to increase tissue or cell-specific distribution of the antiparasitic drugs and the time they spend in contact with them. Their shapes can also be

Table 13.3 Nanoparticle delivery systems for antiparasitic drugs.

S/N	Type of nanoparticles	Parasitic disease	Causative organism
1)	Nano-nitazoxamide	Cryptosporidiosis	*Cryptosporidium parvum*
2)	Albendazole-chitosan micropsheres	Echinococcosis	*Echinococcus multilocularis*
3)	Copper oxide, silver NPs	*Entemoeba*	*Entamoeba histolytica cryptosporidium parvum*
4)	Silver NPs	*Fasciola*	*Fasciola hepatica*
5)	Silver Chitosan Curcumin	Giardiasis	*Giardia lambila*
6)	Gold NPs	Giardiasis	*Giardia lambila*
7)	AmB encapsulated in poly(D,L-lactide-co-glycolide)	Leishmaniasis	*Leishmania*
8)	TiO_2 Ag_2O	Leishmaniasis	*Leishmania*
9)	Chitosan	Leishmaniasis	*Leishmania onfantum*
10)	Gold NPs	Leishmaniasis	*Leishmania major*
11)	Silver NPs	Leishmaniasis	*Leishmania major*
12)	Selenium Silver	Leishmaniasis	*Leishmania major*
13)	Silver	Leishmaniasis	*Leishmania major*
14)	Silver NPs	Leishmaniasis	*Leishmania tropica*
15)	Silver NPs	Leishmaniasis	*Leishmania tropica*
16)	Chitosan-tripolyphosphate conjugated chloroquine	Plasmodia	*Plasmodium berghei*
17)	Curcuminoids-loaded lipid	*Plasmodium*	*Plasmodium berghei*
18)	Copper (II) nanohybrid solids [$LCu(CH_3COO)_2$ & d $LCuCl_2$]	Malaria	*Plasmodium falciparum*
19)	Silver NPs	*Plasmodium*	*Plasmodium falciparum*
20)	Chitosan Silver	Toxoplasmosis	*Toxoplasma gondii*
21)	Chitosan	Trichinellosis	*Trichinella spiralis*
22)	Liposomal amphotericin	Disseminated leishmaniasis	
23)	Liposomal AmB	Indian visceral leishmaniasis	*Leishmania donovani*
24)	Liposomal megluumine antimoniate	Cutaneous leishmaniasis	*Leishmania tropica*
25)	Topical AmB 0.4% liposomal gel	Cutaneous leishmaniasis	*Leishmania tropica*
26)	Leish-111f + MPL-SE vaccine	Mucosal leishmaniasis	*Leishmania braziliensis, L panamensis, or L amazonensis*

Table 13.3 (Continued)

S/N	Type of nanoparticles	Parasitic disease	Causative organism
27)	Artemether nanosuspensions	Malaria	*Plasmodium falciparum*
28)	AmB-polymeric nanoparticles	Granulomatous amoebic encephalitis	*Balamuthia mandrillaris*
39)	Spiramyci-polymeric nanparticles	Toxoplasmosis	*Toxopasma gondii*
30)	Liposomal-praziquantel	Schistosomiasis (bilharzias)	*Trematode worm*
31)	Praziquantel-SLNS	Taeniasis	Tapeworm
32)	Liposomal-fenbendazole	Toxocariasis	*Toxocara canis or T cati*

Source: Pandian et al., 2019; Sun et al., 2019; Tijjani et al., 2022c.

made into spheres, rods or into any other shape that will enhance their cell penetrability. Similarly, their surface charges and hydrophilicity or hydrophobicity must be modified, such that they will promote effective distribution and transport kinetics of the drugs [12]. These nanoparticles properties (size, shape, surface charges and surface hydrophobicity) play crucial roles in the absorption, distribution, metabolism and excretion of nanoparticles-loaded drugs.

Their interplay is necessary for the absorption of drugs either through transcellular transport absorption by endocytosis [86, 87]; through paracellular transport absorption by the opening of a tight junction on cell's top side membrane [88]; via lymphatic transport absorption [12]; or by direct drug molecular absorption [89]. Distribution of nanoparticles via blood circulation is made passively selective to the infection site by nanoparticles' shape, size, surface morphology, constituents, and administration route, and at the end reduces the drug's toxicity [90]. Their metabolism in the liver mainly depends on their composition, which is a reflection of their sizes, shapes, surface charges, and surface hydrophobicity [91], before they are finally excreted through pathway related to the size of nanoparticles and route of administration [92–95]. Also, their excretion rate depends on the charges on them [96].

13.8 Liposomes for Delivery of Antiparasitic Drugs

PEGylated-liposomes doxorubicin became the pioneer nanoformulation authorized for the treatment of cancer diseases because liposomes profoundly enhance drugs delivery as well as reduce their toxicity. Its significance to antiparasitic drugs delivery is portrayed by Khodabandeh et al., [97] in the zeroside effect and speedy curing of the most severe form of leishmaniasis (Kala-azar) that has developed resistance, using liposomes AmB in combination with IFN-γ and allopurinol. In the study above and that of Moosavian et al., [98], liposomes increase the repelling action of the antiparasitic drugs used and lower the degree to which they can damage cells. Earlier, Baruah et al., [5] established that artificial spherical shaped vesicles prepared by thin-film hydration using phosphotidylcholine, phosphotidylethanolamine, cholesterol, and 1, 2-dioeolyl-Sn-glycerol-3-phosphotidylcholine, enhanced the bioavailability and targetability of antiparasitic chemical agents for treating RBCs infected by *Plasmodium*. Berberine-loaded liposomes used for the treatment of leishmaniasis have been shown by Calvo et al., [99] to decrease the cytotoxicity of berberine by enhancing its selectivity index seven fold, while bupervaquone-liposomes (BPQ-LP) effectively eliminated the parasites causing murine leismaniasis through induction of Th1-type protective immunity as a result of elevated amount of a pro-inflamatory cytokine (TNF) and a key chemokine (MCP-1) [100]. Generally, the structure of liposomes is similar to biomembranes and can be biodegraded into non-toxic products, making them beneficial nanocarriers [4, 101–103]. Conjugating them with protein ligands before using them for antiparasitic drugs delivery not only improves their targetability, extends their distribution, and lowers their degree of harm, but their phospholipids component can also directly affect parasites in infected cells or tissues [12, 14, 104].

13.9 Solid Lipid Nanoparticles for Delivery of Antiparasitic Drugs

Solid lipid nanoparticles (SLNs) offer promising advantages over other nanoparticles because of their ability to provide swift release rate and specific targeting due to their better entrapment of hydrophobic drugs, cell penetrability, and biocompatibility [12, 105]. They are superior to other nanoparticles for brain drug delivery because they possess the good properties of lipid nanoemulsions, liposomes and polymeric nanoparticles, and are devoid of their bad properties [14]. For example, albendazole-loaded glyceryl trimyristate SLNs exhibited slower release of albendazole as well as prevent its acid degradation in the stomach, and thereby improved its bioavailabiltity [106, 107]. More recently, 2-hydroxypropyyl-β-cyclodoxtrin, a modified SLNs (m-DDSLNs) prepared by Parvez et al., [108] and used for delivery of PM combined with AmB, significantly decreased levels of *Leishmania donovani* in the liver of infected mice because of their site-specific delivery. H_2ibdtc embedded in SLNs developed by Carneiro et al., [109], to cure American trypanosomiasis, lowers parasitemia in mice 100 times more than normal benznidazole (BNZ) as well as decreases liver and heart inflammation and lesions resulting in survival of all the infected mice. Dihydroartemisinin-SLNs (DHA-SLNs) prepared through single-emulsion solvent evaporation technique using stearic acid as the solid lipid and used against *Plasmodium* demonstrated 13.9% drug loading, 62.3%

encapsulation efficiency, 0.16 PDI, 90 days storage stability and drug sustained release of over 20 hours, all of which resulted in an increased efficacy of 24.12% and enhancement of the survival time compared to free DHA [105].

13.10 Polymeric Nanoparticles for Delivery of Antiparasitic Drugs

Polymeric nanoparticles amplify the entry of drugs into cells, their retention within it, precision to targeted cell and protection against enzymatic inactivation. Hence, they are continuously evaluated for the delivery of drugs like AmB against leishmaniasis and chloroquine or artemisinin (ART) against intracellular *Plasmodium* [91, 110–113]. Polymeric nanoparticles modified with ligands exhibit stronger targeting thereby reducing damage to other tissues [114, 115]. For instance, paromomycin encapsulated in a mannose-conjugated chitosan nanoparticles prepared from dextran, affects the amastigote and promastigote stages of parasites [116]. Earlier, Kumar et al., [113] embedded AmB into iron oxide nanoparticles [Fe(GlNPs)] coated with glycine, and used it against visceral leishmaniasis. The AmB profoundly increased bioavailability, specific targeting, and produced effects twice that of free AmB. Another promising property of polymeric nanoparticles is in their recognition and binding to overexpressed surface receptors on macrophages [117] that can ameliorate the development of resistance against antiparasitic agents, loss of their physiochemical and therapeutic properties as well reducing the degree of harm [118]. A good case in point is rifampicin encapsulated in mannose-conjugated chitosan nanoparticles reported by Jain et al., [119]. The mannose promotes selective delivery to macrophages and eventually the drug-loaded nanocarrier enters and discharge rifampicin inside it. Other advantages are that polymeric nanoparticles are generally stable during gastrointestinal tract transit, and can be used for delivery of all kinds of loads because of the variation in their methods of preparations and the versatility of their ligands which could be natural, semi-synthetic or synthetic [85]. Oyeyemi et al., [120] coentrapped curcumin-artesunate (CA) within copolymer of poly (lactic acid) and poly (glycolic acid) (PLGA) in their attempt towards solving pharmacokinetics impairment caused by instability of antiparasitic drugs. The CA-PLGA nanoparticles used against *Plasmodium berghei*-infected mice suppressed parasitemia five days after oral administration of 5 mg/kg of the compound, better than the free ATS. Presently, azithromycin combined with chloroquine (CQ) that is known to exhibit synergistic effect against parasite growth, is being considered for encapsulation into PLGA nanoparticles with a view to overcoming the problem of drug resistance through intercellular targeting [121].

13.11 Nanosuspensions for Delivery of Antiparasitic Drugs

Nanosuspensions are submicron colloidal dispersions of nanosized drug particles stabilized by surfactants [122, 123]. They are used to enhance the bioavailability [124]. They are inexpensive to produce and have been in the market since the year 2000 when rapamune VR was first released for the prophylaxis of organ rejection. However, anti-parasitic nanosuspensions have not yet come into the market even though studies have been conducted for some time [12]. An example is ivermectin nanosuspensions developed by Starkloff et al., [125], with nanoparticles and amorphous, which improve ivermectin's solubility by four times. Nanosuspensions also have the advantage of improving drug brain delivery as demonstrated by Lemke et al., [126], who studied nanosuspensions of AmB coated with polysorbate 80 and sodium cholate, and found them to have significantly increased brain delivery of the drug as well as enhanced parasite inhibition in vitro model. Kayser [127] also investigated the ability of nanosuspensions in macrophages targeting when he prepared aphidicolin nanosuspensions, which he shown that they target macrophages. Meanwhile, in a study to improve the efficacy of melarsoprol, melarsoprol-nanosuspension was prepared by complexing it with a cyclic heptasaccharide called methyl β-cyclodextrin. A burst release of drug was observed, which led to extreme toxicity in mice. No further study was conducted [54].

13.12 Mesoporous Silica for Delivery of Antiparasitic Drugs

Mesoporous silica nanoparticles (MSNs) were first introduced by Beck in 1992 [128] and were first applied as alternative nanocarriers to other nanoparticles by Vallet Regí et al., [129], owing to their tunable pore size or volume, large surface area, nontoxicity, high thermal property and biocompatibility among others, which confers controlled release of a large amount of the drug [130]. Henceforth, only a few studies were done on MSNs as antiparasitic drugs delivery systems. Among these few studies include that of Urban et al., [131], which demonstrated that the presentation of CQ and PQ to *p*RBC using a class of dendrimer called poly(amidoamines) (PAMAM) reduced 96.5% of *Plasmodium falciparium*. In a recent investigation, Amolegbe et al. [132] evaluated the delivery effectiveness of MSNs-loaded quinine (QN) against a known nano-silica encapsulated ATS drug and found that MSNs-loaded quinine shows longer time of release [85].

13.13 Some Obstacles in the Use of Nanoparticles for Delivery of Antiparasitic Drugs

Much progress have been made in the use of nanoparticles for delivering antiparasitic drugs for managing diseases caused by parasites. However, the application of nanoformulations still presents some challenges. Currently, there is no antiparasitic synthetic agent nanoparticle formulations in the market as a result of problems linked to gaps between preclinical and clinical studies [133]. These gaps include the design of safe and efficient nanoformulations. Among the challenges that translate into resistance, poor efficacy and side effects of antiparasitic drugs, is the limited studies on how the chemical and physical properties of nanoparticles decisively control their entry into cells and their retention within. This limits knowledge needed for the formulation of appropriate nanocarriers that will target parasite infected sites, and dictate the discharge its loaded drugs [12]. Another challenge is the lack of systemic study into ensuring that nanoparticles are harmless, affordable, and easily reproduceable at commercial and industrial scales. Also, current researches on the safety of nanoparticles are mainly focused on their biosafety at molecular, cells, tissues, organs and organism levels, leaving aside concerns on their environmental safety that is effects of their production processes on water, soil, etc. [12]. Again, the preparation technology of nanoparticles especially the polymeric nanoparticles are still at the laboratory stage and seem sluggish in the transition into commercial and industrial scale production required for clinical applications. This is because of paucity of knowledge in preparation methods, elimination pathways, and mechanisms of nanoparticles actions [30].

13.14 Conclusion and Future Prospects

For decades, parasitic diseases continued to undermine the public health and economy of developing countries because of the poor efficacy of decade old antiparasitic drugs arising from developed resistance and toxic side effects. The drive to overcome these parasitic burdens necessitated the development of new therapeutic options, and re-formulation of existing conventional antiparasitic drugs into nanoformulation using nanoparticles is being studied with a view to enhancing delivery and efficacy of these drugs. Today, a wide variety of these nanoparticles, including liposomes, polymeric nanoparticles, and nanoemulsions, are being created and researched pre-clinically all over the world, with some of them currently undertaking clinical trials. Fortunately, ongoing formulation designs, preclinical research, and clinical trials have demonstrated the potential of nanoparticles to enhance the absorption, sustained release, target specificity, and intracellular delivery of conventional antiparasitic drugs at sufficient concentrations to achieve desired results. However, there is currently no antiparasitic drug nanoparticle formulation in the market because of some challenges associated with gaps between preclinical and clinical phases of researches required for the production of safe and efficient nanoformulation, among other limitations. Based on these challenges documented in the use of antiparasitic drug nanoparticle systems, huge investment in terms of funding is required to cover the cost for antiparasitic drug nanoparticle formulations, their pre-clinical studies and clinical trials before release to market. Notwithstanding the highlighted challenges in this chapter, nanoparticles present a promising future in advancing the re-formulations of antiparasitic drugs for enhanced absorption, sustained release, intracellular delivery and lower toxicity, if researches are conducted to overcome the challenges. Additionally, developments in green nanotechnology for the manufacturing of liposomes and the creation of tissue-specific nanopores and nanobio-circuits (Tijjani et al., 2022a) foretell a promising future for nanomedicine against various diseases including parasitic infections.

References

1 Pink, R., Hudson, A., Mouries, M.A., and Bendia, M. (2005). Opportunities and challenges in antiparasitic drug delivery. *Nature Reviews Drug Discovery* 4: 727–740. doi: 10.1038/nrd1824.

2 Cao, B. and Guiton, P. (2018). Important human parasites of the tropics. *Frontiers for Young Minds* 6: 58. doi: 10.3389/frym.2018.00058.

3 Alsina, M.D.R., Bereket, M., Chenciner, L. et al. (2019). Parasitic diseases: a challenge to global health. [MSc Thesis]. Department of global public health. Karolinsko Institutet.

4 Balaña-Fouce, R., Reguera, R.M., Cubría, J.C., and Ordóñez, D. (1998). The pharmacology of leishmaniasis. *General Pharmacology* 30 (4): 435–443. doi: 10.1016/s0306-3623(97)00268-1.

5 Baruah, U.K., Gowthamarajan, K., Vanka, R. et al. (2017). Malaria treatment using novel nano-based drug delivery systems. *Journal of Drug Targeting* 25 (7): 567–581. doi: 10.1080/1061186X.2017.1291645.

6 Farouk, S.N., Yahaya, A., and Muhammad, A. (2018). Application of liposomes nanoparticles in the treatment of malaria: a mini review. *Journal of Biotechnology & Bioresearch* 1 (2): JBB.000510.

7 Dennis, E., Peoples, V.A., Johnson, F. et al. (2015). Utilizing nanotechnology to combat malaria. *Journal of Infectious Diseases & Therapy* 3 (229): doi: 10.4172/2332-0877.1000229.

8 McDonald, T.O., Siccardi, M., Moss, D. et al. (2015). *The Application of Nanotechnology to Drug Delivery in Medicine.* In: Nanoengineering: global approach to health and safety issues (ed P.I. Dolez). 173–223. Elsevier.

9 Patra, J.K., Das, G., Fraceto, L.F. et al. (2018). Nano based drug delivery systems: recent developments and future prospects. *Journal of Nanobiotechnology* 16 (71): doi: 10.1186/s12951-018-0392-8.

10 Roper, C., Pearce, R., Bredenkamp, B. et al. (2003). Antifolate antimalarial resistance in Southeast Africa: a population-based analysis. *Lancet* 361 (9364): 1174–1181.

11 Bailey, R. (1990). Parasitic diseases. *Practitioner* 234 (1494): 844–849.

12 Sun, Y., Chen, D., Pan, Y. et al. (2019). Nanoparticles for antiparasitic drug delivery. *Drug Delivery* 26 (1): 1206–1221. doi: 10.1080/10717544.2019.1692968.

13 Zapalski, M.K. and Hubert, B.L.M. (2011). First fossil record of parasitism in Devonian calcareous sponges (stromatoporoids). *Parasitol* 138: 132–138. doi: 10.1017/S0031182010001071.

14 Tijjani, H., Mahmoud, F.A., Abdulkarim, H. et al. (2022c). Nanotherapeutics of phytoconstituents for parasitic diseases: a short review. *Gadau Journal of Pure and Allied* 1 (1): 1–11. doi: 10.54117/gjpas.v1i1.2.

15 Smith Darr, J. and Conn, D.B. (2015). Importation and transmission of parasitic and other infectious diseases associated with international adoptees and refugees immigrating into the United States of America. *BioMed Research International* 2015: 763715. doi: 10.1155/2015/763715.

16 Torgerson, P.R., Devleesschauwer, B., Praet, N. et al. (2010). World Health Organization estimates of the global and regional disease burden of 11 foodborne parasitic diseases. A Data Synthesis. *PLoS Med.* 2015;12 (12): e1001920. doi: 10.1371/journal.pmed.1001920.

17 Leiby, D.A., O'Brien, S.F., Wendel, S. et al. WPTTID subgroup on parasites (2019). International survey on the impact of parasitic infections: frequency of transmission and current mitigation strategies. *Vox Sanguinis* 114 (1): 17–27.

18 Piperaki, E.T. and Tassios, P.T. (2016). Parasitic infections; their positions and impact in the postindustrial world. *Clinical Microbiology and Infection* 22: 469–470.

19 Pisarski, K. (2019). The global burden of disease of zoonotic parasitic diseases: top 5 contenders for priority consideration. *Tropical Medicine and Infectious Disease* 4 (1): 44. doi: 10.3390/tropicalmed4010044.

20 Witkins, B.M. (2003). Drugs for the control parasitic diseases: current status and development. *Trends in Parasitology* 19 (11): 477–478.

21 Marie, C. and Petri, W.A., Jr..(2021). Approach to parasitic infections. MSD manual professional version. (Accessed April 21 2022) from https://www.msdmanuals.com/professional/infectious-diseases/approach-to-parasitic-infections.

22 World Health Organization (2013). *Sustaining the Drive to Overcome the Global Impact of Neglected Tropical Diseases: Second WHO Report on Neglected Diseases.* Geneva, Switzerland.

23 Lu, M., Xiong, D., Sun, W. et al. (2017). Sustained release ivermectin-loaded solid lipid dispersion for subcutaneous delivery: in vitro and in vivo evaluation. *Drug Delivery* 24 (1): 622–631. doi: 10.1080/10717544.2017.1284945.

24 Sarangi, B., Jana, U., Sahoo, J. et al. (2018). Systematic approach for the formulation and optimization of atorvastatin loaded solid lipid nanoparticles using response surface methodology. *Biomed Microdev* 20 (3): 53. doi: 10.1007/s10544-018-0285-5.

25 Ali, M., Diso, S.U., Minjibir, A.A. et al. (2017). Assessment of medicinal plants used in treatment of various ailments in kano city, Nigeria. *Asian Journal of Biotechnology and Bioresource Technology* 2 (4): 1–11.

26 Al-Snafi, A.E. (2017). Nutritional and therapeutic importance of *Daucus carota*- a review. *The IOSR Journal of Pharmacy* 7 (2): 72–88.

27 Tijjani, H., Egbuna, C., and Luka, C.D. (2018). *Biosynthesis of Phytochemicals, in Phytochemistry*, 1. Chapter 2, Fundamentals, modern techniques, and applications. P. Canada: Apple Academic Press Inc. 37–78.

28 Maude, R.J., Pontavornpinyo, W., Saralamba, S. et al. (2009). The last man standing is the most resistant: eliminating artemisinin-resistant malaria in Cambodia. *Malaria Journal* 8: 1.

29 Wissing S.A., Kayser O., Müller, R.H. (2004). Solid lipid nanoparticles for parenteral drug delivery. *Advanced Drug Delivery Reviews* 56 (9): 1257–1272.

30 Volpedo, G., Costa, L., Ryan, N. et al. (2019). Nanoparticulate drug delivery systems for the treatment of neglected tropical protozoan diseases. *The Journal of Venomous Animals and Toxins Including Tropical Diseases* 25: e144118.

31 Safari, J. and Zarnegar, Z. (2014). Advanced drug delivery systems: nanotechnology of health design a review. *Journal of Saudi Chemical Society* 2014 (18): 85–89.

32 Joralemon, M.J., McRae, S., and Emrick, T. (2010). PEGylated polymers for medicine: from conjugation to self-assembled systems. *Chemical Communications (Cambridge, England)* 46: 1377–1393.

33 Li, L. and Huh, K.M. (2014). Polymeric nanocarrier systems for photodynamic therapy. *Biomatrixcs Research* 18: 19.

34 Peer, D., Karp, J.M., Hong, S. et al. (2007). Nanocarriers as an emerging platform for cancer therapy. *Nature Nanotech* 2: 751–760.

35 Singh, R. and Lillard, J.W., Jr. (2009). Nanoparticle-based targeted drug delivery. *Experimental and Molecular Pathology* 86 (3): 215–223.

36 Costa, P.M., Wang, J.T., Morfin, J.F. et al. (2018). Functionalised carbon nanotubes enhance brain delivery of amyloid-targeting pittsburgh compound B (PiB)-derived ligands. *Nanotheranostics* 2 (2): 168–183. doi: 10.7150/ntno.23125.

37 Kalimuthu, K., Lubin, B.C., Bazylwich, A. et al. (2019). Gold nanoparticles stabilize peptide-drug-conjugates for sustained targeted drug delivery in cancer cells. *Journal of Nanobiotechnology* 16 (1): 34.

38 Shen, Y., Cao, B., Snyder, N.R. et al. (2018). ROS responsive resveratrol delivery from LDLR peptide conjugated PLA-coated mesoporous silica nanoparticles across the blood–brain barrier. *Journal of Nanobiotechnology* 16 (1): 13. doi: 10.1186/s12951-018-0340-7. PMID: 29433522; PMCID: PMC5810018.

39 Petros, R.A. and deSimone, J. (2010). Strategies in the design of nanoparticles for therapeutic applications. *Nature Reviews Drug Discovery* 9: 618–627.

40 Bahnson, S.M., Kassam, H.A., Moyer, T.J. et al. (2016). Targeted nitric oxide delivery by supramolecular nanofibers for the prevention of restenosis after arterial injury. *Antioxid Redox Signaling* 24 (8): 401–418. doi: 10.1089/ars.2015.6363.

41 Contreras Lancheros, C.A., Pelegrino, M.T., Kian, D. et al. (2018). Selective antiprotozoal activity of nitric oxide-releasing chitosan nanoparticles against *Trypanosoma cruzi*: toxicity and mechanisms of action. *Current Pharmaceutical Design* 24 (7): 830–839. doi: 10.2174/1381612824666180209105625.

42 Look, M., Bandyopaadhyay, A., Blum, J.S., and Fahmy, T.M. (2010). Application of nanotechnologies for improved response against infectious diseases in the developing world. *Advanced Drug Delivery Reviews* 62: 378–393. doi: 10.1371/journal.pntd.0001582.

43 Mehtani, D., Seth, A., Sharma, P. et al. (2019). Biomaterials and bionanotechnology advances in pharmaceutical product development and research. Academic Press. Cambridge, MA, USA: Chapter 4, Biomaterials for sustained and controlled delivery of small drug molecules. 89–152. doi: 10.1016/B978-0-12-814427-5-00004-4.

44 Kunjiappan, S., Pavadai, D., Vellaichamy, S. et al. (2020). Surface receptor-mediated targeted drug delivery systems for enhanced cancer treatment: a state-of-the-art review. *Drug Development and Research* 82 (3): 309–340. doi: 10.1002/ddr.21758.

45 Shrestha, H., Bala, R., and Arora, S. (2014). Lipid-based drug delivery systems. *Journal in Pharmacology (Cairo)* 2014: 801820. doi: 10.1155/2014/801820.

46 Müller, R.H., Radtke, M., and Wissing, S.A. (2002). Solid lipid nanoparticles (SLN) and nanostructured lipid (NLC) in cosmetic and dermatological preparations. *Advanced Drug Delivery Reviews* 54 (1): 5135–5155. doi: 10.1016/S0169-409X(02)000118-7.

47 Nelson, O.A., Patrick, O.O., and Ndidi, N.C. (2009). Nanotechnology and drug delivery part 1: background and applications. *The Tropical Journal of Pharmaceutical Research* 8 (3): 265–274.

48 Patel, M., Souto, E.B., and Singh, K.K. (2013). Advances in brain drug targeting and delivery: limitations and challenges of solid lipid nanoparticles. *Expert Opinion on Drug Delivery* 10: 889–905.

49 Wilczewska, A.Z., Niemirowicz, K., Markiewicz, K.H., and Car, H. (2012). Nanoparticles as drug delivery systems. *Pharmacological Reports* 64: 1020–1037.

50 Dos Santos Giuberti, C., de Oliveira Reis, E.C., Ribeiro Rocha, T.G. et al. (2011). Study of the pilot production process of long-circulating and pH-sensitive liposomes containing cisplatin. *Journal of Lipid Research* 21: 60–69.

51 Sunderland, C.J., Steiert, M., Talmadge, J.E. et al. (2006). Targeted nanoparticles for detecting and treating cancer. *Drug Development and Research* 67: 70–93.

52 Atrooz, O.M. (2011). Effects of alkylresorcinolic lipids obtained from acetonic extract of Jordanian wheat grains on liposome properties. *The International Journal of Biological and Chemical Sciences* 5 (5): 314–321.

53 Li, M., Du, C., Guo, N. et al. (2019). Composition, design and medical application of liposomes. *European Journal of Medicinal Chemistry* 164: 640–653.

54 Pandian, S.R.K., Panneerselvam, T., Pavadai, P. et al. (2021). Nanobased approach for the treatment of neglected tropical diseases. *Frontiers in Nanotechnology* 3: 665274.

55 Jaiswal, M., Dudhe, R., and Sharma, P.K. (2015). Nanoemulsion: an advanced mode of drug delivery system. *3 Biotech* 5 (2): 123–127.

56 Gurpreet, K. and Singh, S.K. (2018). Review of nanoemulsion formulation and characterization techniques. *Indian Journal of Pharmaceutical Sciences* 80 (5): 781–789. doi: 10.4172/pharmaceutical-sciences-10000422.

57 Ali-Boucetta, H. and Kostarelos, K. (2013). Pharmacology of carbon nanotubes: toxicokinetics, excretion and tissue accumulation. *Advanced Drug Delivery Reviews* 65: 2111–2119.

58 Hormann, K. and Zimmer, A. (2016). Drug delivery and drug targeting with parenteral lipid nanoemulsions A review. *Journal of Controlled Release* 223: 85–98.

59 Andreas, W. and Karola, V.U. (2011). Liposome technology for industrial purposes. *Journal of Drug Delivery* 201: 1–9. doi: 10.1155/2011/591325.

60 Zheng, J., Wan, Y., Elhissi, A. et al. (2014). Targeted paclitaxel delivery to tumors using cleavable PEG-conjugated solid lipid nanoparticles. *Pharmaceutical Research* 31: 2220–2233.

61 Sinha, V.R. and Khosla, L. (1998). Bio-available polymers for implantable therapeutic systems. *Drug Development and Industrial Pharmacy* 24 (12): 1129–1138. doi: 10.3109/03639809108572.

62 Sung, Y.K. and Kim, S.W. (2020). Recent advances in drug delivery systems. *Biomaterials Research* 24 (12): doi: 10.1185/s40824-020-00190-7.

63 Sanchez, A., Vila-jato, J.L., and Alonso, M.J. (2003). Development of biodegradable microsphere and nanospheres for the controlled release of cyclosporine. *American International Journal of Pharmaceutics* 99 (2–3): 263–273. doi: 10.1016/037-5173(93)90369-Q.

64 Verrecchia, T., Huve, P., Bazile, D. et al. (1993). Adsorption/desorption of human serum albumin at the surface of poly (lactic acid) nanoparticles prepared by a solvent evaporation process. *Journal of Biomedical Materials Research* 27 (8): 1019–1028. doi: 10.1002/jbm.820270807.

65 Croy, S.R. and Kwon, G.S. (2006). Polymeric micelles for drug delivery. *Current Pharmaceutical Design* 12: 4669–4684.

66 Kopecek, J. (2013). Polymer-drug conjugates: origins, progress to date and future directions. *Advanced Drug Delivery Reviews* 65: 49–59.

67 Caminade, A.M., Laurent, R., and Majoral, J.P. (2005). Characterization of dendrimers. *Advanced Drug Delivery Reviews* 57 (15): 2130–2146. doi: 10.1016/j.addr.2005.09.011.

68 Svenson, S. and Tomalia, D.A. (2005). Dendrimers in biomedical applications - reflections on the field. *Advanced Drug Delivery Reviews* 57: 2106–2129.

69 D'Emanuele, A. and Attwood, D. (2005). Dendrimer-drug interactions. *Advanced Drug Delivery Reviews* 57: 2147–2162.

70 Menjoge, A.R., Kannan, R.M., and Tomalia, D.A. (2010). Dendrimer-based drug and imaging conjugates: designs considerations for nanomedical applications. *Drug Discovery Today* 15: 171–187.

71 Malik, N., Evagorou, E.G., and Duncan, R. (1999). Dendrimer-platinate: a novel approach to cancer chemotherapy. *Anticancer Drugs* 10 (8): 767–776. PMID: 10573209.

72 Sharifi-Red, J., Quispe, C., Butnariu, M. et al. (2021). Chitosan nanoparticles as a promising tool in nanomedicine with particular emphasis on oncological treatment. *Cancer Cell International* 21: 318. doi: 10.1186/S12935-021-02025-4.

73 Saeed, R.M., Dmour, I., and Taha, M.O. (2020). Stable chitosan-based nanoparticles using polyphosphoric acid or hexametaphosphate for tandem ionotroic/covalent crosslinking and subsequent investigation as novel vehicles for drug delivery. *Front Bioengg Biotech* 8: 4. doi: 10.3389/fbioe.2020.00004.

74 Karomah, A. (2021). What are silica nanoparticles? *AZoNano*. from https://www.azonano.com/article.aspx?ArticleID=5857 (Accessed April 15 2022).

75 Liu, W., Huang, Y.W., Zhou, X.D., and Ma, Y. (2006). In vitro toxicity of silica nanoparticles in human lung cancer cells. *Toxicology and Applied Pharmacology* 217: 252–259.

76 Park, E.J. and Park, K. (2009). Oxidative stress and pro-inflammatory responses induced by silica nanoparticles in vivo and in vitro. *Tetrahedron Letters* 184: 18–25.

77 Yu, K.O., Grabinski, C.M., Schrand, A.M. et al. (2009). Toxicity of amorphous silica nanoparticles in mouse keratinocytes. *Nano Research* 11: 15–24.

78 Liu, X. and Sun, J. (2010). Endothelial cells dysfunction induced by silica nanoparticles through oxidative stress via JNK/P53 and NF-$_{K}$B pathways. *Biomat* 31: 8198–8209.

79 Mody, V.V., Cox, A., Shah, S. et al. (2013). Magnetic nanoparticle drug delivery systems for targeting tumor. *Applied Nanoscience* 4: 385–392. doi: 10.1007/s13204-013-0216-y.

80 Grassi-Schultheis, P.P., Heller, F., and Dobson, J. (1997). Analysis of magnetic material in the human heart, spleen and liver. *Biometals* 10: 351–355.

81 Garg, A. (2017). Therapeutic application of nanomaterials. *IGI Global* 290–420. doi: 10.4018/978-1-5225-0751-2ch015.

82 Cai, X., Luo, Y., Zhang, W. et al. (2016). pH-sensitive ZnO quantum dots-doxorubicin nanoparticles for lung cancer targeted drug delivery. *ACS Applied Materials & Interfaces* 8 (34): 22442–22450. doi: 10.1021/acsami.6b04933.

83 Zhao, M.K. and Zhu, E.J. (2016). The research and application of quantum dots as nano-carriers for targeted drug delivery and cancer therapy. *Nanoscale Res Letters* 11: 207. doi: qo.1186/s11671-016-1394-9.

84 Valizadeh, A., Mikaeili, H., Samiei, M. et al. (2012). Quantum dots: synthesis, bioapplications, and toxicity. *Nanoscale Research Letters* 7: 480. doi: 10.1186/1556-276x-7-480.

85 Tijjani, H., Zangoma, M.H., Olatunde, A. et al. (2022a). Nanotechnology application for effective delivery of antimalarial drugs. In *Applications of Nanotechnology in Drug Discovery and Delivery* (eds C. Egbuna, M.-A. Găman and J. Jeevanandam), Elsevier.

86 Conner, S.D. and Schmid, S.L. (2003). Regulated portals of entry into the cell. *Nature* 422: 37–44.

87 Mayor, S. and Pagano, R.E. (2007). Pathways of clathrin-independent endocytosis. *Nature Reviews Molecular Cell Biology* 8: 603–612.

88 He, J., Hou, S.X., Feng, J.F., and Cai, B.Q. (2005). Effect of particle size on oral absorption of silymarin-loaded solid lipid nanoparticles. *Zhongguo Zhong Yao Za Zhi* 30 (21): 1651–1653. Chinese. PMID: 16400939.

89 Varshosaz, J., Ahmadipour, S., Tabbakhian, M., and Ahmadipour, S. (2018). Nanocrystalization of pioglitazone by precipitation method. *Drug Research* 68 (10): 576–583. doi: 10.1055/a-0591-2506.

90 Dufort, S., Sancey, L., and Coli, J.L. (2012). Physico-chemical parameters that govern nanoparticless fate also distate rules for their molecular evolution. *Advanced Drug Delivery Reviews* 64: 179–189.

91 Asthana, S., Gupta, P.K., Jaiswal, A.K. et al. (2015). Overexpressed macrophage mannose receptor targeted nanocapsules-mediated cargo delivery approach for eradication of resident parasite: in vitro and in vivo studies. *Pharmaceutical Research* 32 (8): 2663–2677. doi: 10.1007/s11095-015-1651-0.

92 He, Q., Zhang, Z., Gao, F. et al. (2011). *In vivo* study of biodistribution and urinary excretion of silica nanoparticles formulation against leishmania in mice model. *Parasite Immunology* 38: 599–608.

93 Simpson, C.A., Salleng, K.J., Cliffel, D.E., and Feldheim, D.L. (2013). *In vivo* toxicity, biodistribution, and clearance of glutathione-coated gold nanoparticles. *Nanomed* 9 (2): 257–263. doi: 10.1016/j.nano.2012.06.002.

94 Sancey, L., Kotb, S., Truillet, C. et al. (2015). Long-term in vivo clearance of gadolinium-based AGuIX nanoparticles and their biocompatibility after systemic injection. *ACS Nano* 9 (3): 2477–2488. doi: 10.1021/acsnano.5b00552.

95 Chen, X.L., Li, J.C., and Huang, Y.Z. (2017). The biodistribution, excretion and potential toxicology of different-sized Pd nanosheets in mice following oral and intraperitoneal administration. *Biomaterials Science* 5: 2448–2455.

96 Souris, J.S., Lee, C.H., Cheng, S.H. et al. (2010 July). Surface charge-mediated rapid hepatobiliary excretion of mesoporous silica nanoparticles. *Biomaterials* 31 (21): 5564–5574. doi: 10.1016/j.biomaterials.2010.03.048.

97 Khodabandeh, M., Rostami, A., Borhani, K. et al. (2019). Treatment of resistant visceral leishmaniasis with interferon gamma in combination with liposomal amphotericin B and allopurinol. *Parasitology International* 72: 101934. doi: 10.1016/j.parint.2019.101934.

98 Moosavian, S.A., Fallah, M., and Jaafari, M.R. (2019). The activity of encapsulated meglumine antimoniate in stearylamine-bearing liposomes against cutaneous leishmaniasis in BALB/c mice. *Exp Para* 200: 30–35.

99 Calvo, A., Moreno, E., Larrea, E. et al. (2010). Berberine-loaded liposomes for the treatment of *Leishmania infantum*-infected BALB/c mice. *Pharmaceutics* 12 (9): 858. doi: 10.8390/pharmaceutics12090858.

100 de Costa-silva, T.A., AJJr, G., Lindoso, J.A. et al. (2017). Nanoliposomal buparvaquone immunomodulates leishmania infantum-infected macrophages and is highly effective in a murine model. *Ant Agents Chem* 61 (4): e02297–e16. 10.1128.AAC.02297-16.

101 She, Y.X., Basang, W.D., Dong, L.D., et al. (2010). Preparation method of small unilamellar vesicle liposome of ivermectin. PAT:CN101664389.

102 Rathore, A., Jain, A., Gulbake, A. et al. (2011). Mannosylated liposomes bearing Amphotericin B for effective management of visceral leishmaniasis. *Journal of Liposome Research* 21 (4): 333–340. doi: 10.3109/08982104.2011.575381. Epub 2011 May 25. PMID: 21612342.

103 Silva, L.D., Arrúa, E.C., Pereira, D.A. et al. (2016). Elucidating the influence of praziquantel nanosuspensions on the in vivo metabolism of *Taenia crassiceps* cysticerci. *Acta Troicap* 161: 100–105. doi: 10.1016/j.actatropica.2016.06.002.

104 Zhang, H.Y., liu, X.X., Xiao, H.B. et al. (2000). Pharmacokinetics of praziquantel liposomes in rabbits. *Chinese Journal of Veterinary Medicine* 26: 55–56.

105 Omwoyo, W.N., Melariri, P., Gathirwa, J.W.F. et al. (2016). Development, characterization and antimalarial efficacy of dihydroartemisinin loaded solid lipid nanoparticles. *Nanomed* 12: 801–809.

106 Anjali, K., Singh, K., Bharkad, G.P., and Singh, K. (2017). solid lipid nanoparticles of albendazole for treatment of *Toxocara canis* infection: in vivo efficacy studies. *Nanoscience & Nanotechnology-Asia* 7: 80–91.

107 Marslin, G., Siram, K., Liu, X. et al. (2017). Solid lipid nanoparticles of albendazole for enhancing cellular uptake and cytotoxicity against U-87 MG glioma cell lines. *Molecules* 22 (11): 2040. doi: 10.3390/molecules22112040.

108 Parvez, S., Yadagiri, G., Gedda, M.R. et al. (2020). Modified sold lipid nanoparticles encapsulated with amphotericin B and paromomycin: an effective oral combination against experimental murine visceral leishmaniasis. *Scientific Reports* 10: 1–14. doi: 10.1038/s41598-020-69276-5.

109 Carneiro, Z.A., da S. Maia, P.I., Sesti-Costa, R. et al. (2014). In vitro and in vivo trypanocidal activity of $h_2$bdtc-loaded solid lipid nanoparticles. *PLOS Neglected Tropical Diseases* 8 (5): e2847. doi: 10.1371/journal.pntd.0002847.

110 Talisuna, A.O., Bloland, P., and D'Alessandro, U. (2004). History, dynamics and public health importance of malaria parasite resistance. *Clinical Microbiology Reviews* 17: 235–254.

111 Afonso, A., Hunt, P., Cheesman, S. et al. (2006). Malaria parasites can develop stable resistance to artemisinin but lack mutations in candidate genes atp6 (encoding the sarcoplasmic and endoplasmic reticulum Ca2+ ATPase), tctp, mdr1, and cg10. *Anti-Cancer Agents Chemistry* 50 (2): 480–489. doi: 10.1128/AAC.50.2.480-489.

112 Tripathy, S., Mahapatra, S.K., Chattopadhyay, S. et al. (2013). A novel chitosan based antimalarial drug delivery against *Plasmodium berghei* infection. *Acta Troicap* 128 (3): 494–503. doi: 10.1016/j.actatropica.2013.07.011.

113 Kumar, R., Pandey, K., Sahoo, G.C. et al. (2017). Development of high efficacy peptide coated iron oxide nanoparticles encapsulated amphotericin B drug delivery system against visceral leishmaniasis. *Materials Science and Engineering C* 75: 1465–1471. doi: 10.1016/j.msec.2017.02.145.

114 Labhasetwar, V.D. and Dorle, A.K. (1990). Nanoparticles- A colloidal drug delivery system for primaquine and metronidazole. *Journal of Controlled Release* 12: 113–119.

115 Mbela, T.K.M., Puopaert, J.H., and Dumont, P. (1992). Poly (diethylmethylidenemalonate) nanoparticles as primaquine delivery system to liver. *International Journal of Pharmaceutics* 79: 29–38.

116 Esfandiari, F., Motazedian, M.H., Asgari, Q. et al. (2019). Paromomycin-loaded mannosylated chitosan nanoparticles: synthesis, characterization and targeted drug delivery against leishmaniasis. *Acta Tropica* 197: 105045. doi: 10.1016/j.actatropica.2019.105045. Erratum in: *Acta Tropica* 2019 Jul 9;105072.

117 Blanco, E., Shen, H., and Ferrari, M. (2015). Principles of nanoparticles design for overcoming biological barriers to drug delivery. *Nature Biotechnology* 33: 941–951.

118 Yasuzai, M., Khan, M., Nadhman, A., and Shahnazi, G. (2013). Drug resistance in leishmaniasis: current drug delivery systems and future perspectives. *Future Medicinal Chemistry* 5: 1877–1888. doi: 10.4155/fmc.13.143.

119 Jain, N.K., Mishra, V., and Mehra, N.K. (2013). Targeted drug delivery to macrophages. *Expert Opinion on Drug Delivery* 10: 353–367. doi: 10.1517/17425247.2013.751370.

120 Oyeyemi, O., Morenkeji, O., Afolayan, F. et al. (2018). Curcumin-artesunate based polymeric nanoparticle; antiplasmodial and toxicological evaluation in murine model. *Pharmaceutical Fronts* 9: 562. doi: 10.3389/fphar.2018.00562.

121 Jawahar, N., Baruah, U.K., and Singh, V. (2019). Co-delivery of chloroquine phosphate and azithromycin nanoparticles to overcome drug resistance in malaria through intracellular targeting. *Journal of Pharmaceutical Sciences and Research* 11 (1): 33–40.

122 Müller, R.H., Jacobs, C., and Kayser, O. (2001). Nanosuspension as particulate drug formulations in therapy rationale for development and what we can expect for the future. *Advanced Drug Delivery Reviews* 47: 3–19.

123 Patel, V.R. and Agrawal, Y.K. (2011). Nanosuspension: an approach to enhance solubility of drugs. *Journal of PharmTech Research* 2 (2): 81–87. doi: 10.4103/2231-4040.82950.

124 Sattar, A., Chen, D., Jiang, L. et al. (2017). Preparation, characterization and pharmacokinetics of cyadox nanosuspension. *Scientific Reports* 7 (1): 2289. doi: 10.1038/s41598-017-02523-4.

125 Starkloff, W.J., Bucalá, V., Palma, S.D., and Gozalez Vidal, N.L. (2016). Design and in vitro characterization of ivermectin nanocrystals liquid formulation based on a top-down approach. *Pharmaceutical Development and Technology* 22: 809–817.

126 Lemke, A., Kiderlen, A.F., Petri, B., and Kayser, O. (2010). Delivery of amphotericin B nanosuspensions to the brain and determination of activity against *Balamuthia mandrillaris* amebas. *Nanomedicine* 6 (4): 597–603. doi: 10.1016/j.nano.2009.12.004.

127 Kayser, O. (2000). Nanosuspensions for the formulation of aphidicolin to improve drug targeting effects against *Leishmania*-infected macrophages. *International Journal of Pharmaceutics* 196: 253–256.

128 Beck, J.S., Vartuli, J.C., Roth, W.J. et al. (1992). A new family of mesoporous molecular sieves prepared with liquid crystal templates. *Journal of the American Chemical Society* 114 (27): 10834–10843. doi: 10.1021/Ja00053a020.

129 Vallet Regí, M., Rámila, A., Del Real, R.P., and Pérez Pariente, J. (2001). A new property of MCM-41: drug delivery system. *Chemistry of Materials* 13: 308–311.

130 Weynom, E.J., Badawi, M.M., Yisa, A. et al. (2019). Nanotechnology for improved anti-malaria efficacy; review update. *Journal of Cell Science & Molecular Biology* 6 (3): 87–93. doi: 10.19080/IJCSMB.2019.06.555690.

131 Urban, P., Valle-Delgado, J.J., Mauro, N. et al. (2014). Use of poly (amidoamine) drug conjugates for the delivery of antimalarial to *Plasmodium*. *Journal of Controlled Release* 177: 84–95.

132 Amolegbe, S.A., Hirano, Y., Adebayo, J.O. et al. (2018). Mesoporous silica nanocarriers encapsulated antimalarials with high therapeutic performance. *ScientificReports* 8: 3078. doi: 10.1038/s41598-018-21351-8.

133 Anamika, J., Nikhar, V., Laxmikant, G. et al. (2020). Nanobiotechnological modules as molecular target tracker for the treatment and prevention of malaria: options and opportunity. *Drug Delivery and Translational Research* 10 (4): 1095–1110. doi: 10.1007/s13346-020-00770-z.

14

Vaccination Against Parasitic Infection

From Past to Current Approaches in the Development of a Vaccine

Mukesh Kr Singh[1,*], *Gurdeep Singh*[2], *Varsha Tiwari*[1], *and Abhishek Tiwari*[1,*]

[1] *Faculty of Pharmacy, IFTM University, Lodhipur-Rajput, Moradabad, Uttar Pradesh, India*
[2] *School of Pharmaceutical Sciences, Lovely Professional University, Phagwara, Punjab, India*
* *Corresponding authors*

14.1 Introduction

Infections with parasites are the root cause of one of the most serious and widespread diseases that can affect both humans and animals. Significant economic losses are connected with infections. These losses can be broken down into two categories: losses in productivity and losses arising from infertility. The management of animal parasitic diseases should have as their primary objective the enhancement of animal productivity and overall mass. As a result, it is necessary to implement control measures that are not only efficient but also cheap and long-lasting. Even though there are many effective treatments available for the treatment of the majority of important diseases, there is still a pressing need for the development of efficient vaccines. The problem of drug resistance among parasites is getting worse, there are no newly developed drugs that are effective, and there are drug residues in milk, goods made from milk, and meat. The invention of immunisations is largely attributable to these circumstances.

In many regions of the world, parasite illnesses provide a substantial challenge to the efficient production of cattle. For the past 55–60 years, antiparasitic drugs have been utilised extensively in an effort to restrict the spread of these illnesses. However, there are some diseases for which there are no effective antiparasitic treatments, and there are others for which chemotherapy is only partially

Parasitic Infections: Immune Responses and Therapeutics, First Edition.
Edited by Abhay Prakash Mishra and Manisha Nigam.

effective due to the relatively late onset of clinical signs and diagnosis following infection. Both of these situations are caused by the parasites that cause the diseases. On the other hand, parasites like ticks, mites, enteric nematodes, and avian coccidia that rapidly multiply, recycle, and disseminate across host populations call for additional continual mass medicine administration through food, drink, or spraying.

Vaccination has long been recognised as the strategy that has proven to be the most effective in lowering the number of illnesses caused by parasites in both humans and animals. Despite the years of study that have gone into it, the few vaccines for parasites that are presently available on the market all have licences to be used against parasites that affect veterinary animals.

Because of the biological complexity of parasites, the creation of vaccines can be a particularly challenging endeavor. Unlike viruses and bacteria, parasites progress through a variety of growth phases, each of which presents the host with a distinct antigenic profile. There are very few instances in which sustaining parasite populations does not necessitate passing through or creating chronic infections in their particular animal host. A significant obstacle to progress is the lack of in vitro methods for the growth of the appropriate stages of many different parasites. The ability of many parasites to alter the immune responses of their hosts in order to prevent or delay the clearance of parasites further complicates the process of vaccine development [1].

Vaccines have had a significant impact on the improvement of public health, particularly since the 1960s when national immunisation programmes were successfully established and structured for the first time. In countries where there is a high level of vaccination programme coverage, many of the diseases that were once responsible for the majority of fatalities have almost completely disappeared. According to the World Health Organization (WHO), the current vaccination schedule saves between two and three million lives annually, which contributes to a reduction in the overall mortality rate among children under the age of five across the globe [2]. The purpose of this chapter is to investigate the methods used in the creation of vaccinations against parasitic infections from the beginning of time up until the present day.

For this reason, it is recommended that a multidisciplinary approach be utilised to tackle parasitic infections. This approach should include chemotherapy, grazing management, biological control, genetic host resistance, and parasite vaccinations. Because so many parasites have developed sophisticated immune evasion mechanisms, it is currently difficult to imagine the development of vaccinations that are actually effective. However, given the rapid advances being made in immunology and cell genetics, these predictions may soon be rendered obsolete [3].

14.2 Different Vaccines against Parasites

14.2.1 Protozoan Vaccines

Because protozoan infections in animals are linked to considerable production losses and because many of these organisms either cause zoonotic diseases or have close ties to human parasites, the importance of protozoa as infection reservoirs for human diseases is increased. Protozoan infections in animals are linked to considerable production losses. Vaccines against protozoa are not yet available for humans, although there are several that are in the process of being developed for animals [4].

14.2.1.1 Live Protozoan Vaccines

In live vaccinations, living organisms are administered to their hosts in the hope of eliciting an immune response and thereby replicating the effects of a disease. Since the immunological processes that are involved in protection and the stages of protozoan parasites that are involved in infections are mostly unknown, the majority of vaccines use live organisms themselves to trigger the necessary protective immune response in protozoan infections. This is because live organisms can trigger the necessary protective immune response in protozoan infections. T-cell driven immune responses can be induced using a live vaccine technique by ensuring that antigens are properly processed and presented within cells, as well as combining these antigens with major histocompatibility complex Class I and Class II antigens. Live vaccination methods more precisely resemble the development of both the innate and adaptive immune responses that would occur in a natural infection. As a result, live immunisation will induce the appropriate inflammatory and regulatory immunological responses in the host animals [5]. It is believed that T-cell responses are necessary to provide adequate protection against intracellular infections (Table 14.1).

14.2.1.2 Killed and Subunit Protozoan Vaccines

Several inactivated vaccines made of indeterminate antigenic structures or, more recently, crude whole organisms have been registered for use on the market for companion animals. These vaccines come in a variety of different formulations. Vaccines like these have the potential to lessen the severity of an illness or stop its transmission, but in general, their efficacy is inferior to that of living organisms. They have the potential to act as the building blocks for the development of recombinant vaccines. The selection of a sufficient parasite antigen and the delivery of that antigen to the immune system in a manner that will allow it to properly trigger innate and adaptive immune responses are two of the most challenging aspects in developing a dead vaccination that is effective [6].

Table 14.1 Different types of vaccine with their advantages, disadvantages and mechanisms.

Type of Vaccine	Invention	Advantages	Disadvantages	Vaccine using current technology	Mechanisms/ property of vaccine	Refences
Live vaccines	Edward Jenner	Single dose, long-term immunity Stimulates immune system naturally Produces strong immunity Elicits both humoral and cellular immune response Multiplication in the host Induces innate and adaptive immune responses Induces CD4+ and CD8+ T cells	Not safe Has side effects Uses whole pathogen Generation of unfavorable immune response	Measles, mumps, rubella, yellow fever, influenza, oral polio, typhoid, Japanese encephalitis, rotavirus, BCG, varicella zoster	Live vaccinations include living, weakened germs that can still reproduce in the host. By inducing a wider variety of immune responses, both humoral (B cells) and cellular (CD8+ and CD4+ T cells), genuine infection is replicated almost precisely.	[46]
Killed vaccines	Almroth Edward Wright	No risk and safer than live vaccines Inexpensive No amplification in the host	Less powerful than live vaccines Requires multiple doses to have significant reactogenicity Increased risk of allergic reactions	Whole-cell pertussis, polio, influenza, Japanese encephalitis, hepatitis A, rabies	Vaccine action is induction of both CD^{4+} and CD^{8+} T-cell responses, which require both endogenous as well as exogenous antigen processing and presentation of antigens with MHC class I and II molecules	[5]
Toxoid	Ramon	Diphtheria, which can be very serious and even deadly, as well as tetanus, and whooping cough . Prevents your child from developing a thick coating in the back of the nose or throat from diphtheria that can make it hard to breathe or swallow.	Vaccines do have some risk of an adverse reaction, the most common being redness and soreness at the injection site or fever and allergic reactions.	Diphtheria, tetanus	The diphtheria vaccine is made by taking the diphtheria toxin and inactivating it with a chemical. The inactivated toxin is called a "toxoid". Once injected, the toxoid causes an immune response to the toxin, but unlike the toxin	[2]

(Continued)

Table 14.1 (Continued)

Type of Vaccine	Invention	Advantages	Disadvantages	Vaccine using current technology	Mechanisms/ property of vaccine	Refences
Recombi-nant protein vaccine	William Rutter	No risk No interference with the maternal immunity and/or other vaccines Induces strong immune response	Needs cold chain for transportation Expensive, no biological activity Needs adjuvant to enhance immunogenicity Contamination with bacteria substances Large-scale preparation is difficult	Pertussis, influenza, hepatitis B, meningococcal, pneumococcal, typhoid, hepatitis A	CD^{4+}T cell and humoral responses. Poor induction of cellular responses, particularly of CD8+ T cells	[5, 59]
DNA vaccines	Enzo Paoletti and Dennis Panicali	Safe to use, no adjuvant needed Stable at room temperature and shipment is inexpensive Easy for manipulation and production Elicits antigen-specific immune responses	Low potency in humans Often weak immunogenicity	Hepatitis B vaccine, HPV vaccine	Induces both CD4+ and CD8+ T cells Elicits both humoral and cellular immune response Due to CpG motifs, are immunomodulators Prime antigen-specific memory T cells	[58]

Live genetically modified vaccines	Maurice Hilleman	Stable Safer than general live vaccines Stimulate immune system as in natural infection Multiplication in the host	Risk of reversion to virulent state Needs cold chain for transportation Unknown memory formation and duration Stimulates a weaker immune system response	Rotavirus vaccine MMR vaccine BCG vaccine against TB	Live vaccines use a weakened form of the germ that causes a disease. Because these vaccines are so similar to the natural infection that they help prevent, they create a strong and long-lasting immune response.	[50]
Live non-pathogenic vaccines	Edward Jenner	Safe to use Life-long immunity No reversion to virulent state	Unknown memory formation and duration Need cold chain for transportation	SARS-CoV-2	Stimulates immune system similar to natural infection Multiplication in the host Live vaccines induce T-cell mediated immune responses and mimic natural infection through correct processing and presentation of antigens in association with MHC class I and II antigens	[5, 59]

Neospora caninum is a common factor in the occurrence of abortions in cattle. A crude *N. caninum* vaccine is available in the United States of America. This vaccine combines inactivated *N. caninum* tachyzoites with an adjuvant in order to help reduce the number of abortions caused by the parasite in healthy pregnant cattle and to prevent it from spreading to calves while they are in gestation. Neogaurd is the brand name of this particular immunisation [7, 8]. *Giardia intestinalis* is a type of enteric parasite that affects a wide variety of animal species and has the potential to cause significant damage to the immune systems of young children and humans. Giardia Vax®, a commercially available vaccination, has been granted authorisation by the government of the United States for use in canines and felines so that it can significantly cut down on the frequency, intensity, and length of cyst shedding. An unprocessed mixture of *G. intestinalis* trophozoites that have been disturbed and artificially produced is contained within the vaccine [9].

Vaccination has been recognised for a very long time as the most effective and long-lasting technique for managing parasite disease in both humans and animals. (Table 14.1) Despite years of research, the few parasite vaccines that are now on the market are all licenced for use against parasites in veterinary species. This is the case despite the fact that there are no vaccines against human parasites. *Toxoplasma gondii*, *Babesia bovis*, *Eimeria* species, *Theileria annulata*, and other live parasites that are often attenuated or partially disabled by irradiation or concurrent antibiotic administration were the only effective vaccinations. Because the bulk of these live vaccines have to be created using parasites that have been extracted from animal hosts, there are significant challenges associated with standardisation, quality control, shelf life, and the costs of production. Despite the fact that live vaccines have traditionally been regarded as unacceptable for use in humans, a recent initiative has been instigated to investigate the use of *Plasmodium falciparum* sporozoites for vaccination. This initiative was motivated by the well-documented challenges encountered in developing subunit vaccines for malaria [1].

14.3 Administration Route

With one important exception, all routinely administered vaccines continue to rely on the injection route as the principal means of presenting the vaccine antigen to the recipient's immune system.

The challenge of meeting the demand for sterile syringes and needles in less developed countries is a primary focus of the work made by the EPI. Even if the pain from the injections is slight, it is something to keep in mind, particularly in light of the numerous vaccines that infants receive frequently. This is especially important to keep in mind in light of the fact that the immunisations can be repeated several times. A basic approach to reducing the total number of injections required is to incorporate multiple antigens into a single vaccination. Consequently, the live vaccines for measles, rubella, and mumps as well as the combination vaccines for diphtheria, tetanus, and pertussis are the kind of vaccines that are utilised most commonly. Additional combinations – of which there are many available on the market – raise concerns about the possibility of decreased immune responses when several antigens are competing for the same piece of equipment as well as the expense. Each new combination will need to go through a series of rigorous clinical tests, which will drive up the overall price of the finished product. This raises significant problems because immunisations should be administered to every child in the world, including and especially the poorest of those children.

The immunisation against live poliovirus, which is given orally, constitutes the exemption that was just discussed. Due to the fact that it replicates in the intestinal epithelium, the attenuated form of the polio virus is able to enter the body on its own. It is believed to have the additional benefit of activating the immune system of the intestinal mucosa, which is the body's natural first line of defence, and creating a defence that is more effective than a vaccine that is administered via injection. However, the vast majority of infections are naturally obtained through mucosae, whether in the respiratory system, the bowels, or the genital tract; as a consequence, local mucosal defences may be the most efficient type of resistance. In point of fact, the mucosae have a highly developed immune surveillance system, which is comprised of specialised cells that assess the antigens that are loading onto them and then convey the information to antigen-presenting dendritic cells and the lymphocytes that lie beneath them. It is a mystery how the various components of food and the microorganisms found in the environment are able to circumvent the immune activation pathway. One explanation for this phenomenon is self-tolerance, which is also sometimes referred to as "oral tolerance". However, it is unknown whether an immune response or the formation of tolerance is responsible for determining which channel is taken [10].

14.4 Trematodes and Helminths Spread through Soil

14.4.1 Schistosomiasis

Schistosoma haematobium and *S. mansoni*, both of which are dimorphic parasitic flatworms that are transmitted by freshwater snails and circulate in the bloodstream of mammalian hosts, are the primary etiological agents of schistosomiasis. *S. haematobium* and *S. mansoni* cause schistosomiasis. At the moment, there are 52 countries that are home to 219 million people who have been exposed to schistosomes, with Africa accounting for 90% of the disease's expected cases in 2003.

Several studies that have been conducted up to this point have uncovered proteins that have the potential to be used in the development of novel diagnostic tools and vaccinations [11].

Proof of concept for a vaccination against *Schistosoma* trematodes was provided by mice and non-human primates inoculated with natural, irradiated cercariae. This vaccine offered about 80% protection against schistosomula challenge [12, 13, 14]. In light of the fact that attenuated preparations were deemed unsuitable for use as a human cercarial vaccine, efforts were made to produce a subunit vaccine [15]. The majority of the targets for the existing lead subunit vaccines were found by using traditional biochemical methods on parasite protein fractions (for instance, *S. mansoni* glutathione S-transferase antigens of 28 kDa and 14 kDa). These methods were used to isolate the target proteins from the parasites [14]. These compounds originate predominantly from the tegumental membranes of migrating schistosomulum stages (which have an effect on parasite invasion) and adult females (which have an effect on parasite survival and fecundity), which are the objectives of the treatment. Recent advancements in high-throughput technology have helped to speed up the process of identifying antigens for important schistosome species. Despite this, there is still a lack of translation from basic research to clinical and field studies. At the moment, there are just a handful of recombinant targets that are in various stages of clinical and commercial development [11] (Figure 14.2).

14.4.2 Hookworm Disease: Helminths that Spread Through Soil

Soil-transmitted helminths include medically significant endoparasites such as hookworms, roundworms, and whipworms. Some examples of these endoparasites include *Necator americanus* and *Ancylostoma duodenale.* Roundworms include *Ascaris lumbricoides*. Whipworms include *Trichuris trichiura*. Hookworms include *Necator americanus* [16]. Since hookworm infections presently account for around two-thirds of the global human illness burden from soil-transmitted nematode infections, a significant amount of work has been put into the search for a vaccine. Approximately 440 million people across Asia, sub-Saharan Africa, the Caribbean, and Latin America are afflicted with human hookworm disease, which is mostly brought on by the parasites *N. americanus* and *A. duodenale*. This condition is a global health concern [17, 18].

14.4.3 Helminthiases: Endoparasite Vaccines

The bulk of disease-causing parasites that are of relevance to medicine and veterinary science are endo-parasites, with helminthiasis being a major contributor to the problem [19]. These parasites are responsible for a range of disorders that can be chronic and/or debilitating, which can interfere with normal physical and cognitive development. Children are particularly sensitive to these diseases because of their developing bodies and brains [19, 20]. Mass anti-helminthic drug administrations (MDAs), such as praziquantel for schistosomiasis and albendazole or mebendazole for soil-transmitted helminthiases, are the primary public health interventions that are recommended to control morbidity in endemic countries. These are the types of helminthiases that are transmitted from the soil to humans (STH). However, because of the risk of developing chemical resistance, treatment is made more difficult [21–23].

14.5 Arthropods that Act as Parasites

14.5.1 Ectoparasite Vaccines

The ectoparasitic arthropods make up the most numerous animal phylum and are becoming increasingly important in the fields of veterinary and medical research [24, 25]. More than twenty percent of all newly discovered infectious diseases that occurred between the years 1940 and 2004 were caused by arthropod-borne illnesses [26] (Figure 14.2).

14.5.2 Flies That Carry Parasites

The haematophagous biting flies, the non-biting annoyance flies, and the flies that induce myiasis, are the three primary groups of flies that are significant to veterinary medicine [27]. Important biting flies for veterinarians include the *Haematobia irritans* (also known as the horn fly), *H. irritansexigua* (also known as the buffalo fly), and *Stomoxys calcitrans* (also known as the stable fly) [28]. (Figure 14.2). Cattle horn flies are a nuisance that can be financially devastating since they cause severe itching, a reduction in milk production, lower body weight, considerable blood loss, and damage to hides [29]. In addition, horn flies are capable of transmitting a variety of illnesses, including several species of *Staphylococcus* that can cause mastitis in dairy heifers [30].

14.6 The Protector Immune Response

The creation of vaccines has been hampered by a lack of knowledge regarding the specific immune effectors that are responsible for the attrition of parasites and the antigens that trigger them. According to [31], the effective development of a schistosomiasis vaccine had been hampered by disagreements over the necessary immune response as well as a lack of understanding of the effector mechanisms that mediate immunity. This was one of the reasons why the disease is so difficult to treat. For natural immunity to be maintained, it is often necessary to have recurrent infections, stage-specific immunological responses, as well as a wide array of antibody classes and T-cell responses. Even though the technology necessary to produce recombinant proteins has been available for around 30 years, it is still rather uncommon to find proteins that have the required level of effectiveness. The most prominent and groundbreaking exceptions are the advances that have been made in tick and cestode vaccines. These are successes that have set a new standard [32, 33].

14.7 Antigenic Differences and Their Relation to Stage-Specific Antigen Expression

When a host is infected with protozoan or metazoan parasites, the immune system of the host is subjected to a wide variety of antigens that are stage-specific and temporally expressed. There will be a connection established between some of them and immune responses that are protective, while for others there will not. The greatest challenge for the vaccine researcher is to identify the components that are essential for the production of protective immunity. An illustration of the enormous antigenic variation that all parasites display is provided by the variant surface glycoprotein (VSG) of trypanosomes. Trypanosomes are the pathogens responsible for sleeping sickness, which is transmitted by tsetse flies and is seen in high prevalence in Africa. The surface of the parasite is covered in VSGs, which act as a physical barrier to protect the underlying proteins from the effectors of the immune system [34]. This surface coat is responsible for inducing a one-of-a-kind trypanocidal immune response, which is then neutralised by antigenic variation. During this process, the trypanosomes switch to the production of a separate VSG, which, if it is antigenically novel, promotes the clonal growth of the switched cells and creates a new parasitaemia peak. Although only one VSG gene is expressed by each trypanosome, these cells have the ability to flip between hundreds of thousands of different VSG genes [35] (Figure 14.2).

14.8 Vaccination as a Preventative

Vaccines are the most efficient, trustworthy, and long-term method of controlling parasite disease, and they are also beneficial in reducing reliance on pharmaceutical drugs and pesticides. Vaccines are also a sustainable method [36]. The use of vaccines has many benefits, some of which include improving animal health and welfare by lowering the incidence of animal infections; improving public health by lowering the incidence of zoonoses and other food-borne pathogens in animals; resolving problems with acaricide, antibiotic, and anthelmintic resistance; preventing the use of chemicals on animals and in the environment; and preserving biodiversity (Figure 14.2). All of these features should raise the economic advantage of animal production and improve its long-term viability [5].

14.8.1 Malaria

Malaria is the most significant tropical parasite sickness that can be found anywhere in the world. The disease is widespread with an estimated yearly mortality toll of approximately 1.1 million deaths. Each year, there are between 350 and 500 million clinical cases of the disease. The costs associated with treatment as well as days missed from work might be significant. The life cycle of the parasite is quite complex, and it all begins when female mosquitoes feed on blood and release sporozoites into the bloodstream. The sporozoites then make their way to the liver and enter hepatocytes there. As they develop into merozoites over the course of about two weeks, tens of thousands of new ones are produced at this location. After leaving the hepatocytes, the merozoites continue their journey through the circulation to infect the erythrocytes (RBCs). There, they multiply and mature within twenty-four to seventy-two hours, until the RBCs lyse and release the merozoites, which then infect new RBCs to finish the cycle. The acute fever episodes and rigors that are characteristic of malaria, which happen every 48 to 72 hours, occur at the same time as the synchronised lysis of infected RBCs, which releases the newly matured merozoites. This is because the synchronised lysis of infected RBCs is triggered by the synchronised release of newly matured merozoites. Some of the merozoites will develop into gametocytes, which is the sexual stage of development. These gametocytes have the ability to reproduce sexually and create new sporozoites if they are swallowed by an anopheline mosquito, which in turn restarts the cycle [37].

The fact that human participants who were given irradiated sporozoites were protected against experimental infection to a high degree (about 90%), showing the feasibility of vaccination. [38] People who live in regions where malaria is common can build a natural immunity to the disease, but it is a slow process that requires constant antigenic stimulation and has an effect on how severely they are affected by the disease. Hyperimmunoglobulins derived from malaria-resistant individuals have been passively transplanted into healthy human volunteers who have never been infected with the disease [38].

There are three different approaches to creating a vaccine against malaria. 1) To provoke potent immune responses that will prevent merozoites from entering the bloodstream; 2) vaccination results in the development of disease-limiting immunity; and 3) transmission of the virus by mosquitoes is prevented by means of population-based immunisation.

Techniques for administering pre-erythrocytic vaccines make an effort to either elicit a cell-mediated immune response that will limit intra-hepatic parasites or an antibody response that will destroy sporozoites and prevent them from penetrating the hepatocyte. Both of these immune responses are intended to limit the number of parasites that can be found within the liver. If a vaccination of this nature existed, it would prevent the spread of clinical sickness. The most promising pre-erythrocytic vaccine option comes from the circumsporozoite protein (CSP), which is an essential part of the surface of the sporozoite. The first evidence that people may be protected from malaria infection with a subunit vaccination was provided by prototype vaccines. These vaccines were created to stimulate antibody responses against the repetitive epitopes of the circumsporozoite protein [39, 40]. This was the first proof that people may be protected from malaria infection with a subunit vaccination. Because the immunogenicity of the prospective vaccines was low, an amalgam of circumsporozoite protein synthesised with unfused hepatitis B surface antigen (HBsAg) was constructed as a vaccine. This vaccine is now in use [41].

Asexual blood-stage (erythrocytic) vaccine techniques are designed to elicit T-cell responses that will prevent the establishment of the parasite in RBCs as well as antibodies that will inactivate merozoites and/or target malarial antigens expressed on the surface of RBCs. Additionally, these techniques are intended to elicit antibody responses that will target malarial antigens expressed on the surface of RBCs. This type of immunisation would primarily serve as a disease-reduction vaccine in endemic nations by slowing down the exponential growth of merozoites, which is the infectious stage of the parasite. Recently, there was a publication of an in-depth analysis of the development of MSP-1 as a potential vaccine candidate [42]. To summarize, MSP1 provided protection against challenge infection with *Plasmodium yoelii* in laboratory mice. Passive vaccination with monoclonal antibodies (mAbs) also provided protection against challenge infection in the same model, highlighting the importance of antibodies for providing protection. Extensive structural research made feasible by gene sequence analysis led to the discovery of a disulphide-rich region of approximately 100 amino acids at the C-terminus of the gene that encodes two epidermal growth factor (EGF) domains. This area is located at the end of the gene [37].

14.8.2 Leishmania

Leishmaniasis is a widespread disease that affects people in many parts of the world, including Africa, Latin America, south and central Asia, the Mediterranean basin, and the Middle East. It is caused by a wide variety of species of flagellated protozoan parasites that belong to the *Leishmania* genus. There are an estimated 12 million new instances of the disease worldwide every year. It can lead to severe disfigurement as well as death . Although the parasite can be passed on to healthy humans, the vast majority of infections are contracted from wild animals who serve as a reservoir for the parasite (small rodents, dogs). Visceral symptoms, mucocutaneous manifestations, and cutaneous manifestations are all possible for this condition. Cases of visceral leishmaniasis occur on a yearly basis at a rate ranging from 1.5 million to 2 million, and epidemics can emerge with high rates of death. In order to treat it, chemotherapy is administered, which is both expensive and risky due to the development of drug resistance. After recovering from an infection, a person develops a resistance to subsequent infections; this demonstrates that it is possible to create an effective vaccine [43]. Vaccines against attenuated parasites have a good chance of producing the same immune response as genuine infections and replicating such infections. However, there are concerns with transitory infections that only deliver moderate amounts of antigen, which elevates the possibility of an inappropriate skew in the Th1 response's bias. This can be a problem for people who suffer from allergies. This method of immunisation will also provide a high number of parasite antigens, which is a potential benefit. This is in contrast to the few parasite antigens that can be obtained by employing subunit or recombinant antigens [44].

According to research conducted on mice, the genetics of the host may have an effect on the protective immune responses elicited by vaccination [43].

DNA vaccination is regarded as a potentially fruitful avenue for the advancement of vaccine research because it often results in Th1 responses [45]. The results of the testing of several DNA vaccine designs have yielded a wide range of outcomes. Sand flies are the insect vectors for the parasite, and a 15kDa protein found in sand flies has showed promise as a means of triggering some defensive response. Despite the many advances that have been made, it is instructive to refer back to the concept of the ideal antileishmanial vaccine that was presented by [43].

14.8.3 Contemporary Approaches to Leishmaniasis Vaccination

Leishmania and the recent discovery of the DNA sequence of the human genome have lately opened up new possibilities for the development of innovative vaccine techniques. Vaccines that are created using recombinant DNA techniques include recombinant subunit vaccines, recombinant DNA vaccines, genetically modified live vaccines, recombinant viral-based vaccines, and recombinant vaccines based on epitopes or peptides. Other types of recombinant vaccines include genetically modified live vaccines.

Prior to the creation of recombinant vaccines, susceptible individuals were protected against leishmaniasis by being immunized with crude antigens. After the development of recombinant DNA technology, numerous components of the parasite, in particular genes that are essential for the parasite's continued existence, were targeted, and the immunological response that was elicited in the host has been characterized. A separate strategy was utilized to generate genetically attenuated live vaccines, and it involved silencing a number of important genes. Vaccines against *Leishmania* can be developed using recombinant techniques with the use of essential information on the genes involved in the parasite's structure, metabolism, and virulence. There is no risk of sickness due to the presence of unfavorable pathogenic qualities when using individual or multiple proteins from one or more species of an organism instead of the entire parasite cell antigens. This is the primary benefit of using individual or multiple proteins from one or more organisms. Since there may be a synergistic effect on the immune response, utilising multiple antigens rather than just one may produce better results than using just one antigen. This is the conclusion drawn from a number of studies. As a direct consequence of this, these vaccinations have a lower risk than the conventional immunisations.

Numerous antigens derived from different species of *Leishmania* have been studied for their potential use as recombinant proteins and DNA vaccines. However, none of them were able to establish a complete level of long-term immunity; rather, they only produced a partial level of protection against the infection. Researchers have created novel methods, such as cocktail vaccines and heterologous prime-boost vaccination, in order to lengthen the duration of the immune response in the host. Specifically, this was done in order to combat the challenge of chronic infections [46]

14.9 The Zoonotic Consequences of Parasitism

There are a wide variety of zoonotic diseases that are brought on by parasites and can affect both metazoan and protozoan organisms. Public health is threatened in many parts of the world by a condition known as hydrated illness, which is brought on by an infection with tapeworms belonging to the *Echinococcus* genus. People become ill from the disease when they swallow the eggs of the organism that causes it, which hatch in the intestines of dogs and other canids. This makes proper sanitation a concern for the disease. The sickness caused by the digenetic trematode *S. japonicum*, which is also a major cause of mortality in sub-Saharan Africa, affects more than 250 million people throughout Southeast Asia [47]. More than 40 million individuals in China's tropical and subtropical regions are affected by this disease [48, 49]. The disease known as fascioliasis is brought on by the liver fluke *Fasciola hepatica*. This widespread infection often strikes bovines and sheep first, but it also has the potential to strike humans. Infection can occur in people when they consume unwashed and raw aquatic plants that are contaminated with encysted larvae. In humans, *Fasciola* spp. can cause severe diseases that can even be fatal; nevertheless, animals can carry enormous worm burdens without becoming ill from the infestation. In most cases, fasciolosis in humans is linked to locally endemic cases of fasciolosis in animals. People, dogs, and cats can all become infected with the protozoan parasite known as *Giardia*, which is responsible for severe gastrointestinal sickness. People whose immune systems are impaired are especially at risk for developing a severe gastrointestinal sickness caused by *Cryptosporidium parvum* [37].

14.10 Current Spectacular Technical Advancements in Vaccines Development

14.10.1 Recombinant Protein Vaccines

Recombinant vaccines rely on the ability of one or more specific antigens to generate immunity against the disease when they are given with additives, when they are created by plasmids, or when they are delivered by non-harmful bacterial or viral vectors during administration. When compared to vaccines that are based on purified macromolecules, recombinant protein vaccines allow for the avoidance of a number of potential problems. These problems include the risk of co-purification of undesirable contaminants and the reversal of the toxoids to their toxigenic forms, for instance, when thinking about diphtheria or tetanus toxoid vaccines. Another significant issue that is alleviated by the utilisation of this technology is the challenge posed by the acquisition of adequate quantities of purified antigenic components. The majority of the vaccines that are now under investigation are derived from highly purified recombinant proteins or components of infectious diseases [50] (Figure 14.2). One well-known example of a recombinant protein vaccine is the hepatitis B vaccination, which is currently administered to human patients. The infection caused by the hepatitis B virus (HBV) is a form of chronic liver disease that affects people all over the world. HBV has a very strong attraction for human liver cells, and part of what contributes to this affinity is a receptor that is produced on the surface of infected cells. In order to produce vaccines against hepatitis B, the hepatitis B surface antigen, often known as HBsAg, is first produced in yeast cells. The formation of virus-like particles (VLPs) by the HBsAg, which are highly immunogenic, contributes significantly to the remarkable efficacy of the HBV vaccination. It is possible that the yeast expression system will secrete the antigen into the culture supernatant, which will make its purification much simpler [51, 52]. In addition, because yeast cells are able to glycosylate proteins, these cells provide some of the eukaryotic cellular machinery that is required for the post-translational modification of proteins. The HBV vaccination is now within reach of the vast majority of developing countries as a direct result of price reductions brought about by increased market competition and the dissemination of production technology to a large number of manufacturers (Table 14.1).

14.10.2 The Creation of Vaccines Using Recombinant DNA Technology

When beginning a new vaccination study, how do you choose which prokaryotic or eukaryotic vector to employ for the production of recombinant proteins? There are a wide variety of prokaryotic and eukaryotic vectors available. This choice will be affected by one's familiarity with the pertinent antigen. Does post-translational modification play a role in the activation of a protective response or is it simply necessary for efficient protein folding? Even if glycans from yeast and insect cells would not be suitable, bacterial cells do not glycosylate their proteins [53]. According to the results of the tests, the majority of the sheep's antibody response to the protective gut antigens from *Haemonchus contortus* was to the glycan component. However, these authors noted that current research suggested that this was unlikely to be a factor in the protection brought on by the vaccine. Since parasite-specific patterns of glycosylation, such as *H. contortus* H11 [54], have been found, the currently available commercial eukaryotic expression systems will not result in the appropriate glycosylation of recombinant antigens. This is because glycosylation is a process in which sugars are attached to proteins (Figure 14.2).

14.10.3 The Use of Monoclonal Antibodies in the Research of Parasitic Diseases

Because of the features of monoclonal antibodies, the utilisation of these reagents has resulted in the provision of a number of advantages during the course of the inquiry into parasite illnesses. Immunotherapy, vaccine development, and diagnosis are the three main application areas that can be generally classified under this umbrella term. The antigens from a parasite of interest, in addition to the antigens from parasites that are closely related to it, can all be screened using monoclonal antibodies as a library for diagnostics in order to identify those that are unique. This can be done in order to find the antigens that are specific to the parasite of interest (Table 14.1).

Using monoclonal antibodies, it is now possible to recognize specific coccidial antigens and analyse their contribution to the immunological regulation of parasitism in vitro. This is achievable thanks to recent advances in the field. In a further step, recombinant parasite antigens cloned into bacteriophage from cDNA libraries have been discovered with the help of monoclonal antibodies and polyclonal immunological serum [55]. Because of this, a protein that was produced by *E. coli* and selected in this manner has been utilized to partially protect chickens from *E. tenella* infection, which is a challenge infection [56].

14.10.4 Utilizing Monoclonal Antibodies

In 1975, Kohler and Milstein outlined a method for combining healthy B lymphocytes with cancerous myeloma cells in a procedure that they called fusion. The hybrids that were produced were endowed with the ability to continue growing in vitro indefinitely and possessed the capacity to secrete antibodies much like their B lymphocyte parent. Clones are characterized by their inherent similarity and their ability to produce IgG with a consistent molecular structure and antigen specificity. Clones can be generated from hybridized cell populations. This is due to the fact that every B cell produces its own immunoglobulin molecule, which has a predetermined level of specificity.

Because of their high level of specificity, monoclonal antibodies are among the most useful tools for research into the antigens of parasites. Parasites are complicated organisms that progress through numerous phases of development, and at each stage of their life cycle, they provide the host with a wide variety of antigens. The use of monoclonal antibodies makes it much simpler to recognize, separate, and examine the various antigens that are produced by these complex organisms.

Also, if the antigens do not contain any repetitive epitopes, the monoclonal antibodies' specificity for a single epitope may preclude their use in certain testing methods, such as precipitation reactions. This is because these methods require the antibodies to recognize multiple epitopes simultaneously. Because a hybrid cell line may only generate one kind of antibody, isotype-dependent properties such as complement fixation and protein A binding may have a restricted applicability for monoclonal antibodies that are otherwise valuable.

The only way to make hybrids is by fusing together cells from different rodents, which has slowed down the process of developing monoclonal antibodies (and recently human cells). In the field of animal illness research, the utilisation and investigation of murine monoclonal antibodies are subject to a number of different limitations [57].

14.10.5 DNA Vaccines

This technique involves delivering the genetically modified DNA of a particular antigen so that immune cells can come into direct contact with the antigen and trigger a variety of reactions to various defence mechanisms. This allows immune cells to better fight off infections. DNA vaccines may have a number of advantages over conventional immunisations, including the ability to evoke a wider spectrum of immune reactions. Vaccination is the most effective method of disease control in both humans and productive animals; however, the development of vaccines to prevent parasitic infections in animals is still in its early stages. Vaccination is the most effective method of disease control in both humans and productive animals [58] (Table 14.1).

Concurrently with the advent of protein vaccines, a new feasible delivery technology known as DNA vaccines that encode one or more immunogenic proteins has been developed and made available. In 1990, RNA and DNA expression vectors were successfully injected into the skeletal muscle of mice for the very first time. These vectors contained reporter genes [59]. DNA vaccines hold a great deal of potential for use in the prevention and treatment of a wide variety of diseases, including leishmaniasis and other conditions for which Th1 responses and cell-mediated immunity are necessary to provide protection. DNA vaccines have been developed over the past 20 years in order to combat infectious microorganisms. This method of immunisation elicits both humoral and cellular immune response against native forms of protein, and antigens are produced in the natural conformation of the host due to post-translational modifications of the expressed proteins inside the cells. Reduce the amount of DNA that is used in order to increase the level of protection produced. The magnitude of the immune response that is induced can, of course, be influenced by a wide range of parameters, including the immunogenicity of the gene that is sought to be expressed and the presence of synthetic oligodeoxynucleotides. By acting as an adjuvant, the immunomodulators that can be found in the CpG patterns of plasmid DNA cause a change in the immune response towards Th1 [60]. In addition, immunisation with DNA alone triggers a protein called TLR9, which results in a considerable amount of IL-12 production and, eventually, a Th1 immune response [45]. DNA vaccines also have the advantage of being more stable throughout shipment and transition than conventional immunisations.

14.11 The Effects of Parasitism on the Entire World

The consumption of antiparasitic drugs is, in all practical respects, the one and only method for mitigating the negative consequences of parasitism in both humans and livestock. Parasites that are drug-resistant are becoming increasingly common in both animal and human populations, which presents a challenge for parasite control. Vaccination being the primary means of disease prevention, it is critical to develop other long-term control strategies. Twenty-five years after the ability to create recombinant parasite proteins was hailed as a major success in vaccine development, only a few recombinant vaccines against livestock parasitic diseases have advanced to the point where they can be released. This is despite the fact that it was possible to create recombinant parasite proteins [37]. The development of the first recombinant vaccination against a parasite that can infect people has not yet been successful. It is anticipated that by the year 2050, more than half of all people living in poor nations would reside in urban and peri-urban regions, the majority of whom will reside in slums, which are ideal environments

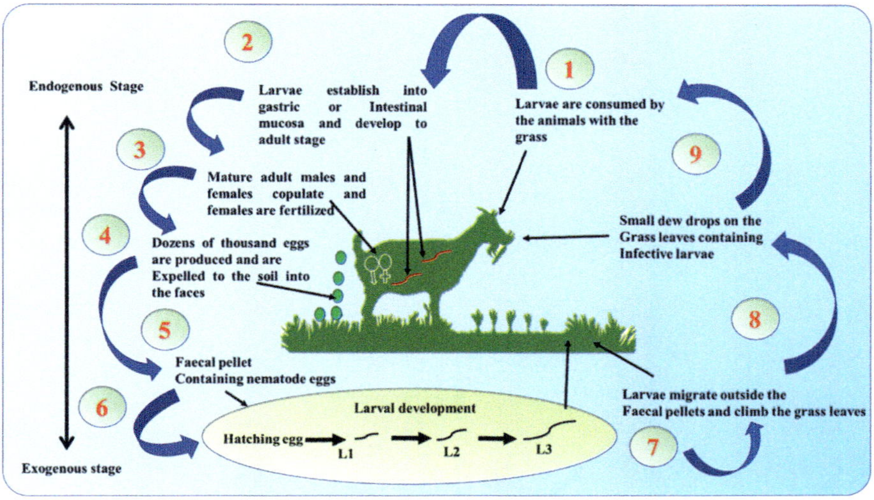

Figure 14.1 Development cycle of *Haemonchus contortus* in goat.

for the transmission of a wide variety of parasites [61]. Parasitic diseases, which can be transmitted by a wide variety of helminths, protozoan parasites, and ectoparasites, also have a substantial impact on the amount of cattle produced around the world. One of the most prominent is the *Haemonchus contortus* nematode, which is a pathogenic blood-feeding nematode parasite that affects sheep and goats. Both large-scale sheep and goat producers and, perhaps more significantly, resource-poor small holders, whose production directly affects family diet and, ultimately, survival, are directly affected by infection and its intensity. This parasite is especially common in tropical and subtropical regions (Figure 14.1).

14.12 Acuity Control

Both for the treatment of clinical disease and for larger-scale programmes aimed at disease control, pharmaceuticals continue to be an indispensable tool. However, drug resistance in the target parasites is already quite common, which presents a challenge for sustainable management, most notably in the context of livestock. Many people have been successfully treating themselves with currently available medicines for years. The development of drug resistance in cattle hosts has been linked to the use of every type of anthelmintic [62]. Nematodes that are resistant to many treatments are becoming more prevalent in certain regions, which places the continued operation of production facilities for small ruminants at risk of

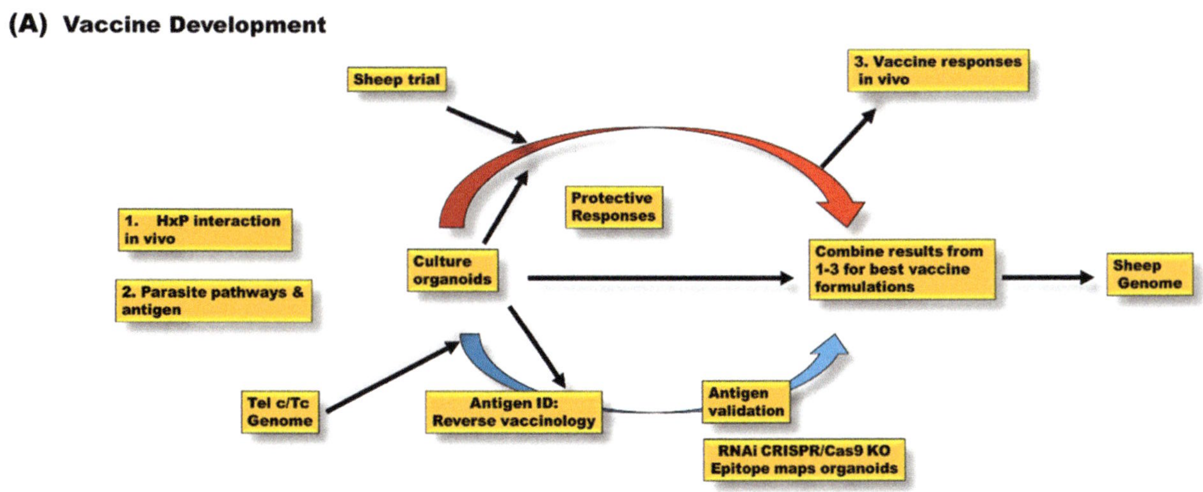

Figure 14.2 Different stages of vaccine development.

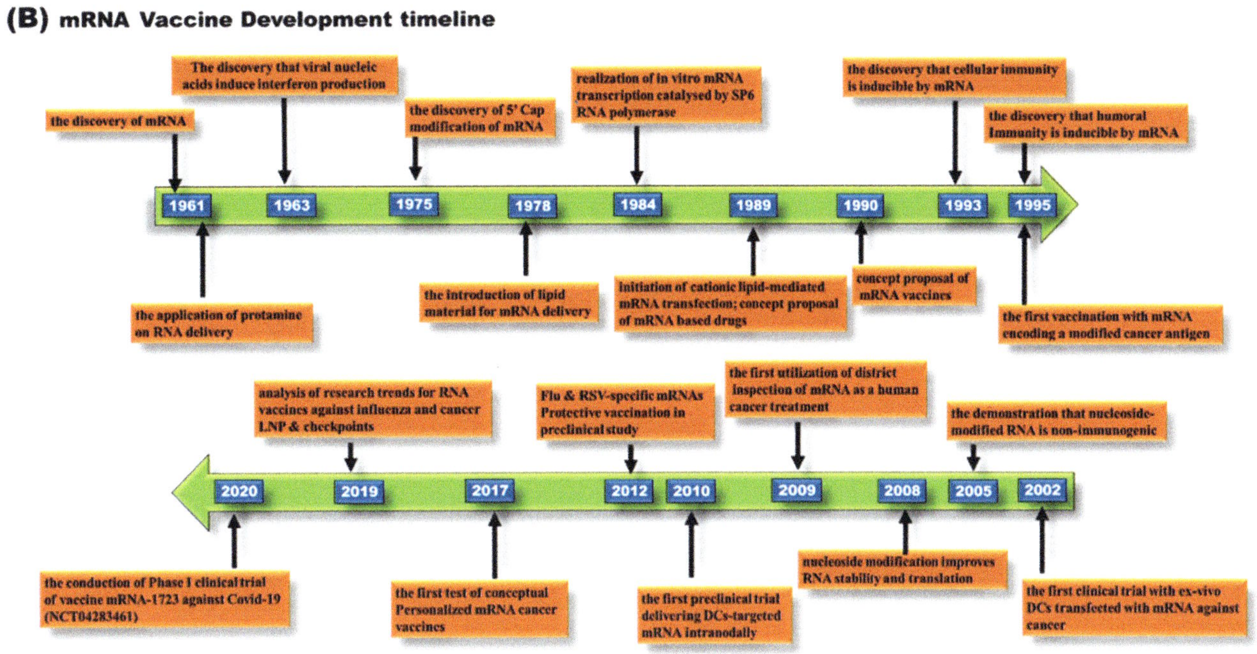

Figure 14.2 (Cont'd)

extinction. In a review of this nature, it is imperative to discuss the widespread problem of ectoparasites, in particular the tick and mite infestations that are frequent in both humans and animals. Some parasitic diseases should become less common over time if public health education, access to clean water, and sanitary living conditions all continue to advance. However, this is not always the case with illnesses that are transmitted by vectors. Malaria is spreading unchecked in many parts of the world, and patients' resistance to treatment is increasing, particularly in sub-Saharan Africa. Every socioeconomic category will be affected by this. The only effective method of mosquito control is to protect oneself from bites by employing measures like using bed nets impregnated with insecticide. The most defenceless people of society are the ones who are affected when there is an outbreak of trypanosomiasis or leishmaniasis. The World Health Organization, along with other international organisations and charities, is coordinating significant efforts to find creative ways to control a number of the major parasites that affect both humans and animals, with the ultimate goal of eradicating as many of them as possible.

14.13 Conclusion

New insights into the immunobiology of the parasites have been provided as a result of the development of new vaccines. This has assisted in the concentration of efforts on the identification of new candidate antigens for vaccine testing and, in some instances, has provided new ideas for the design of vaccines. Vaccines should be able to achieve this goal without causing the disease to spread further since they stimulate an immune response in the animal host that afterwards identifies infectious microorganisms and helps the body fight sickness. Using recombinant DNA technologies, scientists have been able to create live genetically modified organisms, recombinant dead vaccines, and genetic immunisations that do not cause illness but still elicit a powerful immune response. These advancements have been made possible by the creation of genetically modified organisms that can replicate on their own. For the purpose of developing vaccines utilising rDNA technologies, a comprehensive understanding of the infectious agent, in particular the antigens that are absolutely necessary for providing protection and the factors that contribute to the development of the disease, is required.

There is good reason to assume that vaccinations will be developed against some or all of the most significant animal diseases as soon as the necessary financial and technical resources become available. Nevertheless, it is necessary to acknowledge the complexity of the problems that are being discussed at this time. Through immunisation, communities are protected from diseases that historically claimed the lives of millions of people each year, particularly children. According to the United Nations Convention on the Rights of the Child, every child has the right to the highest attainable level of health, and as a direct consequence, every child has the right to get vaccines.

The creation of vaccinations as a method for preventing parasite infections in animals has a promising future and is the strategy that is both the most feasible and the most cost-effective. It is essential to preserve the synergy that exists between parasite vaccinations and anti-parasitic medications. Doing so will reduce the amount of anti-parasitic medication that must be taken in order to get the desired effect and will stop the development of drug resistance. Conventional methods of parasite control, such as chemotherapy, grazing management, and biological control of parasites, are not economically viable in many countries, including India, despite the fact that they can be used in an integrated pattern for effective control. This is true even though they can be employed individually for effective control. Vaccines give long-lasting immunity that can be periodically increased, which is beneficial in the event that animals are exposed to natural diseases.

References

1 Morrison, W. and Tomley, F. (2016). Development of vaccines for parasitic diseases of animals: challenges and opportunities. *Parasite Immunology* 38 (12): 707–708.

2 Pollard, A.J. and Bijker, E.M. (2021). A guide to vaccinology: from basic principles to new developments. *Nature Reviews Immunology* 21 (2): 83–100.

3 Sharma, N., Singh, V., and Shyma, K.P. (2015). Role of parasitic vaccines in integrated control of parasitic diseases in livestock. *Veterinary World* 8 (5): 590–598.

4 Meeusen, E.N., Walker, J., Peters, A. et al. (2007). Current status of veterinary vaccines. *Clinical Microbiology Reviews* 20 (3): 489–510.

5 Innes, E.A., Bartley, P.M., Rocchi, M. et al. (2011). Developing vaccines to control protozoan parasites in ruminants: dead or alive? *Veterinary Parasitology* 180 (1–2): 155–163.

6 Saravanan, B.C., Ray, D.D., and Sankar, M. (2013). Conventional and molecular vaccine against protozoans infecting livestock. *Molecular Biological Approaches for Diagnosis and Control of Parasitic Diseases*. Izatnagar, UP, India: Indian Veterinary Research Institute. 182.

7 Innes, E.A., Wright, S., Bartley, P. et al. (2005). The host–parasite relationship in bovine neosporosis. *Veterinary Immunology and Immunopathology* 108 (1–2): 29–36.

8 Romero, J.J., Perez, E., and Frankena, K. (2004). Effect of a killed whole *Neospora caninum* tachyzoite vaccine on the crude abortion rate of Costa Rican dairy cows under field conditions. *Veterinary Parasitology* 123 (3–4): 149–159.

9 Holroyd, N. and Sanchez-Flores, A. (2012). Producing parasitic helminth reference and draft genomes at the Wellcome Trust Sanger Institute. *Parasite Immunology* 34 (2–3): 100–107.

10 Mäkelä, P.H. (2000). Vaccines, coming of age after 200 years. *FEMS Microbiology Reviews* 24 (1): 9–20.

11 Stutzer, C., Richards, S.A., Ferreira, M. et al. (2018). Metazoan parasite vaccines: present status and future prospects. *Frontiers in Cellular and Infection Microbiology* 8: 67.

12 Coulson, P.S. (1997). The radiation-attenuated vaccine against schistosomes in animal models: paradigm for a human vaccine? *Advances in Parasitology* 39: 271–336.

13 Coulson, P.S. and Wilson, R.A. (1997). Recruitment of lymphocytes to the lung through vaccination enhances the immunity of mice exposed to irradiated schistosomes. *Infection and Immunity* 65 (1): 42–48.

14 Hagan, P., Abath, F.G., and Dunne, D.W. (1995). Prospects for immunological control of schistosomiasis. *The Lancet* 345 (8963): 1488–1492.

15 Molehin, A.J., Rojo, J.U., Siddiqui, S.Z. et al. (2016). Development of a schistosomiasis vaccine. *Expert Review of Vaccines* 15 (5): 619–627.

16 World Health Organization (2016). Schistosomiasis and soil-transmitted helminthiases: number of people treated in 2015. *Weekly Epidemiological Record* 91 (49/50): 585–595.

17 Hotez, P.J., Diemert, D., Bacon, K.M. et al. (2013). The human hookworm vaccine. *Vaccine* 31: B227–32.

18 Pullan, R.L., Smith, J.L., Jasrasaria, R., and Brooker, S.J. (2014). Global numbers of infection and disease burden of soil transmitted helminth infections in 2010. *Parasites & Vectors* 7 (1): 1–9.

19 Hotez, P.J., Pecoul, B., Rijal, S. et al. (2016). Eliminating the neglected tropical diseases: translational science and new technologies. *PLoS Neglected Tropical Diseases* 10 (3): e0003895.

20 Briggs, N., Weatherhead, J., Sastry, K.J., and Hotez, P.J. (2016). The hygiene hypothesis and its inconvenient truths about helminth infections. *PLoS Neglected Tropical Diseases* 10 (9): e0004944.

21 Geurden, T., Chartier, C., Fanke, J. et al. (2015). Anthelmintic resistance to ivermectin and moxidectin in gastrointestinal nematodes of cattle in Europe. *International Journal for Parasitology: Drugs and Drug Resistance* 5 (3): 163–171.

22 Keenan, J.D., Hotez, P.J., Amza, A. et al. (2013). Elimination and eradication of neglected tropical diseases with mass drug administrations: a survey of experts. *PLoS Neglected Tropical Diseases* 7 (12): e2562.

23 Sutherland, I.A. and Leathwick, D.M. (2011). Anthelmintic resistance in nematode parasites of cattle: a global issue? *Trends in Parasitology* 27 (4): 176–181.

24 Goddard, J. (2016). *Physician's Guide to Arthropods of Medical Importance*. CRC press.

25 Mathison, B.A. and Pritt, B.S. (2014). Laboratory identification of arthropod ectoparasites. *Clinical Microbiology Reviews* 27 (1): 48–67.

26 Jones, K.E., Patel, N.G., Levy, M.A. et al. (2008). Global trends in emerging infectious diseases. *Nature* 451 (7181): 990–993.

27 Pape, T., Blagoderov, V., and Mostovski, M.B. (2011). Order Diptera Linnaeus, 1758. In: *Animal Biodiversity: An Outline of Higher-level Classification and Survey of Taxonomic Richness* (ed. Z.-Q. Zhang), 222–229. Zootaxa. 3148(1).

28 Pruett, J.H. (2002). Immunological intervention for the control of ectoparasites of livestock-A review. *Journal of veterinary parasitology* 16 (1): 1–10.

29 Byford, R.L., Craig, M.E., and Crosby, B.L. (1992). A review of ectoparasites and their effect on cattle production. *Journal of Animal Science* 70 (2): 597–602.

30 Hibler, C.P. (1966). Development of *Stephanofilaria stilesi* in the horn fly. *The Journal of Parasitology* 52 (5):890–898.

31 Wynn, T.A. and Hoffmann, K.F. (2000). Defining a schistosomiasis vaccination strategy–is it really Th1 versus Th2? *Parasitology Today* 16 (11): 497–501.

32 Lightowlers, M.W. (2006). Cestode vaccines: origins, current status and future prospects. *Parasitology* 133 (S2): S27–42.

33 Willadsen, P. (2006). Vaccination against ectoparasites. *Parasitology* 133 (S2): S9–25.

34 Cross, G.A. (1996). Antigenic variation in trypansosomes: secrets surface slowly. *Bioessays* 18 (4): 283–291.

35 Barry, J.D. and McCulloch, R. (2001). Antigenic variation in trypanosomes: enhanced phenotypic variation in a eukaryotic parasite. *Adv Parasitol* 49: 1–70.

36 Ramaswamy, K. (2015). Role of parasite vaccines in sustained animal health and production. *Journal of Veterinary Parasitology* 29 (2): 73–83.

37 Knox, D.P. (2010). Parasite vaccines: recent progress in, and problems associated with their development. *The Open Infectious Diseases Journal* 4 (1).

38 McGregor, I.A. (1964). The passive transfer of human malarial immunity. *The American Journal of Tropical Medicine and Hygiene* 13 (1-Part-2): 237–239.

39 Ballou, W.R., Hoffman, S.L., Sherwood, J.A. et al. (1987). Safety and efficacy of a recombinant DNA *Plasmodium falciparum* sporozoite vaccine. *The Lancet* 329 (8545): 1277–1281.

40 Hoffman, S.L., Wistar, R., Jr, Ballou, W.R. et al. (1986). Immunity to malaria and naturally acquired antibodies to the circumsporozoite protein of *Plasmodium falciparum*. *New England Journal of Medicine* 315 (10): 601–606.

41 Stoute, J.A., Slaoui, M., Heppner, D.G. et al. (1997). A preliminary evaluation of a recombinant circumsporozoite protein vaccine against *Plasmodium falciparum* malaria. *New England Journal of Medicine* 336 (2): 86–91.

42 Holder, A.A. (2009). The carboxy-terminus of merozoite surface protein 1: structure, specific antibodies and immunity to malaria. *Parasitology* 136 (12): 1445–1456.

43 Kedzierski, L., Zhu, Y., and Handman, E. (2006). *Leishmania* vaccines: progress and problems. *Parasitology* 133 (S2): S87–112.

44 Handman, E. (2001). Leishmaniasis: current status of vaccine development. *Clinical Microbiology Reviews* 14 (2): 229–243.

45 Gurunathan, S., Klinman, D.M., and Seder, R.A. (2000). DNA vaccines: immunology, application, and optimization. *Annual Review of Immunology* 18 (1): 927–974.

46 Taheri, T. and Rafati, S. (2013). Leishmaniasis: recombinant DNA vaccination and different approaches for vaccine development. *Clinical Investigation* 3 (11): 1023–1044.

47 Van der Werf, M.J., de Vlas, S.J., Brooker, S. et al. (2003). Quantification of clinical morbidity associated with schistosome infection in sub-Saharan Africa. *Acta Tropica* 86 (2–3): 125–139.

48 McManus, D.P. and Dalton, J.P. (2006). Vaccines against the zoonotic trematodes *Schistosoma japonicum*, *Fasciola hepatica* and *Fasciola gigantica*. *Parasitology* 133 (S2): S43–61.

49 McManus, D.P., Li, Y., Gray, D.J., and Ross, A.G. (2009). Conquering 'snail fever': schistosomiasis and its control in China. *Expert Review of Anti-infective Therapy* 7 (4): 473–485.

50 Nascimento, I.P. and Leite, L.C.C. (2012). Recombinant vaccines and the development of new vaccine strategies. *Brazilian Journal of Medical and Biological Research* 45: 1102–1111.

51 Adkins, J.C. and Wagstaff, A.J. (1998). Recombinant hepatitis B vaccine. *BioDrugs* 10 (2): 137–158.

52 Dertzbaugh, M.T. (1998). Genetically engineered vaccines: an overview. *Plasmid* 39 (2): 100–113.

53 Knox, D.P. and Redmond, D.L. (2006). Parasite vaccines–recent progress and problems associated with their development. *Parasitology* 133 (S2): S1–8.

54 Haslam, S.M., Coles, G.C., Munn, E.A. et al. (1996). *Haemonchus contortus* glycoproteins contain N-linked oligosaccharides with novel highly fucosylated core structures. *Journal of Biological Chemistry* 271 (48): 30561–30570.

55 Danforth, H.D., McCandliss, R., Libel, M. et al. (1985). Development of an avian coccidial antigen by recombinant DNA technology. *Poultry Science* 64: 85.

56 Danforth, H.D. and Augustine, P.C. (1986). Use of hybridoma antibodies and recombinant DNA technology in protozoan vaccine development. *Avian Diseases* 30(1): 37–42.

57 Gamble, H.R. (1987). Monoclonal antibody technology in the development of vaccines for livestock parasites. *Journal of Animal Science* 64 (1): 328–336.

58 Zhao, G., Yan, R., Muleke, C.I. et al. (2012). Vaccination of goats with DNA vaccines encoding H11 and IL-2 induces partial protection against *Haemonchus contortus* infection. *The Veterinary Journal* 191 (1): 94–100.

59 Wolff, J.A., Malone, R.W., Williams, P. et al. (1990). Direct gene transfer into mouse muscle in vivo. *Science* 247 (4949): 1465–1468.

60 Bode, C., Zhao, G., Steinhagen, F. et al. (2011). CpG DNA as a vaccine adjuvant. *Expert Review of Vaccines* 10 (4): 499–511.

61 Sibley, C.H. and Hunt, S.Y. (2003). Drug resistance in parasites: can we stay ahead of the evolutionary curve? *TRENDS in Parasitology* 19 (11): 532–537.

62 Kaplan, R.M. (2004). Drug resistance in nematodes of veterinary importance: a status report. *Trends in Parasitology* 20 (10): 477–481.

63 Ceri H, M.E.O. and Morck, D.W. (2000). *Giardia* vaccination. *Parasitology Today* 16: 213–217.

64 Murray, C.J., Vos, T., Lozano, R. et al. (2012). Disability-adjusted life years (DALYs) for 291diseases and injuries in 21 regions, 1990–2010: a systematic analysis for the Global Burden of Disease Study 2010. *The Lancet* 380 (9859): 2197–2223.

65 Stewart, V.A. and Coppel, R. (2009). Issues in malaria vaccine development. *Parasite Immunology* 31: 489–491.

15

Current Trends in Parasitic Diseases and Precautionary Measures

Nisha Singh

Hemvati Nandan Bahuguna Garhwal University (A Central University), Srinagar (Garhwal), Uttarakhand, India

15.1 Introduction

Parasitic infections, such as malaria, schistosomiasis, and soil-transmitted helminth infections, pose a significant threat to public health globally, affecting over 1.5 billion people and causing significant morbidity and mortality, particularly in developing countries. One of the major challenges in treating these infections is the rise of antibiotic resistance, making it increasingly difficult to effectively treat parasitic infections caused by protozoan parasites [1]. Climate change also plays a role in the spread of these diseases, as changes in temperature and rainfall patterns can alter the habitats of vectors that transmit parasitic infections. In addition to these challenges, there is a need for new and innovative solutions for controlling parasitic infections, including genetic engineering and the use of genetically modified organisms to control vectors that transmit parasitic diseases [2]. Despite these challenges, simple measures such as promoting sanitation and personal hygiene, community education, and awareness campaigns can greatly reduce the transmission of parasitic infections. However, diagnosing and treating parasitic infections in developing countries remains a significant challenge, and alternative therapies and traditional medicine may offer additional treatment options, although further research is required to adequately comprehend its efficacy [3]. The WHO has set a goal to reduce the number of deaths from parasitic

Parasitic Infections: Immune Responses and Therapeutics, First Edition.
Edited by Abhay Prakash Mishra and Manisha Nigam.

diseases, making it crucial to continue to invest in the research and development of new treatments and control measures for parasitic infections [4]. The objective of this chapter is to provide a general overview of the current state of parasitic infections, the necessity of adopting precautions to stop and control their spread and the obstacles in efficiently managing them as a consequence of issues like antibiotic resistance and climate change. In order to limit transmission, the chapter highlights the importance of fostering good hygiene, education, and research into both conventional and alternative treatments, especially in developing countries. The importance of ongoing research and the development of novel therapies, with genetic engineering as a potential remedy, has also been emphasized.

15.2 Trends in Parasitic Diseases: An Overview of Current Developments

Public health experts and researchers are particularly concerned about a few contemporary trends in parasite illnesses. The prevalence of drug-resistant parasites that are becoming more difficult to treat is one trend. Additionally, climate change and increased international travel have led to the spread of parasitic diseases to new geographic regions, making them a global health concern. Another trend is the growing burden of parasitic diseases in low- and middle-income countries, where access to diagnosis and treatment is often limited. Finally, there is increasing recognition of the social and economic impact of parasitic diseases, which can perpetuate poverty and impede economic development.

15.2.1 The Emergence of New Parasitic Diseases

New parasite diseases may appear for several reasons, including changes in environment, human behavior, and the development of drug resistance in existing parasites [5, 6]. Climate change, deforestation, and urbanization can all contribute to the spread of new parasitic diseases by altering the habitats and behaviors of the host organisms and the parasites that infect them [7]. Additionally, increased global travel and trade can also facilitate the spread of parasitic diseases to new geographic regions [8, 9]. Chagas disease also known as "American trypanosomiasis", has emerged in recent years due to changes in the environment [10]. It is caused by the parasite *Trypanosoma cruzi*, which is spread to humans through the bites of infected triatomine bugs. These bugs are commonly found in rural areas of Latin America, and they typically live in the cracks and crevices of poorly constructed homes made from mud, adobe, or thatch [11]. Deforestation and urbanization have led to an increase in the number of people living in close proximity to the natural habitats of triatomine bugs, resulting in an enhanced number of people infected with Chagas disease. Additionally, climate change has also affected the distribution of triatomine bugs and the risk of Chagas disease, as warmer temperatures and changes in rainfall patterns have led to the expansion of their range [11]. Malaria which is caused by *Plasmodium* sp., is transmitted to humans through the bites of infected *Anopheles* mosquitoes. Climate change has been shown to affect the distribution and abundance of mosquitoes and the transmission of malaria. Warmer temperatures and changes in rainfall patterns can lead to an increase in the number of mosquitoes and the transmission of malaria in areas where the disease was previously not a major concern [12].

Drug resistance is also a major factor contributing to the emergence of new parasitic diseases [13]. For example, malaria caused by *Plasmodium falciparum* has developed resistance to many of the drugs that were previously used to treat it, including chloroquine and sulfadoxine-pyrimethamine. This has led to the emergence of new strains of the parasite that are more difficult to treat, and has made it more challenging to control and eliminate malaria in affected areas [14]. Another recent example is that there has been a significant increase in the number of cases of drug-resistant strains of the parasite that causes leishmaniasis. Leishmaniasis is a parasitic disease that is transmitted to humans through the bites of infected sand flies, and it can cause a wide range of symptoms, including skin sores, fever, and weight loss. The emergence of drug-resistant strains of the parasite has made it more difficult to treat and control leishmaniasis in affected areas [15–17].

15.2.2 The Rise of Antibiotic/Drug Resistance in Parasitic Infections

The rise of antibiotic/drug resistance in parasitic infections is a growing concern in the field of parasitic disease control. It occurs when parasites become resistant to the drugs used to treat them, making it more difficult to treat and control the spread of these infections.

One of the most well-known examples of resistance in parasitic infections is the rise of drug-resistant malaria. The *Plasmodium* parasite has developed resistance to a number of drugs, including chloroquine and sulfadoxine-pyrimethamine [18]. This has made it more difficult to treat and control the spread of malaria, particularly in regions where the parasite has become resistant to multiple drugs. The development of resistance to artemisinin-based combination therapies (ACTs), the recommended first-line treatment for malaria, has also become a major concern in Southeast Asia and Africa [19]. In addition, resistance to multiple classes of antimalarials has been reported in some areas, which poses a significant threat to malaria control efforts [20].

Another example of drug resistance in parasitic infections is the rise of drug-resistant soil-transmitted helminths such as hookworm. Soil-transmitted helminths are a group of parasitic worms that infect people living in areas with poor sanitation. These parasites have developed resistance to drugs such as albendazole and mebendazole, making it more difficult to treat and control the spread of these infections. The emergence of resistance to these drugs can lead to treatment failures and may impede progress towards achieving the WHO's goal of reducing the burden of soil-transmitted helminths [21].

Drug resistance in parasitic infections can also occur as a result of overuse of drugs. It can occur in both the host and the vector. Additionally, the use of suboptimal drug regimens, such as using a single drug when multiple drugs are required, can also contribute to the development of resistance [13].

15.2.2.1 Mechanism of Drug Resistance

Drug/antibiotic resistance in parasitic diseases can occur through a variety of mechanisms. Some of the most common mechanisms are discussed here.

15.2.2.1.1 Alteration of Drug Targets

Mutations in the genes that produce the enzymes or receptors that the drugs are designed to target can occur in parasites [22]. For example, *Plasmodium falciparum* has developed resistance to a number of drugs by mutating the gene that codes for the PfCRT protein, which is the target of chloroquine [23]. This has led to changes in the structure of the protein, making it less susceptible to the drug.

15.2.2.1.2 Efflux Pumps

Some parasites can develop efflux pumps, which are proteins that pump drugs out of the cell before they can have an effect. For example, *Plasmodium falciparum* has developed resistance to a number of drugs by expressing efflux pumps such as Pgh1 and PfMDR1, which pump drugs out of the cell before they can have an effect [24].

15.2.2.1.3 Drug Modification

Some parasites can develop enzymes that modify the drugs, making them less effective. For example, *Leishmania donovani*, the parasite that causes leishmaniasis, has developed resistance to a number of drugs by expressing enzymes that modify the drugs [15].

15.2.2.1.4 Gene Amplification

Some parasites can amplify genes that code for enzymes or receptors that the drugs target. For example, *Plasmodium falciparum* has developed resistance to a number of drugs by amplifying the gene that codes for the PfMDR1 protein, which is the target of a number of antimalarial drugs. In another example, in arsenite-resistant *Leishmania tarentolae* gamma-glutamylcysteine synthetase gene gsh1 and of the ABC transporter gene pgpA expression is increased [22]. This can lead to an overproduction of the protein, making the parasite less susceptible to the drug.

15.2.2.1.5 Reduced Drug Uptake
Some parasites can reduce the uptake of drugs, making it more difficult for the drugs to reach their targets. For example, *Trypanosoma brucei*, the parasite that causes sleeping sickness, has developed resistance to a number of drugs by reducing the uptake of drugs through changes in the expression of drug transporters [25].

15.2.2.1.6 Increased Repair Mechanisms
Some parasites can increase the repair of damage caused by drugs, making it more difficult for the drugs to have an effect. For example, *Plasmodium falciparum* has developed resistance to a number of drugs by upregulating repair mechanisms such as the NHEJ pathway, which repairs double-stranded breaks in DNA caused by drugs [26].

15.2.2.1.7 Development of Resistance Through Horizontal Gene Transfer
Some parasites can acquire resistance genes through horizontal gene transfer, which occurs when genetic material is transferred from one organism to another [27]. For example, *Plasmodium falciparum* has acquired resistance to a number of drugs through horizontal gene transfer of resistance genes from other *Plasmodium* species. Understanding these pathways is essential for the design of successful tactics to combat antibiotic resistance in parasitic diseases [28].

The emergence of antibiotic resistance in parasitic infections is a serious public health concern. It can lead to increased morbidity and mortality, as well as increased healthcare costs. It also complicates the control and elimination of these diseases. To combat the rise of antibiotic resistance in parasitic infections, it is important to use drugs in a responsible and rational manner. This includes implementing appropriate prescribing practices, promoting rational drug use, and implementing effective surveillance systems to monitor the emergence and spread of resistance [29]. Additionally, investing in research and development of new drugs and treatment regimens is crucial.

15.2.3 The Impact of Environment and Climate Change on the Spread of Parasitic Diseases

The environment has a significant impact on how parasitic infections are transmitted and disseminated. Changes in land use, deforestation, water management, and other human activities can all have a significant impact on the distribution and transmission of parasites and the organisms that act as vectors for their transmission [6, 30, 31]. Understanding the potential impacts of the environment on parasitic diseases is crucial for designing and implementing effective control measures.

One of the most notable examples is the impact of deforestation on the transmission of malaria. Deforestation can create new habitats for the *Anopheles* mosquito, the primary vector of malaria, leading to an increase in mosquito populations and an increased risk of transmission. This is because the *Anopheles* mosquito prefers to breed in small, stagnant water bodies, which are often created as a result of deforestation. In addition, deforestation can also increase the number of human–mosquito interactions by bringing people into closer contact with mosquitoes in the newly created habitats [32].

Water management is another important factor in the transmission of parasitic diseases. Alterations in water flow and the creation of new water bodies can have a significant impact on the breeding and survival of vectors and intermediate hosts. For example, the construction of dams and irrigation systems can create new habitats for snails, the intermediate host of *Schistosoma* parasites, leading to an increased risk of schistosomiasis transmission. Similarly, changes in water flow can also affect the breeding and survival of mosquitoes, leading to an increased risk of malaria transmission [33, 34].

Land use changes such as urbanization and agriculture can also have a significant impact on the transmission of parasitic diseases. Urbanization can lead to the destruction of natural habitats, which can reduce the number of vectors and intermediate hosts, thus reducing the risk of disease transmission. However, it can also lead to an increased risk of transmission by bringing people into closer contact with infected vectors or hosts, as well as creating conditions that are favorable for the breeding and survival of vectors [35, 36]. Agriculture can also have a significant impact on the transmission of parasitic diseases. For example, the use of pesticides and other chemicals can reduce the number of vectors and intermediate hosts, thus reducing the risk of disease transmission. However, it can also lead to an increased risk of transmission by creating new habitats for vectors and intermediate hosts and by bringing people into closer contact with infected vectors or hosts [37].

Climate change is also a major environmental factor that can affect the spread of parasitic diseases. Warmer temperatures, changes in precipitation patterns, and sea level rise can all affect the distribution and transmission of parasites and the organisms that act as vectors for their transmission [38]. Understanding the potential impacts of climate change on parasitic diseases is crucial for designing and implementing effective control measures [39].

One of the most well-known examples is the potential for climate change to increase the range and transmission of malaria. Warmer temperatures can accelerate the development of the *Plasmodium* parasite in the *Anopheles* mosquito, leading to an increased number of mosquitoes and an increased risk of transmission to humans. This can potentially extend the geographic range of malaria-endemic areas, exposing new populations to the disease. In addition, warmer temperatures can also increase the rate of mosquito biting, leading to an increased transmission of the disease. Changes in precipitation patterns can also affect the breeding and survival of mosquitoes, leading to an increased risk of malaria transmission. For example, changes in the timing and frequency of rainfall can create more favorable conditions for mosquito breeding, leading to an increase in mosquito populations [40].

Climate change can also affect the distribution and transmission of other parasitic diseases. For example, changes in precipitation patterns can affect the breeding and survival of snail intermediate hosts, leading to an increased risk of schistosomiasis transmission [41, 42]. Similarly, sea level rise can raise the liklehood of infection with soil-transmitted helminths such as hookworm by increasing the flooding of areas with contaminated soil. This can increase the potential for human exposure to contaminated soil and increase the risk of infection [43].

Climate change can also have a negative impact on the effectiveness of current control measures. For example, higher temperatures can reduce the effectiveness of pesticides used to control mosquitoes, and changes in precipitation patterns can affect the breeding and survival of mosquitoes and snail intermediate hosts, making it more difficult to control the transmission of parasitic diseases. This can make it harder to control the transmission of these diseases, even when using the same strategies and interventions that have been effective in the past.

In addition to the direct impacts of climate change on parasitic diseases, there are also indirect impacts to consider. For example, climate change can lead to changes in land use and agricultural practices, which can in turn affect the distribution and abundance of vectors and intermediate hosts. This can further increase the risk of disease transmission. Climate change can also lead to increased human migration, as people move to escape the effects of extreme weather events and rising sea levels. This can increase the risk of disease transmission by exposing new populations to infected vectors or hosts [44]. In conclusion, parasite infections are transmitted and spread by the environment, which is a key aspect to note.

15.2.4 Application of Genetic Engineering in Parasitic Disease Control

A significant global health concern is parasitic infections, which can be caused on by a variety of organisms such as protozoa, helminths, and arthropods. They affect over 1.5 billion people, primarily in developing countries, and are responsible for significant morbidity and mortality. Malaria, caused by the *Plasmodium* parasite, is one of the leading causes of death in sub-Saharan Africa, while schistosomiasis, caused by the *Schistosoma* parasite, affects over 200 million people worldwide [45, 46].

Genetic engineering, a set of technologies that allow for the manipulation of an organism's genetic makeup, becomes a potentially effective tool in the fight against parasitic diseases. The ability to target specific genes in the parasite or its vector has the potential to greatly reduce or even eradicate the transmission of these diseases. The earliest forms of genetic engineering can be traced back to the 1970s, with the advent of recombinant DNA technology. As a result, methods like CRISPR-Cas9 genome editing and RNA interference have been created, which allow for precise manipulation of the genetic code [47]. In addition, the creation of transgenic organisms and the use of gene drive have expanded the possibilities of genetic engineering in controlling parasitic diseases.

One of the most promising applications of genetic engineering in parasitic disease control is the targeting of the vector or intermediate host of the parasite. In malaria the *Plasmodium* parasite is transmitted to humans by the *Anopheles* mosquito. Scientists are using genetic engineering techniques such as CRISPR-Cas9 and RNA interference to target specific genes in the mosquito that are essential for the transmission of the parasite. For example, scientists have successfully edited genes in the mosquito that control its ability to sense and locate human hosts, thereby reducing its ability to transmit the parasite [48, 49]. Schistosomiasis, caused by the *Schistosoma* parasite, is transmitted to humans through contact with freshwater infested with infected snail intermediate hosts. Scientists are using genetic engineering to target the snail intermediate host by introducing a gene that makes them resistant to the parasite. This can drastically reduce the transmission of the disease by breaking the lifecycle of the parasite [50, 51]. Toxoplasmosis, caused by *Toxoplasma gondii*, can be transmitted to humans through contact with contaminated food and water or by ingesting undercooked meat. Genetic engineering has been used to create transgenic mice that are resistant to the parasite, which can be used to study the disease and to find potential treatments [52]. Lymphatic filariasis, caused by *Wuchereria bancrofti*, is transmitted to humans by mosquitoes. Genetic engineering has been used to create transgenic mosquitoes that are resistant to the parasite, which can drastically reduce the transmission of the disease [53]. For trypanosomiasis, also known as sleeping sickness, researchers have created genetically modified trypanosomes and schistosomes respectively, that are unable to reproduce, which can help to reduce the spread of the disease [54]. In leishmaniasis scientists are working on using genetic engineering to create genetically modified *Leishmania* parasites that are unable to infect humans and animals, which can help to reduce the spread of the disease [55].

15.2.5 Big Data and Artificial Intelligence Applications for Control and Prevention Parasitic Diseases

Big data and artificial intelligence (AI) are increasingly being used in the field of parasitic diseases to improve our understanding of these diseases and to develop new tools and strategies for their control and prevention. One key area where big data and AI are being used is in the analysis of epidemiological data. This includes the use of large datasets to identify patterns and trends in the distribution, transmission, and risk factors for parasitic diseases. This information can be used to inform the development of more effective control and prevention strategies, such as identifying high-risk areas and populations and targeting control efforts to these areas and populations [56]. The spread of parasite illnesses can also be predicted and forecast using big data analytics and machine learning approaches, which can help identify possible outbreaks before they occur and respond rapidly to contain them [57, 58].

Another area where big data and AI are being used is in the development of new diagnostic tools. For example, big data analytics can be used to identify new biomarkers for parasitic diseases, which can be used to develop more accurate diagnostic tests. AI algorithms can also be used to analyze large datasets of images and other medical data, which can be used to improve the accuracy of diagnostic tools such as microscopy and radiology [59, 60].

Big data and AI are also being used in drug discovery and development. AI-based approaches can be used to identify new drug targets, as well as to screen large numbers of compounds for their potential efficacy against parasitic diseases. Big data analytics can also be used to analyze large datasets of genetic information, which can be used to identify new drug targets and to understand the molecular mechanisms of parasitic diseases [61, 62].

However, there are also some challenges associated with the use of big data and AI in parasitic disease research. The necessity to combine data from various sources and the lack of high-quality data are two major obstacles. Additionally, there may be ethical and privacy concerns related to the use of big data and AI, especially when working with sensitive information such as medical data. Therefore, it is important for researchers to ensure that data is collected and handled responsibly, and that appropriate measures are in place to protect the privacy of individuals [63–65].

In conclusion, big data and AI are playing a crucial role in parasitic disease research by providing new insights, tools and strategies for control and prevention. These technologies are being used to analyze epidemiological data, develop new diagnostic tools, and discover new drugs and treatments. However, there are also challenges associated with the use of big data and AI in parasitic disease research, such as lack of quality data, and ethical and privacy concerns. Therefore, it is important to address these challenges in order to fully leverage the potential of big data and AI in parasitic disease research (Figure 15.1).

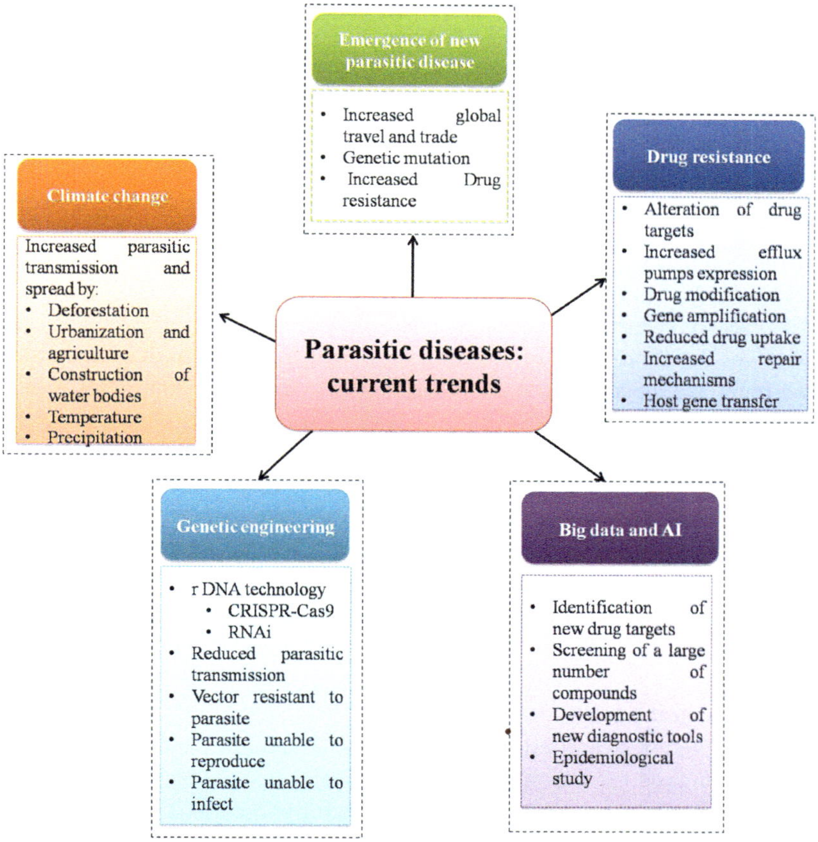

Figure 15.1 Current trends in parasitic diseases.

15.3 Combating Parasitic Diseases: Control Strategies and Preventive Measures

Combating parasitic diseases requires a multifaceted approach that includes both control strategies and preventive measures. Identifying and treating vulnerable individuals as well as taking steps to stop the parasite from spreading are all part of control strategies. Vaccination, public awareness campaigns on the consequences of parasitic diseases, and taking steps to reduce the number of parasite-transmitting vectors are all examples of preventive measures.

The World Health Organization (WHO) recommends a variety of control measures for parasitic diseases that will be discussed next.

15.3.1 Vector Control

Vector control is a critical strategy in the control and prevention of parasitic diseases. This approach involves reducing or eliminating the populations of insects or other organisms that transmit parasites to humans, such as mosquitoes, snails, and flies. Vector control is often implemented in conjunction with other control measures such as preventive chemotherapy, improved sanitation and hygiene, and health education [66]. There are several methods used in vector control to reduce the populations of vectors that transmit parasitic diseases.

15.3.1.1 Source Reduction

This involves the elimination of breeding sites of vectors, such as stagnant water sources, to reduce the number of vectors in an area. This is particularly important in controlling malaria and dengue, as both diseases are transmitted by mosquitoes that breed in standing water.

15.3.1.2 Chemical Control

This involves the use of insecticides to control adult vectors, as well as the use of larvicides to control immature stages of vectors. This method is effective in reducing vector populations but has potential negative impacts on the environment and non-target organisms.

15.3.1.3 Biological Control

This involves the use of natural predators such as dragonflies and fish to control vectors. This method is environmentally friendly and sustainable, but it may not be as effective as chemical control in reducing vector populations.

15.3.1.4 Physical Control

This involves the use of barriers such as screens and nets to prevent vectors from entering homes and other buildings. This is an effective method to prevent vectors from biting people and transmitting disease.

15.3.1.5 Genetic Control

This involves the release of genetically modified mosquitoes that are unable to transmit the disease or have reduced ability to reproduce to reduce the population of vectors.

15.3.1.6 Health Education

This involves educating individuals and communities about the transmission, symptoms, and prevention of parasitic diseases, and how to reduce exposure to vectors. Community participation and buy-in is crucial for the success of vector control efforts.

Vector control is particularly important in areas where parasitic diseases are prevalent and traditional diagnostic tools are not readily available or feasible. For example, in sub-Saharan Africa, malaria is a major public health problem and vector control is an essential component of malaria control efforts. The use of insecticide-treated bed nets and indoor residual spraying are effective methods of reducing vector populations and preventing transmission of the disease.

However, vector control is not without its limitations. For example, the use of chemical control measures such as insecticides can have negative impacts on the environment and on non-target organisms. Additionally, a significant issue that needs to be resolved is the development of pesticide resistance in vectors. Community participation and buy-in is crucial for the success of vector control efforts, and it can be challenging to engage communities in these efforts.

15.3.2 Preventive Chemotherapy

Preventive chemotherapy is a strategy that involves administering drugs to individuals or populations at risk of infection to prevent the establishment of parasitic infections. This approach is particularly useful for parasitic diseases that are difficult to diagnose or for which traditional diagnostic tools are not readily available or feasible. Preventive chemotherapy is an important tool in the control and prevention of parasitic diseases such as malaria, schistosomiasis, and soil-transmitted helminths [67].

Mass drug administration (MDA) is a widely used form of preventive chemotherapy that involves distributing drugs to entire communities or populations, regardless of their individual infection status, in order to target parasitic diseases that are highly prevalent in a given area. MDA with anti-malarial drugs is recommended by the World Health Organization in areas where the prevalence of malaria is greater than 20%. This strategy is effective in minimizing the spread of the disease as well as the number of infections and fatalities linked to malaria. Similarly, MDA with drugs such as praziquantel is effective in reducing the burden of schistosomiasis in affected communities [68, 69]. MDA is also used to control soil-transmitted helminths. However, MDA and other preventive chemotherapy strategies have their limitations, such as the difficulty of reaching remote or hard-to-reach areas, the cost, and the potential development of drug resistance among parasites [70]. Intermittent preventive treatment (IPT) and the use of chemoprophylaxis are other strategies that can be used to control parasitic diseases [71].

15.3.3 Improved Sanitation and Hygiene

Improved sanitation and hygiene are essential strategies for controlling and preventing parasitic diseases. The supply of safe and clean drinking water is a crucial component of improved sanitation since it can stop the spread of parasite diseases including schistosomiasis and soil-transmitted helminths [72]. Proper waste management, including the collection, transport, and disposal of solid and liquid waste in a way that minimizes the risk of disease transmission, is another important aspect of improved sanitation. Personal and community hygiene practices such as handwashing with soap and clean water, proper food preparation and handling, and maintaining personal cleanliness are also essential for preventing the transmission of parasitic diseases [4]. However, improving sanitation and hygiene can be challenging in low-income or resource-limited settings and may face cultural or societal barriers. Therefore, it is important to work closely with communities to understand their specific needs and to involve them in the design and implementation of sanitation and hygiene interventions [73].

15.3.4 Health Education

A key strategy in the management and prevention of parasite diseases is health education. It involves educating individuals and communities about the transmission of parasitic diseases, including information on how parasitic infections are acquired, the vectors that transmit the parasites, and the conditions that promote the transmission of the disease. Another important aspect is educating individuals and communities about the symptoms of parasitic diseases, including common signs and symptoms, and when to seek medical attention. Prevention is also an important aspect of health education, including information on how to reduce exposure to vectors and the use of preventive chemotherapy measures such as MDA and IPT [74].

One key example of a community education based parasitic disease eradication program is the Global Schistosomiasis Elimination Program [75]. Schistosomiasis, also known as snail fever, is a parasitic infection caused by a group of trematode worms of the genus *Schistosoma*. It is transmitted through contact with contaminated water, and it affects over 200 million people worldwide, mostly in poor and rural areas. WHO launched the Global Schistosomiasis Elimination Program in 2002, with the goal of controlling and eliminating the disease as a public health problem [76].

The program focuses on providing access to safe water and sanitation, as well as mass drug administration with the drug praziquantel. The success of the programme is greatly dependent on the community's understanding of the disease's transmission mechanisms, the value of having access to safe water, and the significance of taking part in MDA. Community mobilization and involvement in the program have been essential in increasing community participation in MDA, as well as in identifying and treating affected individuals. As a result of this program, several countries have reported a significant decline in the prevalence of schistosomiasis. For example, Egypt has reported a decline in the prevalence of *Schistosoma mansoni* infection from 66% in the 1980s to less than 5% in the 2010s [77]. Similarly, China has reported a decline in the prevalence of *Schistosoma japonicum* infection from over 12% in the 1990s to less than 1% in the 2010s. These success stories demonstrate the importance of community education and awareness in reducing parasitic disease transmission and achieving the goal of parasitic disease elimination [78].

15.3.5 Surveillance and Monitoring

Monitoring and surveillance are crucial tactics for the prevention and control of parasitic infections. This approach involves tracking the incidence and distribution of parasitic diseases to identify high-risk areas and populations. Collection of data on the incidence of parasitic diseases, including data on the number of new cases of the disease and data on the number of deaths caused by the disease, is key aspect of surveillance and monitoring. This information can be used to target control and prevention efforts to high-risk areas or populations [79]. Collection of data on the distribution of parasitic diseases, including data on where the disease is present and the specific populations that are affected, is also important aspect of surveillance and monitoring. Monitoring the effectiveness of control efforts is also an important aspect of surveillance and monitoring. However, surveillance and monitoring can be challenging in low-income or resource-limited settings and may face cultural or societal barriers. Therefore, it is important to work closely with communities to understand their specific needs and to involve them in the design and implementation of surveillance and monitoring efforts [80].

15.3.6 Research and Development

An important approach in the management and prevention of parasite diseases is research and development. To treat and prevent parasite infections, new diagnostic techniques, drugs, and vaccinations must be developed [81]. The creation of new diagnostic tools, such as rapid diagnostic tests (RDTs) for malaria and copro-antigen-based diagnostic assays for soil-transmitted helminths, is a crucial component of research and development. The creation of novel medications and therapies for parasite infections, such as artemisinin-based combination therapies (ACTs) for malaria and tribendimidine for soil-transmitted helminths, is another crucial factor. A component of research and development is the development of fresh vaccinations to ward off parasite diseases like malaria and schistosomiasis [82]. However, research and development is a complex and time-consuming process that requires significant financial and human resources and has ethical considerations. To overcome these challenges, it is important to continue to invest in research and development to improve the detection and treatment of parasitic diseases and ensure that the benefits of these new tools and technologies are accessible to all individuals and communities.

15.3.7 Strong Health Systems

Strong health systems are essential for the control and prevention of parasitic diseases. This involves building health systems that can detect and respond to outbreaks of parasitic diseases, and that can provide access to quality care for individuals affected by these diseases. One key aspect of strong health systems is the ability to detect and respond to outbreaks of parasitic diseases through systems for surveillance and monitoring of parasitic diseases, as well as systems for rapid response to outbreaks [83]. Another important aspect is the ability to provide access to quality care for individuals affected by parasitic diseases through systems for the diagnosis and treatment of parasitic diseases, as well as systems for follow-up care and support. However, building strong health systems that can detect and respond to outbreaks of parasitic diseases is challenging in low-income or resource-limited settings, where there may be limited access to diagnostic tools or limited capacity to detect and respond to outbreaks [3].

15.3.8 Integrated Approach

An integrated approach is crucial for the most effective control of parasitic diseases. It involves combining different measures and strategies such as vector control, preventive chemotherapy, improved sanitation and hygiene, health education, surveillance and monitoring, research and development, and strong health systems in a coordinated and holistic way. It ensures that all aspects of parasitic disease control are addressed and that all sectors of society are involved in the effort. Coordination of different control measures, involving all sectors of society and addressing challenges such as limited capacity, cultural and societal barriers are key to implementing an integrated approach to parasitic disease control [84].

15.4 The Challenges of Diagnosing and Treating Parasitic Infections in Developing Countries

Parasitic infections are a significant public health problem in developing countries, with a range of challenges that impede effective diagnosis and treatment. These challenges include a lack of access to accurate diagnostic tools, a shortage of trained healthcare professionals, limited availability of affordable treatments, poor sanitation and lack of clean water, limited surveillance and monitoring systems, socio-economic factors, stigma and discrimination, and limited government funding. These issues can result in misdiagnosis, inadequate treatment, and increased morbidity and mortality for diseases such as malaria, schistosomiasis, and leishmaniasis. Addressing these challenges requires a multi-pronged approach that addresses underlying issues such as lack of sanitation and access to clean water, and improves access to diagnostic tools and trained healthcare professionals, and increases the availability of affordable treatment [85, 86].

15.4.1 Lack of Access to Accurate Diagnostic Tools

Many developing countries do not have the resources or infrastructure to purchase and maintain expensive diagnostic equipment, such as microscopes or PCR machines. For example, in many rural areas in sub-Saharan Africa, healthcare facilities may only have basic laboratory equipment, such as microscope, and lack the ability to perform more advanced diagnostic tests like PCR. This can result in misdiagnosis and inadequate treatment of parasitic infections such as malaria, schistosomiasis, and leishmaniasis.

15.4.2 Lack of Trained Healthcare Professionals

In many developing countries, there is a shortage of trained healthcare workers who are able to properly diagnose and treat parasitic infections. For example, in India a large proportion of healthcare providers in rural areas do not have formal medical training and may not be able to identify and properly treat parasitic infections such as elephantiasis, Chagas disease, and toxoplasmosis.

15.4.3 Lack of Available and Affordable Treatments

Many parasitic infections are treated with drugs that are not readily available or are too expensive for people living in these countries. For example, in many parts of Africa, antimalarial drugs are often in short supply and can be unaffordable for many people. This lack of access to treatment can lead to chronic infections and increased morbidity and mortality for diseases such as malaria, schistosomiasis, and leishmaniasis.

15.4.4 Lack of Sanitation and Clean Water

The lack of sanitation and clean water in many developing countries makes the population more susceptible to parasitic infections such as *Ascaris lumbricoides*, *Giardia intestinalis*, and *Dracunculus medinensis* and makes it more difficult to control and eliminate these infections.

15.4.5 Limited Surveillance and Monitoring Systems

It is challenging to understand the full scope of the issue and properly target therapies in many underdeveloped countries due to the absence of tools for tracking and monitoring parasite infections. This can make it difficult to control the spread of diseases such as Guinea worm disease, and Chagas disease.

15.4.6 Socio-Economic Factors

Poverty, poor education, and poor living conditions contribute to the spread of parasitic infections and make it more difficult for people to access healthcare and treatment. For example, people living in slums or refugee camps may be more likely to contract parasitic infections such as hookworm and schistosomiasis because of poor sanitation and limited access to clean water.

15.4.7 Stigma and Discrimination

Many people in developing countries may be reluctant to seek help for parasitic infections due to stigma and discrimination related to the infection. For example, people with elephantiasis may be shunned by their communities, making it difficult for them to access healthcare and treatment.

15.4.8 Limited Government Funding

In many developing countries, the government may not have the financial resources to invest in healthcare infrastructure, including diagnostic tools and treatment options, which exacerbates the challenges of diagnosing and treating parasitic infections. This could lead to lack of access to diagnosis or treatment for diseases such as loiasis or trypanosomiasis.

15.5 The Use of Traditional Medicine in the Treatment of Parasitic Diseases

Traditional medicine, also known as indigenous or folk medicine, is a system of medicine that has been passed down through generations within a particular culture or community. It is often based on the use of natural substances and therapies, and is widely used in many parts of the world for the treatment of various diseases, including parasitic infections [87].

One traditional medicine that is used for the treatment of parasitic diseases is herbal medicine. Herbal medicine uses plants or plant extracts to treat parasitic infections. Many plants have been found to possess anti-parasitic properties, and have been used for centuries to treat parasitic infections such as malaria, schistosomiasis, and leishmaniasis. Some examples of herbs that are used to treat parasitic infections include *Artemisia annua*, which is used to treat malaria, and praziquantel, which is used to treat schistosomiasis [88].

Acupuncture is a traditional Chinese medicinal practice that involves the insertion of thin needles into specific points on the body to treat a variety of conditions, including parasitic infections. It is believed that the insertion of needles into these specific points can help to balance the body's energy, known as Qi, and in turn, help to combat the parasitic infection [89].

In addition, traditional healers and practitioners also use other methods such as bloodletting, cupping, massage, and moxibustion to help combat parasitic infections. They also often use a combination of natural remedies and spiritual or psychological treatments to help people to cope with the physical and emotional symptoms of parasitic infections.

Furthermore, traditional medicine also includes the use of diet and nutrition as a form of treatment. In some cultures, certain foods are believed to have anti-parasitic properties and are used to prevent and treat parasitic infections. For example, in some African cultures, the seeds of the *Moringa oleifera* tree are believed to have anti-parasitic properties and are used to treat malaria [90].

It is important to note that while traditional medicine can be effective in the treatment of parasitic infections, it should be used in conjunction with conventional medical treatments and under the guidance of a qualified healthcare professional. Also, it is important to note that traditional medicine should be used with caution as some traditional remedies could be harmful or interact with conventional treatments, and, in some cases, may not be effective.

15.6 The Need for Continued Research and Development of New Treatments for Parasitic Infections

Parasitic infections are a major public health concern worldwide, affecting over a billion people, primarily in developing countries. These infections are caused by a diverse group of pathogens, including protozoan parasites (such as those that cause malaria, leishmaniasis, and sleeping sickness) and helminthic parasites (such as those that cause schistosomiasis and soil-transmitted intestinal worms). These infections can have a significant impact on the health, well-being and productivity of those affected [91].

One of the major challenges in treating parasitic infections is the emergence and spread of drug-resistant parasites. Many of the current treatments for parasitic infections are based on a limited number of drugs, and the overuse or improper use of these drugs has led to the development of drug-resistant strains of parasites [92]. For example, *Plasmodium falciparum*, the parasite that causes the deadliest form of malaria, has developed resistance to multiple classes of antimalarials. As a result, the effectiveness of some drugs has significantly decreased, which has made treating malaria more complicated.

Another challenge is the significant unmet need for new treatments for parasitic infections, particularly for neglected tropical diseases (NTDs), which affects the poorest communities in the world. Many of the existing treatments for NTDs are not suitable for use in these communities due to factors such as cost, availability, and difficulty of administration. For example, some drugs require refrigeration, which is not always possible in remote rural areas, whereas others are too costly for the poor population to afford. Moreover, some of the drugs available for the treatment of NTDs have significant side effects, which can make them difficult to administer to vulnerable populations [93].

To address these challenges, there is a need for continued research and development of new treatments for parasitic infections. This involves both the development of novel drugs that target various phases of the parasitic life cycle and the discovery of innovative therapeutic targets within the parasite [94, 95]. In addition, novel drug delivery technologies, such nanoparticles and microneedles, are being created to boost the efficacy and security of already existing therapies. Additionally, a promising field for the creation of novel medications is the discovery and development of new natural substances from plants, marine organisms, and microbes with antiparasitic action.

In conclusion, continued research and development of new treatments for parasitic infections is crucial to addressing the significant global health problem caused by these infections and the emergence of drug-resistant parasites, as well as meeting the unmet needs of the poorest communities affected by neglected tropical diseases.

15.7 Conclusion

Parasitic infections are a serious global health issue that affects millions of people causing significant morbidity and mortality, particularly in developing countries. The rise of antibiotic resistance in parasitic infections, impact of climate change on the spread of parasitic diseases, and the need for new and innovative solutions for the control of parasitic infections, such as genetic engineering, are major challenges facing the fight against parasitic diseases. However, simple measures such as promoting sanitation and personal hygiene, community education and awareness campaigns, can also greatly reduce the transmission of parasitic infections. The challenges of diagnosing and treating parasitic infections in developing countries are significant, and alternative therapies and traditional medicine may offer additional treatment options, but more research is needed to fully understand their effectiveness. Therefore, it is crucial to continue to invest in research and development of new treatments and control measures for parasitic infections to achieve the goal set by WHO to reduce the number of deaths from parasitic diseases.

References

1 Bhutta, Z.A., Sommerfeld, J., Lassi, Z.S. et al. (2014). Global burden, distribution, and interventions for infectious diseases of poverty. *Infectious Diseases of Poverty* 3 (1): 1–7.

2 El-Sayed, A. and Kamel, M. (2020). Climatic changes and their role in emergence and re-emergence of diseases. *Environmental Science and Pollution Research* 27 (18): 22336–22352.

3 Despommier, D.D., Gwadz, R.W., and Hotez, P.J. (2012). *Parasitic Diseases*. Springer Science & Business Media.

4 World Health Organization. (2015). Water sanitation & hygiene for accelerating and sustaining progress on neglected tropical diseases: *a global strategy 2015–2020*. World Health Organization.

5 Short, E.E., Caminade, C., and Thomas, B.N. (2017). Climate change contribution to the emergence or re-emergence of parasitic diseases. *Infectious Diseases: Research and Treatment* 10: 1178633617732296.

6 Patz, J.A., Graczyk, T.K., Geller, N., and Vittor, A.Y. (2000). Effects of environmental change on emerging parasitic diseases. *International Journal for Parasitology* 30 (12–13): 1395–1405.

7 Coppens, C.M., de Boer, S.F., and Koolhaas, J.M. (2010). Coping styles and behavioural flexibility: towards underlying mechanisms. *Philosophical Transactions of the Royal Society B: Biological Sciences* 365 (1560): 4021–4028.

8 Baker, R.E., Mahmud, A.S., Miller, I.F. et al. (2022). Infectious disease in an era of global change. *Nature Reviews Microbiology* 20 (4): 193–205.

9 Capela, R., Moreira, R., and Lopes, F. (2019). An overview of drug resistance in protozoal diseases. *International Journal of Molecular Sciences* 20 (22): 5748.

10 Melo, R.D., Guarneri, A.A., and Silber, A.M. (2020). The influence of environmental cues on the development of *Trypanosoma cruzi* in triatominae vector. *Frontiers in Cellular and Infection Microbiology* 10: 27.

11 Parra-Henao, G., Quirós-Gómez, O., Jaramillo-O, N., and Cardona, Á.S. (2016). Environmental determinants of the distribution of Chagas disease vector *Triatoma dimidiata* in Colombia. *The American Journal of Tropical Medicine and Hygiene* 94 (4): 767.

12 Jetten, T.H., Martens, W.J., and Takken, W. (1996). Model simulations to estimate malaria risk under climate change. *Journal of Medical Entomology* 33 (3): 361–371.

13 Davies, J. and Davies, D. (2010). Origins and evolution of antibiotic resistance. *Microbiology and Molecular Biology Reviews* 74 (3): 417–433.

14 Wongsrichanalai, C., Pickard, A.L., Wernsdorfer, W.H., and Meshnick, S.R. (2002). Epidemiology of drug-resistant malaria. *The Lancet Infectious Diseases* 2 (4): 209–218.

15 Croft, S.L., Sundar, S., and Fairlamb, A.H. (2006). Drug resistance in leishmaniasis. *Clin Microbiol Rev* 19 (1): 111–126. doi: 10.1128/CMR.19.1.111-126.2006.

16 Singh, N., Bajpai, S., Kumar, V. et al. (2013). Identification and functional characterization of *Leishmania donovani* secretory peroxidase: delineating its role in NRAMP1 regulation. *PLoS ONE* 8 (1): e53442.

17 Vora, N. (2008). Impact of anthropogenic environmental alterations on vector-borne diseases. *The Medscape Journal of Medicine* 10 (10): 238.

18 Mita, T., Tanabe, K., and Kita, K. (2009). Spread and evolution of *Plasmodium falciparum* drug resistance. *Parasitology International* 58 (3): 201–209.

19 Ouji, M., Augereau, J.M., Paloque, L., and Benoit-Vical, F. (2018). *Plasmodium falciparum* resistance to artemisinin-based combination therapies: a sword of Damocles in the path toward malaria elimination. *Parasite* 25: 24.

20 Eastman, R.T. and Fidock, D.A. (2009). Artemisinin-based combination therapies: a vital tool in efforts to eliminate malaria. *Nature Reviews Microbiology* 7 (12): 864–874.

21 Bethony, J., Brooker, S., Albonico, M. et al. (2006). Soil-transmitted helminth infections: ascariasis, trichuriasis, and hookworm. *The Lancet* 367 (9521): 1521–1532.

22 Blasco, B., Leroy, D., and Fidock, D.A. (2017). Antimalarial drug resistance: linking *Plasmodium falciparum* parasite biology to the clinic. *Nature Medicine* 23 (8): 917–928.

23 Ibraheem, Z.O., Abd Majid, R., Noor, S.M. et al. (2014). Role of different Pfcrt and Pfmdr-1 mutations in conferring resistance to antimalaria drugs in *Plasmodium falciparum*. *Malaria Research and Treatment* 2014: 950424.

24 Le Bras, J. and Durand, R. (2003). The mechanisms of resistance to antimalarial drugs in *Plasmodium falciparum*. *Fundamental & Clinical Pharmacology* 17 (2): 147–153.

25 Matovu, E., Seebeck, T., Enyaru, J.C., and Kaminsky, R. (2001). Drug resistance in *Trypanosoma brucei* spp., the causative agents of sleeping sickness in man and nagana in cattle. *Microbes and Infection* 3 (9): 763–770.

26 Gupta, D.K., Patra, A.T., Zhu, L. et al. (2016). DNA damage regulation and its role in drug-related phenotypes in the malaria parasites. *Scientific Reports* 6 (1): 1–5.

27 Andam, C.P., Fournier, G.P., and Gogarten, J.P. (2011). Multilevel populations and the evolution of antibiotic resistance through horizontal gene transfer. *FEMS Microbiology Reviews* 35 (5): 756–767.

28 Painter, H.J., Morrisey, J.M., Mather, M.W., and Vaidya, A.B. (2007). Specific role of mitochondrial electron transport in blood-stage *Plasmodium falciparum*. *Nature* 446 (7131): 88–91.

29 Majumder, M.A., Rahman, S., Cohall, D. et al. (2020). Antimicrobial stewardship: fighting antimicrobial resistance and protecting global public health. *Infection and Drug Resistance* 13: 4713.

30 Caminade, C., McIntyre, K.M., and Jones, A.E. (2019). Impact of recent and future climate change on vector-borne diseases. *Annals of the New York Academy of Sciences* 1436 (1): 157–173.

31 Walsh, J.F., Molyneux, D.H., and Birley, M.H. (1993). Deforestation: effects on vector-borne disease. *Parasitology* 106 (S1): S55–S75.

32 Ortiz, D.I., Piche-Ovares, M., Romero-Vega, L.M. et al. (2022). The impact of deforestation, urbanization, and changing land use patterns on the ecology of mosquito and tick-borne diseases in Central America. *Insects* 13 (1): 20.

33 De Leo, G.A., Stensgaard, A.S., Sokolow, S.H. et al. (2020). Schistosomiasis and climate change. *British Medical Journal* 371: m4324.

34 Lund, A.J., Sam, M.M., Sy, A.B. et al. (2019). Unavoidable risks: local perspectives on water contact behavior and implications for schistosomiasis control in an agricultural region of northern Senegal. *The American Journal of Tropical Medicine and Hygiene* 101 (4): 837.

35 White, R.J. and Razgour, O. (2020). Emerging zoonotic diseases originating in mammals: a systematic review of effects of anthropogenic land-use change. *Mammal Review* 50 (4): 336–352.

36 Hassell, J.M., Begon, M., Ward, M.J., and Fèvre, E.M. (2017). Urbanization and disease emergence: dynamics at the wildlife–livestock–human interface. *Trends in Ecology & Evolution* 32 (1): 55–67.

37 Andreazza, F., Oliveira, E.E., and Martins, G.F. (2021). Implications of sublethal insecticide exposure and the development of resistance on mosquito physiology, behavior, and pathogen transmission. *Insects* 12 (10): 917.

38 Blum, A.J. and Hotez, P.J. (2018). Global "worming": climate change and its projected general impact on human helminth infections. *PLoS Neglected Tropical Diseases* 12 (7): e0006370.

39 Lata, K., Das, G., Verma, R., and Baghel, R.P. (2018). *Impact of climate variability on occurrence and distribution of vector and vector-borne parasitic diseases. Journal of Entomology and Zoology Studies* 6: 1388–1393.

40 Liu, Z., Wang, S., Zhang, Y. et al. (2021). Effect of temperature and its interactions with relative humidity and rainfall on malaria in a temperate city Suzhou, China. *Environmental Science and Pollution Research* 28 (13): 16830–16842.

41 Kalinda, C., Chimbari, M., and Mukaratirwa, S. (2017). Implications of changing temperatures on the growth, fecundity and survival of intermediate host snails of schistosomiasis: a systematic review. *International Journal of Environmental Research and Public Health* 14 (1): 80.

42 Studer, A., Thieltges, D.W., and Poulin, R. (2010). Parasites and global warming: net effects of temperature on an intertidal host–parasite system. *Marine Ecology Progress Series* 415: 11–22.

43 Kowalczyk, K. and Klapec, T. (2020). Contamination of soil with eggs of geohelminths *Ascaris* spp., *Trichuris* spp., *Toxocara* spp. in Poland–potential source of health risk in farmers. *Annals of Parasitology* 66 (4): 433–440.

44 Braide, W., Justice-Alucho, C.H., Ohabughiro, N., and Adeleye, S.A. (2020). Global climate change and changes in disease distribution: a review in retrospect. *International Journal of Advanced Research in Biological Sciences* 7 (2): 32–46.

45 Varo, R., Chaccour, C., and Bassat, Q. (2020). Update on malaria. *Medicina Clínica (English Edition)* 155 (9): 395–402.

46 Verjee, M.A. (2019). Schistosomiasis: still a cause of significant morbidity and mortality. *Research and Reports in Tropical Medicine* 10: 153.

47 Barrangou, R., Birmingham, A., Wiemann, S. et al. (2015). Advances in CRISPR-Cas9 genome engineering: lessons learned from RNA interference. *Nucleic Acids Research* 43 (7): 3407–3419.

48 Wickramasinghe, P.D., Silva, G.N., Silva Gunawardene, Y.I., and Dassanayake, R.S. (2021). Advances in *Aedes* mosquito vector control strategies using CRISPR/Cas9. In: *Genetically Modified and Other Innovative Vector Control Technologies (ed B.K. Tyagi)*. 67–87. Singapore: Springer.

49 Gantz, V.M., Jasinskiene, N., Tatarenkova, O. et al. (2015). Highly efficient Cas9-mediated gene drive for population modification of the malaria vector mosquito *Anopheles stephensi*. *Proceedings of the National Academy of Sciences* 112 (49): E6736–43.

50 Grewelle, R.E., Perez-Saez, J., Tycko, J. et al. (2022). Modeling the efficacy of CRISPR gene drive for snail immunity on schistosomiasis control. *PLOS Neglected Tropical Diseases* 16 (10): e0010894.

51 Du, X., McManus, D.P., French, J.D. et al. (2021). CRISPR/Cas9: a new tool for the study and control of helminth parasites. *Bioessays* 43 (1): 2000185.

52 Winiger, R.R. and Hehl, A.B. (2020). A streamlined CRISPR/Cas9 approach for fast genome editing in *Toxoplasma gondii* and *Besnoitia besnoiti*. *Journal of Biological Methods* 7 (4): e140.

53 Wilke, A.B. and Marrelli, M.T. (2015). Paratransgenesis: a promising new strategy for mosquito vector control. *Parasites & Vectors* 8 (1): 1–9.

54 Sánchez-Valdéz, F.J., Pérez Brandán, C., Ferreira, A., and Basombrío, M.Á. (2015). Gene-deleted live-attenuated *Trypanosoma cruzi* parasites as vaccines to protect against Chagas disease. *Expert Review of Vaccines* 14 (5): 681–697.

55 Pandey, S.C., Kumar, A., and Samant, M. (2020). Genetically modified live attenuated vaccine: a potential strategy to combat visceral leishmaniasis. *Parasite Immunology* 42 (9): e12732.

56 Xia, S., Zheng, J.X., Wang, X.Y. et al. (2020). Epidemiological big data and analytical tools applied in the control programmes on parasitic diseases in China: NIPD's sustained contributions in 70 years. *Advances in Parasitology* 110: 319–347.

57 Agrebi, S. and Larbi, A. (2020). Use of artificial intelligence in infectious diseases. In: *Artificial Intelligence in Precision Health* (ed. D. Barh). 415–438. Academic Press.

58 Kang, S. and Lee, D. (2022). Use of artificial intelligence for predicting infectious disease. In: *Big Data Analytics for Healthcare* (ed. P. Keikhosrokiani), 153–163. Academic Press.

59 Peiffer-Smadja, N., Dellière, S., Rodriguez, C. et al. (2020). Machine learning in the clinical microbiology laboratory: has the time come for routine practice? *Clinical Microbiology and Infection* 26 (10): 1300–1309.

60 Huang, K., Jiang, Z., Li, Y. et al. (2021). The classification of six common skin diseases based on Xiangya-Derm: development of a Chinese database for Artificial Intelligence. *Journal of Medical Internet Research* 23 (9): e26025.

61 Zhavoronkov, A., Vanhaelen, Q., and Oprea, T.I. (2020). Will artificial intelligence for drug discovery impact clinical pharmacology? *Clinical Pharmacology & Therapeutics* 107 (4): 780–785.

62 Winkler, D.A. (2021). Use of artificial intelligence and machine learning for discovery of drugs for neglected tropical diseases. *Frontiers in Chemistry* 9: 614073.

63 Villalta, F. and Rachakonda, G. (2019). Advances in preclinical approaches to Chagas disease drug discovery. *Expert Opinion on Drug Discovery* 14 (11): 1161–1174.

64 Ghosh, B. and Choudhuri, S. (2021). Drug design for Malaria with Artificial Intelligence (AI). In: *Plasmodium Species and Drug Resistance* (ed. R.K. Tyagi). IntechOpen.

65 Moreira-Filho, J.T., Silva, A.C., Dantas, R.F. et al. (2021). Schistosomiasis drug discovery in the era of automation and artificial intelligence. *Frontiers in Immunology* 12: 642383.

66 Rozendaal, J.A. (1997). *Vector Control: Methods for Use by Individuals and Communities*. World Health Organization.

67 Bockarie, M.J., Kelly-Hope, L.A., Rebollo, M., and Molyneux, D.H. (2013). Preventive chemotherapy as a strategy for elimination of neglected tropical parasitic diseases: endgame challenges. *Philosophical Transactions of the Royal Society B: Biological Sciences* 368 (1623): 20120144.

68 Keys, H., Ureña, K., Reyes, J. et al. (2021). Rapid ethnographic assessment for potential anti-malarial mass drug administration in an outbreak area of Santo Domingo, Dominican Republic. *Malaria Journal* 20 (1): 1–3.

69 Eisele, T.P. (2019). Mass drug administration can be a valuable addition to the malaria elimination toolbox. *Malaria Journal* 18 (1): 1–5.

70 Farrell, S.H., Coffeng, L.E., Truscott, J.E. et al. (2018). Investigating the effectiveness of current and modified World Health Organization guidelines for the control of soil-transmitted helminth infections. *Clinical Infectious Diseases* 66 (suppl_4): S253–S259.

71 Gosling, R.D., Cairns, M.E., Chico, R.M., and Chandramohan, D. (2010). Intermittent preventive treatment against malaria: an update. *Expert Review of Anti-infective Therapy* 8 (5): 589–606.

72 Grimes, J.E., Croll, D., Harrison, W.E. et al. (2015). The roles of water, sanitation and hygiene in reducing schistosomiasis: a review. *Parasites & Vectors* 8 (1): 1–6.

73 Omarova, A., Tussupova, K., Berndtsson, R. et al. (2018). Protozoan parasites in drinking water: a system approach for improved water, sanitation and hygiene in developing countries. *International Journal of Environmental Research and Public Health* 15 (3): 495.

74 Bieri, F.A., Gray, D.J., Williams, G.M. et al. (2013). Health-education package to prevent worm infections in Chinese schoolchildren. *New England Journal of Medicine* 368 (17): 1603–1612.

75 King, C.H. (2009). Toward the elimination of schistosomiasis. *New England Journal of Medicine* 360 (2): 106–109.

76 World Health Organization. (2002). *Prevention and control of schistosomiasis and soil-transmitted helminthiasis: report of a WHO expert committee*. World Health Organization.

77 Abruzzi, A. and Fried, B. (2011). Coinfection of *Schistosoma* (Trematoda) with bacteria, protozoa and helminths. *Advances in Parasitology* 77: 1–85.

78 Torgerson, P.R. (2013). One world health: socioeconomic burden and parasitic disease control priorities. *Veterinary Parasitology* 195 (3–4): 223–232.

79 Goutard, F.L., Binot, A., Duboz, R. et al. (2015). How to reach the poor? Surveillance in low-income countries, lessons from experiences in Cambodia and Madagascar. *Preventive Veterinary Medicine* 120 (1): 12–26.

80 Singh, G., Angwafor, S.A., Njamnshi, A.K. et al. (2020). Zoonotic and vector-borne parasites and epilepsy in low-income and middle-income countries. *Nature Reviews Neurology* 16 (6): 333–345.

81 Bahk, Y.Y., Shin, E.H., Cho, S.H. et al. (2018). Prevention and control strategies for parasitic infections in the Korea centers for disease control and prevention. *The Korean Journal of Parasitology* 56 (5): 401.

82 Mubanga, C., Mwape, K.E., Phiri, I.K. et al. (2019). Progress on the development of rapid diagnostic tests for foodborne neglected zoonotic helminthiases: a systematic review. *Acta Tropica* 194: 135–147.

83 Nema, S., Verma, A.K., Tiwari, A., and Bharti, P.K. (2021). Digital health care services to control and eliminate malaria in India. *Trends in Parasitology* 37 (2): 96–99.

84 Sharma, N., Singh, V., and Shyma, K.P. (2015). Role of parasitic vaccines in integrated control of parasitic diseases in livestock. *Veterinary World* 8 (5): 590.

85 Mabey, D., Peeling, R.W., Ustianowski, A., and Perkins, M.D. (2004). Diagnostics for the developing world. *Nature Reviews Microbiology* 2 (3): 231–240.

86 Berger, S.A. and Marr, J.S. (2006). *Human Parasitic Diseases Sourcebook*. Jones & Bartlett Learning.

87 Monzon, R.B. (1995). Traditional medicine in the treatment of parasitic diseases in the Philippines. *Southeast Asian Journal of Tropical Medicine and Public Health* 26 (3): 421–428.

88 Singh, A., Mishra, A., Chaudhary, R., and Kumar, V. (2020). Role of herbal plants in prevention and treatment of parasitic diseases. *Journal of Scientific Research* 64: 50–58.

89 Singh, N., Mishra, B.B., Bajpai, S. et al. (2014). Natural product based leads to fight against leishmaniasis. *Bioorganic & Medicinal Chemistry* 22 (1): 18–45.

90 Mwangi, V.I., Mumo, R.M., Nyachieo, A., and Onkoba, N. (2017). Herbal medicine in the treatment of poverty associated parasitic diseases: a case of sub-Saharan Africa. *Journal of Herbal Medicine* 10: 1–7.

91 Secor, W.E., Meites, E., Starr, M.C., and Workowski, K.A. (2014). Neglected parasitic infections in the United States: trichomoniasis. *The American Journal of Tropical Medicine and Hygiene* 90 (5): 800.

92 Singh, N., Kumar, M., and Singh, R.K. (2012). Leishmaniasis: current status of available drugs and new potential drug targets. *Asian Pacific Journal of Tropical Medicine* 5 (6): 485–497.

93 Norman, F.F., Comeche, B., Chamorro, S., and López-Vélez, R. (2020). Overcoming challenges in the diagnosis and treatment of parasitic infectious diseases in migrants. *Expert Review of Anti-infective Therapy* 18 (2): 127–143.

94 Singh, R., Anand, A., Rawat, A.K. et al. (2022). CD300a receptor blocking enhances early clearance of leishmania donovani from its mammalian host through modulation of effector functions of phagocytic and antigen experienced T cells. *Frontiers in Immunology* 12: 5960.

95 Srivastava, A., Singh, N., Mishra, M. et al. (2012). Identification of TLR inducing Th1-responsive *Leishmania donovani* amastigote-specific antigens. *Molecular and Cellular Biochemistry* 359: 359–368.

Index

N.B. Page numbers in **bold** refer to items in tables and number in *italics* refer to figures.

Parasitic Infections: Immune Responses and Therapeutics, First Edition.
Edited by Abhay Prakash Mishra and Manisha Nigam.

h

o

q

r

s

u

Printed and bound by CPI Group (UK) Ltd, Croydon, CR0 4YY
17/09/2026
14964253-0004